P9-CBD-881

The Basics of Earth Science

The Basics of Earth Science

Robert E. Krebs

Basics of the Hard Sciences

GREENWOOD PRESS
Westport, Connecticut • London

SWOSU At Sayre
Library

Library of Congress Cataloging-in-Publication Data

Krebs, Robert E., 1922–
The basics of earth science / Robert E. Krebs.
p. cm—(Basics of the hard sciences)
Includes bibliographical references and index.
ISBN 0–313–31930–8 (alk. paper)
1. Earth sciences. I. Title. II. Series.
QE26.2.K74 2003
550—dc21 2002032075

British Library Cataloguing in Publication Data is available.

Copyright © 2003 by Robert E. Krebs

All rights reserved. No portion of this book may be reproduced, by any process or technique, without the express written consent of the publisher.

Library of Congress Catalog Card Number: 2002032075
ISBN: 0–313–31930–8

First published in 2003

Greenwood Press, 88 Post Road West, Westport, CT 06881
An imprint of Greenwood Publishing Group, Inc.
www.greenwood.com

Printed in the United States of America

The paper used in this book complies with the Permanent Paper Standard issued by the National Information Standards Organization (Z39.48–1984).

10 9 8 7 6 5 4 3 2 1

Unless otherwise noted, all tables and figures are courtesy of Rae Déjur.

550
K871b

To my father, Leslie C. Krebs (1892–1974), professor of geography and nature study, who introduced me to earth science.

13463

Contents

Introduction

One of the most beautiful and informative photographs is *Earth Rise,* taken by a NASA spaceship while orbiting the moon. The image provides a scene that most of us will never see first hand—a view of Earth rising over the moon's horizon. More important, it tells us a great deal about our planet. The blueness, the cloud cover, and its spherical shape are very evident, making it difficult for modern-day believers in the flat earth theory to claim that Earth is not round. The planet Earth is very similar in many ways to its three sister planets (namely, Mercury, Mars, and Venus), insofar as the formation of their basic solid structures. All the planets came into existence ~4.6 billion years ago, after the nuclearsynthetic activation of the sun. As far as we know, however, the one significant difference between the Earth and the other planets in the solar system is Earth's biosphere—the area (sphere) where carbon/oxygen-based life exists.

The Earth (land) was sometimes referred to as Gaia, or Mother Earth, by ancient philosophers/scientists. The Greek noun for "land" is *ge* or *ga,* thus the goddess of Earth, who was born in chaos in the void of nothingness, was called Gaia. In Greek mythology, Gaia gave birth to the sea, mountains, and sky without the benefit of male involvement. More recently, James Lovelock, in his book *The Ages of Gaia,* developed the Gaia hypothesis as a concept postulating that the areas where life exists on Earth, namely, the biosphere (air, soil, rocks, and water), as well as all living matter, *is* one giant living entity and that Gaia exhibits a symbiotic (living together, sharing) relationship between the organic and inorganic constituents of the biosphere.[1] Although a series of complex forces, processes, and systems, including **feedback** mechanisms, continues to change the biosphere, as well as other concentric spheres of the Earth, this does not fulfill the requirements of a living organism, for example, reproduction. The Gaia hypothesis is not the current accepted geological theory. It has been replaced by a global unifying concept involving many forces, processes, and systems related to plate tectonics. These complexities are still not fully understood, and their concepts, including the importance of the biosphere to all life, challenge present-day scientists. As far as we know, our sister planets have no such biosphere and, thus, no life similar to what is found on Earth.

The Basics of Earth Science is organized around the forces responsible for the formation of the concentric spherical structure of the Earth. Although most planets exhibit a concentric spherical structure, there are significant differences between them and Earth. This reference book explores the Earth's forces and spheres and the multidisciplinary nature of the sciences that make up earth science in a manner appropriate for students of earth science as well as for the general public interested in learning more about our Earth.

Technical and important words that are not explained in the text appear in boldface type when first used and are included in the glossary. Simple projects, experiments, and exercises related to material presented in the text and suitable for students' use at home or school are presented in the Appendix. A selected bibliography and index also are included.

Abbreviations and Conventions Used in This Book

B.C.E. is used to denote dates for events "before the common era" or "before the Christian era."
C.E. represents dates during or after the "common era" or "Christian era."
c. (for circa) is used when an exact date is not known.
B.P. means years "before the present" time or year.
mya stands for "millions of years ago."
The symbol "~" is placed in front of an estimated quantity.

Background

Some Earth Facts

The Earth, of course, is not flat, even though when the surface is viewed up close or from space it may seem flat. The Earth was considered flat by ancient people, and during the Greek/Roman period (500 B.C.E. to 500 C.E.) it was thought to be a disk enclosed by several crystal-like hemispheres that contained the stars and so forth. During the Middle Ages (500–1500), Earth was assumed to be a perfect sphere. It is neither flat nor a perfect sphere. Instead, it is an oblate spheroid, or a globe that is somewhat flattened at its poles and bulging at its equator due to centrifugal force generated by Earth's rotation on its axis and the gravitational effects of the moon and sun. The distance measured around the Earth at its equatorial circumference is 40,075 kilometers (km) (24,902 miles [mi]), and the distance around the Earth's polar circumference is 40,008 km (24,860 mi). This difference of 67 km (42 mi) is about 4 times greater than the height of Mount Everest (8.8 km or about 11 mi).

- Earth's equatorial diameter is 12,756.3 km (7,926.6 mi).
- Its diameter from pole to pole is 12,713.6 km (7,900.1 mi).
- The difference between the equatorial and polar diameters is 26.5 mi, more than 4 times the highest landmass on Earth.
- Earth's mean distance from the sun is 149,500,000 km (92,86,000 mi). The Earth is closer to the sun in January than in June; therefore, an average distance is calculated.
- Its rotation period on its axis is 23 hr. 56 min. 4.1 sec. (one day).
- It is inclined ~23.5° to its equatorial plane (one cause for the Earth's four seasons).
- Its atmosphere is composed of 78.09% nitrogen, 20.94% oxygen, 0.03+% carbon dioxide, and less than 0.0001% of trace gases (neon, argon, ozone, etc.).

- Earth's land surface area is ~150,000,000 km^2 (~57,500,000 mi^2).
- The highest point above sea level is Mount Everest, 8,872 meters (m) (29,108 feet [ft] or 5.513 mi).
- Earth's water surface area is ~361,000,000 km^2 (~139,383,000 mi^2), which covers about 71% of its surface.
- Its greatest depth is the Mariana Trench in the Pacific, ~11,034 m (~36,200 ft), deeper than Mount Everest is high.
- Earth's mass is 5.976 $\times$ 10^{24} kilograms (37 million, billion, billion pounds).
- Its average density is 5.518 times the density of water (density varies).

History of the Spherical Nature of the Universe

The ancient Greek Pythagoras (c.580–c.500 B.C.E.) founded a brotherhood or cult of philosophers and mathematicians who greatly influenced western civilization. Today, they are best known for developing the Pythagorean theorem for determining the value of the **hypotenuse** of a right-angle triangle (i.e., $a^2 + b^2 = c^2$). They also conceived the idea that the sphere is a perfect circle and that all the planets, including the Earth, are perfect spheres. Philolaus of Tarentum (c.480–c.400 B.C.E.), a Pythagorean, was most likely the first person to express the idea that the Earth was not the center of the universe and that, like the other planets, Earth revolved around a central fire. As most philosopher/mathematicians of his day, Pythagoras believed that there must be a balance in all things, so he developed the concept of a "counter-earth," which enabled him to construct 10 spheres beyond the Earth's surface (10 being a sacred number) for all the moveable bodies in the heavens, including a sphere of stars.[2]

One of the Pythagorean students was Eudoxus of Cnidus (c.490–c.350 B.C.E.), who is sometimes known as the father of observational **cosmology**. He made an almost correct measurement of the length of a year as 365 days and 6 hours. In addition, he proposed that all heavenly bodies move in a series of concentric celestial spheres with the Earth—itself a solid sphere—at the center. He also observed some irregularities in the motions of planets, so he designed a separate celestial sphere for each known planet. His model had all 27 concentric spheres attached at their axes to a larger sphere that rotated. Later, he added more spheres to arrive at a total of 34.[3] Aristotle (384–322 B.C.E.), a student of Plato, was influenced by the early Pythagoreans' beliefs that the circle and sphere are perfect figures and, therefore, that the Earth, being a sphere, must be a perfect globe. This belief led Aristotle to construct an elaborate model of celestial spheres based on Eudoxus's model. Aristotle considered the Earth's air as a separate sphere (namely, atmosphere). He then continued outward into space with a series of concentric geocentric spheres based on their material composition of less and less densities.[4] Aristotle ended up with 55 concentric celestial spheres. Much later, in 1538 C.E., Giordano Fracastoro (1478–1553) designed a more elegant model of 77 nonintegrated spheres of the cosmos with the Earth still at the center. Aristotle's model was popular until the rise of the Copernican concept of a heliocentric solar system in the mid-sixteenth century.[5] The concept of celestial spheres carried over to the belief that the Earth is composed of layers or regions of matter consisting of differing densities.

A Short History of Earth Science

Modern earth science is a relatively young multidisciplinary science that uses overlapping areas of other, older sciences, principally geology, oceanography, meteorology, astronomy, geophysics, geochemistry, and geography. Each of these fields of study is subdivided into specialties.[6] In addition, earth scientists employ the basic sciences of physics, chemistry, biology, and mathematics.

In the early 1800s, William Smith (1769–1839), a blacksmith's son and a canal digger, traveled extensively in England, Scotland, and Wales. He also spent his life collecting fossils and observing the various strata of rocks and other outcroppings that resulted from his construction work. From his excavations he realized that rocks were arranged in layers and that fossils found in the various layers of rocks were different. He also believed that there was a relationship between the age of the fossils and the strata of rocks. Later in life he created the first geological map based on his many notes and observations. His unique, somewhat three-dimensional colored map depicts the underground geology rather than merely the topographical surface. Many consider Smith to be the father of modern earth science.[7]

By the late nineteenth century, after the publication of Smith's map, an individual scientist could not possibly be regarded as an expert in all of the very broad fields related to earth science. Therefore, a number of scientific specialties emerged, each concerned with some of the many aspects of the Earth and its inhabitants. Some of these specialties are given in the following list:

- *Geochemistry* is the study of the chemical composition of the Earth. Included in this field are mineralogy and petrology (the study of rocks).
- *Geophysics* is the study of the Earth's surface and its interior physical properties. It includes forces that affect structural geology and plate tectonics, as well as volcanoes and earthquakes.
- *Geomorphology* and geography are the study of the Earth's physical surfaces and landforms.
- *Mineralogy* is the study of the chemical and physical nature of minerals and rocks.
- *Stratigraphy* is the branch of geology that studies the layers or strata of rocks.
- *Seismology* is the study of the composition of the inner spheres by tracking the shock waves that travel through the Earth. These shock waves can be natural results of earthquakes or produced by explosive charges near the surface of the Earth.
- *Paleontology* is the study of plant and animal fossil remains, evolution, and Earth's history.
- *Magnetometry* is using a magnetometer to measure the Earth's magnetic field. Measurements at or beneath ground level record *vertical* magnetic intensities, while *total* intensities of the Earth's magnetic field are measured from the air and sea.
- *Oceanography* is the study of the seas and ocean floors.
- *Meteorology* is the study of the atmosphere, climate, weather, and their effects.
- *Economic geology* (and geography) is the study of Earth's resources, for example, mining and gas/petroleum geology.
- *Biology* is the study of the life forms, functions, and habitats of the Earth.

Earth science and, in particular, geology could not have developed as it did in the past 100 years without the invention and development of new instruments (e.g., computers, modern microscopes, chemical analytical instruments, seismographs) for observing, collecting data, and measuring many of the forces and phenomena related to the Earth. The second big twentieth-century boost for earth science came with the development of modern theories of earth dynamics, including the concepts of continental drift that led to the global geological theory of plate tectonics. The twenty-first century promises to offer even more knowledge and insight into the basic geology of the Earth and how our lives will ultimately be affected.

Notes

1. James Lovelock, *The Ages of Gaia: A Biography of Our Living Earth* (New York: Bantam Books, 1988).
2. Charles Singer, *A History of Scientific Ideas: From the Dawn of Man to the Twentieth Century* (New York: Barnes & Noble, 1959), 26–27.
3. Singer, *A History of Scientific Ideas,* 43–44.
4. Singer, *A History of Scientific Ideas,* 52–53.
5. Robert E. Krebs, *Scientific Development and Misconceptions through the Ages* (Westport, Conn.: Greenwood Press, 1999), 164.
6. Sybil P. Parker, ed., *McGraw-Hill Concise Encyclopedia of Science and Technology,* 3d ed. (New York: McGraw-Hill, 1994), 633.
7. Simon Winchester, *The Map That Changed The World: William Smith and the Birth of Modern Geology* (New York: Harper Collins, 2001).

Geological Time

A Short History of the Earth

How old is the Earth? How was it formed? How did the stuff that makes up the Earth get here? How is it different from its sister planets? Why is it structured in layers or spheres? What are some of its major features? What forces, processes, and systems are involved in Earth's well-being? When and how did life form on Earth? What is the importance of the biosphere? Do other planets have biospheres? If not, why not? What are some of the biological and social implications of human interaction with the biosphere? From the beginning of human existence, people have asked these questions and too often have provided mythological answers. Not until early scientists, mainly astronomers and geologists, learned how to both ask and answer questions for which empirical (observable) data could be gathered, measured, and analyzed did we begin to understand the nature of our Earth. Answers to the preceding questions, and many others, are provided in this book. For now, let's review a few facts about the Earth.

The age of the Earth is indirectly determined by several methods. The most recent is radiometric dating of meteorites that have landed on Earth that indicate an age of about 4.6 billion years. (The use of radioactive decay and other dating methods are discussed in a subsequent section.) Other systems for estimating the Earth's age were developed over the years. One is the **uniformitarianism** principle (also known as "gradualism"), which involves estimating changes over long periods of time with the assumption that scientific laws and geological processes and forces that operate today also operated in the past. This method is sometimes referred to as "The present is the key to the past."[1] One early example of its use was the attempt to determine the amount of mineral salts that are added to the oceans from land-based sources each year, then calculate the number of years required for the oceans of today to contain 3.45% salts (dissolved **ions**).[2] This is about the same salinity as two-thirds of a teaspoon of table salt dissolved in a glass of water. Historically, this method yielded a variety of ages for the Earth. One estimate for the age of the oceans was 1,000 million years. Others who used the salt method estimated the age of the Earth as 30 to 90 million years. All of these ages are underestimates, but they are more accurate than the estimates of 6,000 or 10,000 years based on biblical documents. Just as do all living things, the Earth gradually changes over long periods of time, enabling geologists to measure the present in relation to the past by establishing rates of change. Other systems use this concept, which works by calculating the rate of a specific, slow change that occurred in the past, followed by calculating the total amount of change, then dividing the latter (total) figure into the former. Using this method, it is possible to estimate the following: (1) the rate that soil particles form sediments in rivers and ocean beds and are in turn formed into rock layers; (2) the time of the arrival of fossils on the Earth (estimated to be 500 million years ago). A more accurate method utilizes the decay process (fission) based on the half-life of radioactive elements. Using several elements with different half-lives, scientists calculate the current figure for the Earth's age to be about 4.6 billion years.[3] (See subsequent sections in this chapter for information on dating methods.) It is generally accepted that the planets, including Earth, were formed shortly after the sun became a viable thermonuclear fusion star. By the process of accretion the leftover star material formed into approximate sphere-shaped globular bodies as they revolved around the sun. As they settled down to become planets, their gravity continued to attract meteors, comets, and other cosmic debris to their surfaces, a process that continues today. Alternatively, a similar process may have occurred when a massive star passed our sun, resulting in a streaming of material from these bodies that formed the planets as it proceeded to revolve about the sun.

When the chunks of matter, dust, and so forth were pulled together by gravity to form a new planet, this space debris was loosely held together and not completely solidified. Thus, as the Earth continued to spin on its axis, physical forces acted on this material in a manner similar to a cream separator (centrifuge). **Centripetal force** and gravity held heavy materials near the new planet's center, whereas the opposing **centrifugal force** pushed the lighter (in weight), less dense materials to form the outer layers of the geosphere. The inner, heavier materials formed the mantles and the denser metallic core. The same forces formed other planets, although they evolved differently from Earth for a number of reasons. Some were at a greater distance from the sun, and some were closer to the sun than the Earth. In both cases the development and support of life such as that found on Earth was impossible.

The Earth's atmosphere developed in several stages and was based on the availability of about a dozen elements derived from stellar masses. The most important were gases: hydrogen, helium, water (vapor), neon, ammonia, methane, hydrogen sulfide, and argon. Helium, neon, and argon are inactive and thus were not of major importance in the formation of the ancient atmosphere. Earth's gravitational pull was great enough to hold some of these gases, but not all.

The early motions of Earth were the same as several measurable movements that the Earth experiences today. The two major movements of the Earth that directly affect our lives are (1) its rotation on its axis that causes night and day, and (2) its movement as it revolves about the sun that, along with the Earth's inclination to its equatorial plane toward the sun (ecliptic), is responsible for the change of seasons. At the equator the speed of rotation is 1,600 kilometers (km) per hour (about 1,000 miles per hour [mph]). The velocity of the second motion as we revolve around the sun is 105,000 km per hour (65,247 mph). Three additional motions are less apparent. Earth has a slight wobble as it spins on its axis (**precession**), as well as another wobble as it proceeds in its orbit around the sun (**perturbation**) caused by the moon-earth mutual gravitational system. Both of these wobbles are partially the result of weak gravitational forces between the Earth and other nearby planets in the solar system. Recent research indicates that these slight changes in the Earth's motions may have long-term climate effects. A less noticeable motion is the relatively slow spinning of the Milky Way galaxy. Our entire solar system is located in one of the spiral arms of the Milky Way galaxy, *not* in its center. We are located about 30,000 light years to one side of the galaxy's center. It takes the Milky Way about 200 million years to complete one rotation.[4] Amazingly, we have little feeling or awareness of all these movements.

Basic physical laws and principles are universal. As far as we know, they have always affected the Earth, from its birth through its growth and expansion, along with the universe and all of its myriad forces that affect both the form and functions of our planet and all other celestial entities.

Definitions of Time

Most likely, our primitive ancestors needed only to react to daily and seasonal changes, without considering time as we know it. As civilization progressed, however, it was natural for people to try to find ways to track time as it related to changing events. This requires some sort of unitary standard and a means of measuring the passage of time. The most likely and easiest methods of estimating time were available in nature, such as the passing of the sun across the sky during daylight, phases of the moon, motion of the planets and stars, and changing seasons. All of these are caused by the natural motions of the Earth revolving around the sun and rotating on its axis. It was not until ~5000 to ~6000 years ago that Sumerian and Egyptian civilizations devised methods to account for the passage of extended time periods, such as the periodicities of the moon and the seasons, that is, calendars. They also invented ways to divide the passage of time into shorter periods, that is, clocks. It is assumed that the need for and ways of keeping time were related to either religious or bureaucratic activities and, of course, for agricultural purposes. There is a long history of the development of timekeeping methods and devices dating back to prehistoric days. Rather than dwelling on ancient time recording devices, let's look at what is meant by more technically accurate, scientific concepts of time.

Absolute Time

If people believe in the reality of motion, they must also believe in absolute time. Aristotle, Galileo, and Descartes all believed that if an object was already moving in a straight line, and that object was proceeding at a constant speed, then it would have to cover equal distances in equal units of time.[5] Sir Isaac Newton formalized this concept as one of his laws of motion. He also believed that time is absolute, in the sense that it is possible to measure the interval of time that passes between two different events. Moreover, this time interval would be the same for all people who repeated the measurements for the same events, assuming that they used the same or similar accurate measuring device, that is, a clock. Of course, the two measurements must also use the same units for the passage of time or have some means of converting one unit used to measure time to another and/or different unit.

We do know that it takes approximately 365.25 days for the Earth to circle the sun, and one day is the period required for the Earth to make a complete rotation on its axis. These, and the phases of the moon and longer periods of planetary cycles, were natural motions used as time standards by ancient people. But are there any natural units we can use to measure the passing of time during the length of one day? Unlike the revolution and rotation of the Earth, the decisions to divide one day into 24 units of time (hours), and the hour into 60 units (minutes), and the minute into 60 units (seconds), are somewhat arbitrary and are based on the ancient Babylonian numbering system. In ~1800 B.C.E. the Babylonians developed a system based on 60 possibly because 60 can be divided by 2, 3, 4, 5, 6, 10, 12, 15, and 30, thus eliminating the need for fractions. In addition, $6 \times 60 = 360$, which was used as the number of units (degrees) assigned to a circle and which also may have been related to what they believed were the number of days in a year. We still use this system for timekeeping.[6] Scientists searched until they found an atom with the appropriate oscillating frequency that could be used as an accurate standard for measuring time. (Note: The current standard for time is defined by the oscillations of the cesium atom, which is exactly 9,192,631,770 oscillations [cycles] per second. This was determined arbitrarily after it was established that a second was one-sixtieth of a minute.) If the ancient Sumerians and Babylonians had based their numbering system on 10 rather than 60, current daily and hourly time periods could just as easily be stated in metric units as 100 hours in a day, 100 minutes in a hour, and 100 seconds in a minute (or the units could be 10 hr/day, 10 min/hr, and 10 sec/min). Because we now have a very accurate definition for the second, the oscillations of the cesium atom, we can very accurately calculate the passing of absolute time.

Most of us still think of both time and space as absolute—as did scientists from Aristotle to Newton. That is a common-sense way of looking at things since our experiences are Earth-based, whereas time and space can be thought of as being independent of, but related to, each other. The concept of an ideal absolute time as a straight-line progression of time passing from then, to now, and on to the future does not answer such questions as When did time start? and What existed before that time? and If time is absolute, was there a beginning or is there an end to it, or is time (and space) infinite? The concepts of absolute time and space have changed and may not be as absolute as once believed.

Relative Time

If time is relative rather than absolute it can only be defined in terms of changes, but is there a time in the future when change, and thus time, stops? This is defined as a possible stable state for the universe, when entropy has reached a maximum of disorganization and objects can be neither hotter nor colder than anything else (**absolute zero**) and, thus, change and time have stopped.

Scientists began to understand that preexisting or absolute time did not explain the changes in time as an aspect of the relationships between many events, particularly events one would view in motion from different positions (perspectives) in the vast realm of space, as well as at the quantum level when viewing or measuring the very small spaces related to subnuclear particles. The concepts of absolute time and absolute space changed when the speed of light was determined. Scientists contemplated what would happen to time if and when an object with mass approached the speed of light. With the introduction of Einstein's theories of relativity the common-sense notion of Newtonian absolute time was changed forever. Just as with Einstein's concept of the relativity of motion, his space and time are also relative.

When one is on Earth doing earthly things, Newton's laws of motion still rule, as the senses of humans make it rational for us to consider absolute time and absolute space. For instance, we know that it takes *x* minutes to go from here to there on Earth and that we have covered an exact space from the start to the end of our movement in a specific time period. In the larger context of the universe, however, this absolute nature of time and space no longer applies. The theory of relativity refuted the concept of absolute time in the sense that each observer traveling in space must have his or her own measure of time as recorded by his or her own unique clocks. Observers moving relative to each other in space-time will assign different times and locations to the same event. Even though all these measurements are different, they are all correct, and any one observer can tell exactly, in time and space, when another's observations occurred—if their relative velocities are known. In other words, each observer has a unique frame of reference in both time and space. This may seem somewhat complicated because it does not involve ordinary, everyday experiences. A typical earthly example (but not a very good one) of relativity on Earth is when a train passes a station. An occupant in the train, from his or her unique frame of reference in space (point of view), sees the station moving past the train, while the person standing on the platform, from a different frame of reference, sees the train moving. Another oddity is that as velocity increases, not only does mass increase but time slows down. This relationship may be difficult to measure at the relatively low speeds attained during everyday life on Earth.[7]

Scientists have long accepted that time is not completely separated from an independent space (or distance) but is now a new entity called "space-time" and leads to a fourth dimension in cosmological measurements (e.g., height, depth, width, and now time). The use of the oscillations of the cesium atom as an agreed -upon standard for the second has enabled more exact measurements of distances in space. Distance can now be defined in terms of time and the speed of light, known as "light-seconds." Therefore, the meter is defined as the distance light travels in 0.000000003335640952 seconds. Even though this is the scientific definition of a meter, it is usually thought of as 100 centimeters or just slightly longer than a yard.

What does this have to do with earth science? Einstein's space-time makes no distinction between past and future; time can proceed in both directions. This relative space-time is naturally related to the origin of the universe and the cosmology in general and is an indispensable concept for modern scientific research of the cosmos. But for the study of the geological evolution of the Earth the use of an absolute, directional arrow of time is indispensable.

Chronological Time

Chronological time is linear absolute time. It organizes data as to the sequence of events over periods that have a beginning, continue or progress in a direction, and may at some point reach an end. This is often referred to as the "arrow of time," which can be presented as three arrows of time that describe the past □ to the present □ and into the future. These arrows are (1) the *thermodynamic* arrow, which describes the concept of entropy in the universe where order decreases and disorder increases until **equilibrium** is reached; (2) the *psychological* arrow, which is demonstrated by the fact that humans can remember and keep a record of the past but not the future; and (3) the *cosmological* arrow, which traces the origin of the universe and its expansion rather than its contraction. Scientists consider all three arrows of time as establishing a chronology of geological historical events into distinct periods for their study of the history of the Earth. A variety of linear chronological schemes are assigned to scientific, historic, and political periods. Some examples of these schemes are

1. *Astrological time,* which is based on the chronological arrow for celestial phenomena (events) and related scientific laws. Most of the dates for past astrological events are worked out backward using mathematical computations based on astronomical observations and records. The temperature just after the **Big Bang** is estimated to be a thousand million degrees Celsius. By measuring radiation of specific wavelengths (microwaves) from space and determining the residual heat in space left over from the Big Bang (2.73° above absolute zero cosmic microwave radiation), astrophysicists have determined that the origin of the universe was about 13 to 15 billion years ago. As the universe expanded it cooled, and as it continues to inflate it becomes less crowded, with energy spread out more in space, thus becoming less organized and cooler.[8] Astronomers are also able to describe what occurred in the first 3 minutes after the Big Bang and the formation of hydrogen and helium, followed by the formation of stars, heavier elements, and planets.[9]
2. *Geological time,* which covers many events and periods of the entire Earth's history from its origin about 10 billion years after the Big Bang, to the past 4.6 billion years of its history, to the present. In the past, geologists were somewhat handicapped in their ability to estimate accurately the chronology of the Earth's geology. Even today, some people believe that the Earth is only a few thousand years old and claim that humans were created in recent historic times, perhaps as recently as 6,000 years ago. Geological time tells a different story. Examining fossils and the sediment or rock strata where fossils were found has enabled geologists to develop a chronology of the Earth. By comparing different rock strata, fossils, and the relative distribution of various chemical isotopes, geologists were able to devise a time scale that was related to different age periods of the Earth. The discovery of radiometric and other techniques for dating fossils, minerals, and rocks led to an absolute time scale of the Earth. More information on geochronology will be presented in a subsequent section in this chapter.
3. *Archaeological time,* which is the chronology of ancient cultures established by observing, recording, and analyzing the differing layers of soil that contain human artifacts. Archaeological chronology is not based on records kept by humans but instead by relics of life. The science of stratigraphy is based on the "law of superposition," which is the principle that, if undisturbed, the younger, more recent strata of soil and rock will be located above the older and, thus historically earlier, layers. Archaeological and geological time and dating systems both use the stratigraphic method, and since both parallel each other, they also provide a means for comparison and increase the validity of their chronologies.
4. *Historical time,* which is also the chronology of politics that is determined by a particular sequence of events recorded by humans. Naturally, these time periods are of shorter duration than are astrological, geological, or archaeological chronologies. Historical time usually begins with evidence of written historical records and may be divided into arbitrary periods based on human interpretation of events. For instance, the Mesopotamian era, which includes several civilizations in the Middle East, was one of the first cultures to record events.[10]

Geochronology—Geological Time

Geology is the story of the Earth's age, and time is the narrator.

—author unknown

Introduction and History

Geochronology is the clock system used to date and uncover the age of rocks, meteorites, and fossils found on or near the Earth's surface. Ancient philosopher-scientists contemplated the origin, structure, and age of the Earth based on their observations of geological formations. In the fifth century B.C.E., Herodotus discovered shells in inland parts of North Africa. From these sea relics he surmised correctly that the Mediterranean Sea had extended south to cover the northern sections of Egypt and Libya. Eratosthenes, Aristotle, and other ancient Greeks made similar observations when they found fossils on elevated dry land. It wasn't until the eighteenth century that other theories actually connected these observed geological changes to the Earth's history. James Hutton (1726–1797), John Playfair (1748–1819), and William Smith (1769–1839) were among early geologists who supported the basic principles of superposition. All agreed with the premise that different strata of sediments and/or rocks were laid down over different periods of times, but they failed to fully develop their theory. It was Sir Charles Lyell (1797–1875) who explicitly stated that the same scientific laws and geological processes apply in the past, present, and future and, therefore, are responsible for geological changes. His theory was called "uniformitarianism." Lyell believed that erosion was responsible for the formation of strata (layers) of sediments and rocks. He also proposed that strata closest to Earth's surface that contained fossils similar to contemporary living things were the youngest layers, and that as one dug deeper the strata and fossils were older. He also divided rock formations into geological periods designated as Eocene, Miocene, and Pliocene, terms that are still in use today.[11]

Lyell's concept, now used to relate different geological formations from different places on Earth, provides a correlation between the history of specimens and geographic locations and, thus, a means to establish a geological time scale. This concept of similar geographic sediments and fossils was also applied to match up coastlines for continents that moved due to tectonic plate activity and continental drift (see chapter 2). There are really two different time scales that determine the age of the Earth. These are known as the chronostratigraphic and chronometric scales.

The chronostratigraphic scale is a relative scale based on Hutton's, Playfair's, Smith's, and Lyell's theories regarding the horizontal sequences of sediment. The early geologists asserted that upper, newer strata were formed over older strata. Therefore, the top strata were geologically much younger than underlying strata and rock layers. There are specific boundaries between the strata. Types of fossils located in the layers of sediment also indicated units of time. The relative order of the boundaries of the fossils essentially determines the major divisions of geological time. When the chronostratigraphic scale is referred to as being "relative," it does not mean relative in the same sense as Einstein's relativity. Instead, it infers that geologists can relate or correlate data for rock formations and horizontal strata and fossil boundaries from different geographic locations of the Earth to the same geological phenomena and time period.

The chronometric time scale is an absolute scale in the sense that direct measurements are made of both geological and biological specimens (rocks and fossils) to determine their ages.

Early geologists determined that if the age of the Earth was to be clearly understood, there was a need for a time scale that could, at least to some degree of accuracy, identify specific periods of the Earth's history. In other words, a scale was needed that could identify geological events that could be correlated with absolute historical data in human-related time. This was accomplished by separating both geological strata and biological boundaries into smaller and more accurately determined units of time. These units are, from the largest (and oldest) time spans to the shorter (and younger) units, **eons**, eras, periods, and epochs, as related to biological ages.

The Geological Time Scale

Figure 1.1 provides an example of the modern geological time scale that divides the age of the Earth into units relating its physical history of geology with the archaeological history of biology. The following outline is a summary of the major boundaries within the geological time scale and some distinguishing features between the geological and biological periods. Note that the time scale is organized from the time of Earth's origin to the present and that the abbreviation "mya" means "millions of years ago." If the meanings of the terms used by geologists are known, Figure 1.1 is easier to understand. The suffix "-zoic" refers to animal life. The prefix "paleo-" refers to the most ancient or oldest era, "meso-" means middle, and "ceno-" means the more recent era. Therefore, the term "Paleozoic" refers to an era of "ancient life." Also note that not all geologists use the same dates for their time scales. When a date is questionable, the abbreviation "c." (circa), which implies approximation, is used.

1. *Hadean era,* c.4,600–3,800 mya. This represents the time of the formation of the Earth. One theory states that about 4.6 billion years ago gases of a giant nebula started to spin and form the sun. This was followed by leftover smaller, then larger, particles being attracted to even larger rocks by mutual gravitational forces to form the planets by **accretion**. This was the period when the Earth was going through massive changes and before life evolved. The Hadean era is sometimes included along with the Archean and Proterozoic epochs as subdivisions of the Precambrian era
2. *Precambrian era,* c.3,800–570 mya. More than 80% of the Earth's existence is within the Precambrian era. Rocks from the Precambrian era that form most of the crust are old igneous and deformed metamorphic rocks. During this era, as well as in later periods, there was tremendous movement of the crust that resulted in folding and faulting. Not many fossils were formed at this time in history because most organisms were either single cells or small multicelled or other primitive kinds of life, mostly with soft bodies. Accurate time spans for both the Hadean and Precambrian eras were not determined until the development of **radiometric** dating techniques.

   A. *Archean epoch,* c.3,800–2,500 mya. This represents the age of volcanoes and when the first primitive unicellular (single-celled) organisms appeared. Prokaryote bacterial cells lacking a membrane-bounded nuclei were the first living cells on Earth, followed by protista, which are unicellular and early multicellular organisms that contained a defined nucleus. This was also the age of the oldest microfossils.

   B. *Proterozoic epoch,* c.2,500–544 mya. This was the age of early marine life, algae, and bacteria. It represents development of the first complex multicellular organisms and marine life. Fossils from this epoch indicate that some examples of early life were not yet divided into representatives of the plant or animal kingdoms.

Figure 1.1
Geological Time Scale. Note: Not all charts depicting the geological time scale follow exactly the same divisions for time periods.

GEOLOGICAL TIME

TIME (of beginning) in millions of years ago	ERA	PERIOD	EPOCH	MAIN EVENTS
4500	Hadean			Formation of Earth
3800	Precambrian		Archaean	Earliest life: algae, bacteria
2500			Proterozoic	Early marine animals
	Phanerozoic			
544	Paleozoic	Cambrian		Marine animals
505		Ordovician		Early fishes
440		Silurian		Land plants and animals
410		Devonian		Amphibians
360		Carboniferous		Amphibians, reptiles insects, club mosses horsetails, ferns
286		Permian		Mammal-like reptiles
245	Mesozoic	Triassic		Early mammals, cycads, conifers, dinosaurs
208		Jurassic		Height of dinosaurs, birds
146		Cretaceous		Flowering plants, bees
65	Cenozoic		Paleocene	Mammals, extinction of dinosaurs
54			Eocene	Early horses
38			Oligocene	Grasses, grazing mammals
23			Miocene	Increase in mammals
5		Tertiary	Pliocene	Late in era hominids
1.8		Quaternary	Pleistocene	*Homo sapiens*
11,000 (not in millions)			Holocene	Human civilizations

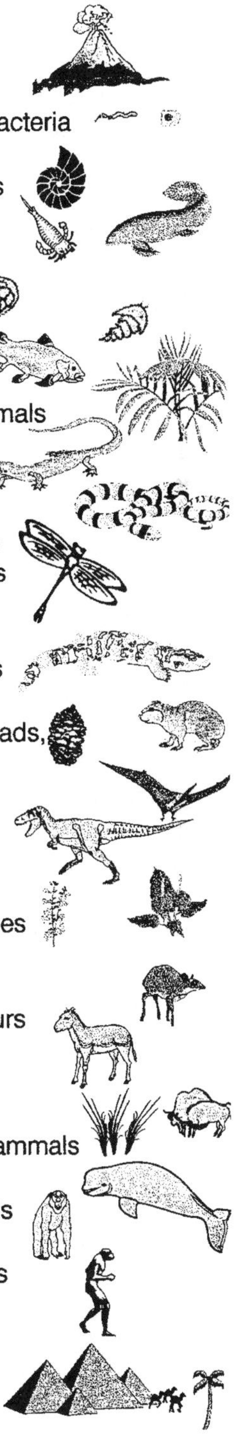

3. *Phanerozoic eon,* c.544 mya to the present. This comprises rocks younger and less complex than those in the Precambrian era. The term "eon" means "visible life" in Greek, which aptly describes the division between the Precambrian and Phanerozoic eons. Unlike the older eon, the Phanerozoic eon is characterized by many similar types of rocks spread throughout the world that are rich in fossils of various extinct life forms.

The Phanerozoic eon is divided into three main age-related units and represents the shortest eon, time-wise. Its three main divisions include many subunits of shorter spans of time that, in turn, are divided into many time-related sub-subunits that are referred to as "periods." The geological time scale can be further subdivided by geologists into shorter and shorter time spans to include epochs according to types of rocks, as well as evolutionary ages established by biologists for ancient plants and animals. The following subunits for the Phanerozoic eon are consolidated for clarity.

A. *Paleozoic era,* c.570–245 mya (ancient life). This era is divided into six main periods representing a span of about 325 million years. During this relatively short period of time the evolution of animals progressed rapidly from simple multicellular forms to more complex marine organisms, including primitive fish, amphibians, reptiles, insects, and small mammals. Plants also evolved from simple plants to ferns and mosses.

(1) *Cambrian period,* c.570–505 mya. The Cambrian period was named after *Cambria,* the Roman name for an area in Wales, where complex forms of life were first found in rocks of this age. This is the period when the first marine shell animals appeared.

(2) *Ordovician period,* c.505–440 mya. The Ordovician period was named after a Celtic tribe called the *Ordovices.* It is sometimes referred to as the age when marine invertebrates evolved. This is also the period when the first fish appeared.

(3) *Silurian period,* c.440–410 mya. The Silurian period is possibly named after a Celtic tribe known as the *Silures* and represents a classification of rocks from this age. This is also the age of simple fish, and fossils of primitive land plants and animals were found in rocks dating from this period.

(4) *Devonian period,* c.410–360 mya. This is the age of more complex fish. It is named after Devonshire, England, where the rocks from this period were first identified. There is some fossil evidence that the very first primitive amphibians appeared about this time.

(5) *Carboniferous period,* c.360–280 mya. The Carboniferous is considered to be a single period in Europe. In the United States, however, this period is divided into lower and upper periods represented as the Mississippian (c.360–320 mya), characterized by a corresponding system of well-exposed rocks found in the Mississippi Valley, and the Pennsylvanian (c.320–280 mya), which consists of somewhat younger rocks and coal found in the state of Pennsylvania. This is the age of reptiles, amphibians, insects, and club mosses, and the period during which (carboniferous) coal was formed.

(6) *Permian period,* c.280–245 mya. This period is named after the province of Perm, in the former Soviet Union, where rocks of this age were first located. The span of years for the Permian period varies greatly among geologists primarily because it is an intermediate period between the older Carboniferous and younger Triassic periods that are also the division between two major eras, the older Paleozoic and the more recent Mesozoic. The Permian period has its own corre-

sponding system of rocks. While it is the end of the age of the trilobites, extinct marine arthropods, it is also the period of continued evolution of amphibians and reptiles.

B. *Mesozoic era,* c.245–65 mya (middle life). This era consists of three periods in a unique geological time span that separates the older Paleozoic period from the more recent Cenozoic period. These three periods, the Triassic, Jurassic, and Cretaceous, represent a time when metamorphic rocks were formed at an intermediate depth at temperatures of about 300–500°C. The rocks also exhibit some shearing due to stress. This is the era that provides many fossils of reptiles and more recent invertebrates.

(1) *Triassic period,* c.245–210 mya (some geologists use a different time span for this period, from c.225 to 190 mya). The name for this period was derived from the word "*tries*" due to the threefold division of a particular formation of rocks found in Germany. It is the age of the first (early) mammals, conifers (pine) trees, and reptiles (dinosaurs).

(2) *Jurassic period,* c.210–145 mya (or c.190–135 mya). This period is named after a corresponding system of rocks found in the Jura Mountains located on the border between France and Switzerland. The Jurassic period is the age when dinosaurs ruled and birds and ferns evolved. Today, it is best known for the fictional book and movie *Jurassic Park,* written by Michael Crichton, where cloned dinosaurs rampaged out of control.

(3) *Cretaceous period,* c.145–65 mya. This period received its name from the Latin word *creta,* meaning "chalk." Chalk rock beds are found in England where **outcroppings** of chalk formed the "White Cliffs of Dover." This is the age of flowering plants and the evolution of insects, including bees and other species that formed colonies.

C. *Cenozoic era,* c.65 mya to the present day (recent life). The Cenozoic is the youngest era on the geological time scale. It consists of two main periods, the Tertiary and the Quaternary. Since it is younger, geologists and other scientists have been able to learn more about the boundary layers of sediment, rocks, fossils, and more recent organisms than those of other periods. Therefore, these periods are divided into shorter time spans called "epochs." More modern-type mammals and other vertebrates, as well as more modern plant fossils, are found in strata representing this era.

(1) *Tertiary period,* c.65–2 mya. The Tertiary is divided into two subperiods.

(a) *Paleogene subperiod,* c.65–25 mya. The Paleogene is again divided into three epochs, that is, the *Paleocene, Eocene,* and *Oligocene.* These epochs, from the oldest to the youngest, represent events that span the extinction of the dinosaurs, about 65 mya, and the evolution of larger grazing mammals, primates, early horses, and hominids.

(b) *Neogene subperiod,* c.25–2 mya. The Neogene is divided into two epochs, the *Miocene* and the *Pliocene.* These epochs represent more recent sediment and rock formations and an increase in both the number and size of mammals (whales). These two epochs also represent the evolution of hominids to premodern *Homo sapiens.*

(2) *Quaternary period,* c.2 mya (or c.1.8 mya). The Quaternary is divided into two uneven (time-wise) epochs.

(a) The *Pleistocene epoch* covers the period beginning about 2 or more million years ago to just 8,000–10,000 years ago. The name also represents the corresponding series of minerals, rocks, and other deposits formed during this epoch. It is the time during which *Homo sapiens* migrated out of Africa to Asia, the Middle East, and later to Europe. The later part of this epoch encompassed the end of the last great Ice Age.

(b) The *Holocene epoch* covers the most recent time span from about 8–10 thousand years ago to the present time. The Egyptian pyramids were built midway through this epoch, and it is the time when human civilization, as we know it, developed.[12–17]

Geological Dating Methods

For many years, geologists and other scientists pursued methods and processes to measure the age of the Earth. Many realized the various strata or layers of sediments, minerals, and rocks are exposed as outcroppings as the Earth heaved and settled over many thousands of millennia. Some of the early techniques such as reading the growth rings of the cross-section cuts of trees are somewhat useful for determining climate changes, but they only indicated periods of a few hundred to a few thousand years of the Earth's history. More accurate techniques have since been devised to read further back into geological time, so that events can be arranged in the order of their occurrences (age) and for specific spans of time. Modern geologists have combined some older concepts into a relative time scale, while developing newer techniques that provide an absolute time scale.

Relative Time Scale

The relative time scale (chronostratigraphic) is based on the sequence of horizontal layering of sediments and rocks with the more recent strata laid down on top of older sediments. From geological studies of the origins of rocks (petrology) and the layering into strata (stratigraphy) of minerals and rocks, it was possible to determine the sequence of events that shaped the outer sphere of the Earth. Geologists noted that land was raised to form mountains and plateaus, while at other locations the land subsided, and what was once dry land was now covered by seas. By studying the layering of sediments and rock formations, particularly outcroppings found in different geographical locations, geologists divided Earth's history into relative time periods. The result was a broadband historical time scale comprised of geological phenomena that covered only a few millions of years and with only rough estimates of the time/event boundaries between broad historical periods. Over the past 200 years, this method of relating similar events and different geological phenomena to different regions led to new geological knowledge. But at the same time, this technique was not exact enough for understanding the significance of the timing of past geological events.

Absolute Time Scale

A more absolute time scale (chronometric) involves direct measurements between sediments, rocks, and biological artifacts and events. It is based on specific units of duration and on the numerical ages assigned to boundaries between different events. Although it is more precise than relative time determination for the ages of the Earth, it continues to be developed as a science, and improvements tend to fine tune published geological time scale charts such as presented in Figure 1.1.

The following list describes some of the older absolute methods.

1. *Dendrochronology.* The reading of tree rings provides a dating method for events and

conditions over a limited period of time. It is based on the number, width, and density of annual rings of older trees that have been cut into cross sections. Scientists used old Douglas fir and white pine trees to establish a master tree index that enables experts using this method to date, rather accurately, both events and climatic conditions over the past 3–4 thousand years.

2. *Varve Analysis.* The term "varve" comes from the Swedish word *vary,* which means "layer." There is a natural tendency for fine thin layers of sediment to form on bottoms of slow-flowing lakes fed by glacial melt. In the summer, fine light-colored silt forms on these lakebeds, followed by a thinner winter layer of more dark-colored, claylike organic deposits. These layers can be analyzed to determine the age of glacial deposits of the Pleistocene epoch (10,000 years ago to 2 mya). By comparing the rate of sedimentation to the number of units of sediment deposited per year, a rather accurate age can be established for some geological glacial events.
3. *Obsidian Hydration Dating.* This method is based on the natural process of water vapor slowly diffusing into the freshly exposed surface of obsidian rock. Obsidian is a glasslike, black to dark-colored material produced by the high heat of volcanic activity. Historically, its shiny, glassy surface made it popular as jewelry. It also fractures rather easily, leaving sharp edges, which made it an excellent cutting tool for early humans. When the fresh surface of chipped obsidian is exposed to water vapor, the hydration process results in formation of a crust (referred to as the "hydration rind") over the exposed surface. The age of the chipped obsidian can be ascertained by determining the thickness of the rind on an ancient tool or for a broken piece of volcanic obsidian glass. This method can be used to date chipped or broken obsidian up to 200,000 years.

Radiometric Dating

Radiometric dating is the most recently developed and accurate absolute dating technology developed to measure geological ages. It makes use of the concept of the natural decay of radioactive elements and their isotopes. The constant half-life for the decay of fissionable elements, when used for dating techniques, is sometimes referred to as an "atomic clock." Uranium, the first radioactive element to be identified, was serendipitously discovered in 1896 by Henry Becquerel (1852–1908) when he accidentally stored some uranium salts on top of an unexposed photo plate. When he developed the photo plate, he saw the image formed by the chunk of uranium ore. Marie Curie (1867–1934) coined the term "radioactive," and in 1898 she and her husband, Pierre Curie (1859–1906), announced their discovery of polonium and radium, both extremely radioactive. But it was not until the 1950s that precise radiometric dating was perfected.

The heavier atoms of elements with higher numbers of **neutrons** also have greater atomic masses (atomic weights), the result of which is that some of their nuclei are unstable. Unstable atoms and **isotopes** can actually break down (transmutate) into more stable atoms of elements with lower atomic weights. During this process, these transmuting atoms emit several types of short wavelength electromagnetic radiation. Alpha rays (helium nuclei) are less penetrating than the other two basic forms of radioactivity; beta rays (high energy electrons) and gamma rays (similar to high energy x-rays and cosmic rays, and have the highest energy, shortest wavelength, and are the most penetrating of all). In 1903, Ernest Rutherford (1871–1937) and Frederick Soddy (1877–1956) proposed the theory that atoms of radioactive elements break down (transmutate) in successive steps in differing and random amounts of time. They also determined that although the process of radioactive decay for individual atoms is random, it is controlled by the average time

Table 1.1
Examples of Elements Used for Radiometric Dating Methods

Radioactive Parent	Half-Life of Parent	Stable Daughter
Samarium-147	106 billion years	Neodymium-143
Rubidium-87	48.8 billion years	Strontium-87
Thorium-232	14 billion years	Lead-208
Uranium-238	4.5 billion years	Lead-206
Potassium-40	1.25 billion years	Argon-40
Uranium-235	705 million years	Lead-207
Carbon-14	5,730 years	Nitrogen-14
Hydrogen-3	12.3 years	Helium-3

in which just half of the element's atoms would be expected to decay. Rutherford coined the word "half-life" to explain the phenomenon that only half of the radioactive atoms (randomly) decay over a specified period of time, and then half of what remains decays in the next period of time, and so on, until all of the radioactive element has been transmuted into more stable elements with lower atomic weights. Most radioactive isotopes have a very short half-life, some less than a microsecond, but others have a half-life of billions of years. Therefore, many of the original elements formed at the early stages of Earth's evolution that have a very long half-life are, thus, still found on the Earth. This concept of half-life was later developed into a fine tuned, very accurate atomic clock for dating artifacts. Radioactive isotopes of some elements, even in elements with lower atomic weights than unstable uranium, exist in extremely small ratios to the major, more abundant stable isotopes of the same element. If this ratio and/or the half-life of this less abundant isotope is known, and is compared to the abundance of the regular atoms found in minerals and ancient artifacts, it can be used as a timing clock.

This process also can be explained as the amount of the stable so-called daughter isotope produced as a result of radioactive decay that is divided by the amount of the unstable so-called parent radioactive isotope. Thus, if the half-life of the parent is known, the age of the sample can be determined. Examples of parent and daughter isotopes, as shown in Table 1.1, may assist in understanding the power of using the half-life of certain elements to determine the age of ancient rocks.

A number of additional radiometric dating techniques were developed in the latter half of the twentieth century. Four examples are given in the following list.

1. The *lead-alpha particle method* uses a **spectrometer** to find the total lead content related to the alpha-particle radiation emitted by either uranium-235 or thorium-232 concentrations as compared with their daughters' lead products. This method can be used to estimate the age of fossils, minerals, and strata of sediments and rocks. The half-life of these isotopes provides accurate dates for material found in the Precambrian era (c.3800–570 mya). Using uranium-238/lead-206 for the alpha decay technique, or rubidium-87/strontium-87 for the beta decay technique, rocks can be dated back to almost the beginning of the Earth, 4.6 billion years ago.
2. The *uranium-thorium reaction,* a parent-to-daughter decay reaction, is the current stan-

dard technique for geological radioactive dating. It uses the concept of the layering of deep-sea sediments on the ocean floor, as well as the fact that seawater dissolves elements that were incorporated into ancient shells and coral. The age of shells and coral can be estimated by the content of this isotope when compared with the different layers of their growth and sediments. This system can determine the age of sediments and fossils up to about 300,000 years.

3. The *rubidium-strontium technique* is used to date igneous and metamorphic rocks found on both the Earth and the Moon. It compares the ratio of the beta decay of rubidium-87 to strontium-87. It is also used to confirm dates established by the potassium-argon radiometric dating methods. The samarium-147 to neodymium-143 method is also used to date Precambrian rocks, as is the lead-alpha particle method.
4. The *radiocarbon-14 dating method* was developed in 1947 by Willard E. Libby (1908–1980) and is considered a reliable technique for determining the age of organic artifacts as far back as 50,000 and sometimes to 70,000 years in the past. In addition to being a useful tool for geology, petrology, oceanography, and climatology, it is also used in the fields of archaeology and anthropology. This method is based on the metabolic uptake (food) of regular carbon-12, which also contains a small portion of the radioactive isotope carbon-14 (^{14}C). Upon an organism's death, the radioactive ^{14}C begins to decay at a specific rate, but with a rather short half-life compared to uranium. Since the radioactive ^{14}C in the dead organic matter is no longer replaced with atmospheric ^{14}C, it can be compared with the stable ^{12}C to determine the date of death of the organism. One of the problems of using ^{14}C as a dating method is that the production of this isotope of carbon in the atmosphere has not been constant over the centuries. It seems that cosmic rays and the Earth's magnetic field affect the rate of production of ^{14}C. Even so, it is a useful method because the error only varies between 1% and 4%, making it about as accurate as other short-time-span dating methods.[18–22]

Radiometric methods are constantly being improved while new techniques are being developed. The fission-track method uses the damage caused to minerals by radiation tracks from ^{238}U; thus, the calculation of the age by spontaneous fission-track density can be made. Also, other radioactive elements are used to assist in estimating the age of geological and biological specimens. A listing of a few samples and their ages as determined by radiometric dating methods is provided in Table 1.2.

Fossils and Evolution

By using radiometric clocks, geologists have determined that the Earth is about 4.6 billion years old. The Precambrian era (3,800–570 mya) is the largest span of time (80%) of all of Earth's geological history. Rocks deposited since that time in the Phanerozoic eon (570 mya to today) are correlated with fossils, as well as with other geological events. Even so, exact time span divisions between boundaries of geological events are not always possible. By augmenting radiometric methods with a study of fossils and the evolution of life, geologists have greatly improved their ability to measure the time span of events and increase our understanding of the geology of the Earth over the past 500 million years.

As mentioned previously, the Earth consists of a complex of spheres and interdependent chemical and physical systems continually in a state of flux. The progression of living organisms is also a complexity of systems in a state of flux and is now referred to as "evolution." In a sense, the geology of the Earth might be referred to as evolution as well—if we consider evolution a process of change. Both proceed over

Table 1.2
Examples of Materials Dated by Radiometric Atomic Clock Methods

Sample	Source	Approx. Age in Years
Cloth	Mummy from Egyptian pyramid, (C-14)	3,000 +
Charcoal	Oldest inhabited site in USA, (C-14)	10,130
Wood	Forest bed near Milwaukee, Wis., (C-14)	11,640
Tuff	Pyroclasitc rocks, pumice	700,000
Ash	African volcanic ash, hominids	1,750,000
Monzonite	Copper ore in Utah	37,500,000
Granite	White Mts. of New Hampshire	180,000,000
Granite	Pike's Peak of Colorado	1,030,000,000
Gneiss	Finland and Baltic region	2,700,000,000
Granite	South Africa	3,200,000,000
Granite	Oldest in USA, Minnesota	3,600,000,000

much greater time periods than that of a human lifetime. Both inorganic and organic evolution are slow processes, with intermittent relatively rapid periods of change. Geologists correlate the progress and products of these two forms of evolution when constructing the geological/biological time scale of the Earth and life. The two types of evolution augment each other as a means of assisting in the determination of the boundaries of geological and biological time spans of the past 3.5 billion or more years, with particular emphasis over the past 2 million years or so (see Figure 1.1).

Definitions and History of Fossils

The Earth's crusts are composed mainly of plutonic and metamorphic rocks, with an overlay of sediments created by the aggregation of weathered and eroded surface materials. Internal forces create outcroppings and cliffs that often contain the preserved remains of former living plants and animals that became known as fossils. As we understand the nature of a fossil today, it is evidence indicative of an organism having lived more than about 10,000 years ago, and it is remains that are either the original or a replacement of the original. Fossilization occurs when organic matter of the original specimen, particularly shells and some bones, is replaced by calcium carbonate or silica from ground water. For instance, structures of both bone and wood can, over long periods of time, be replaced with molecules of minerals, resulting in a stonelike replica of the original structure. An imprint, track, or trail of the original living organism made in mud or other substance may also be preserved in the same fashion and become a fossil of the original impression made by a living organism.

Ancient people observed and were, no doubt, puzzled by the fossilized remains of marine and land animals and plants found in broken and exposed rocks. They may have been even more puzzled by finding shells of sea animals on high ground far from the oceans. These quandaries resulted in a number of ancient beliefs and theories concerning the deposition of fossil imprints of plants and seashells in rocks found high in mountains and on dry land far from the oceans. At one time it was proposed that fossils were natural carvings in stones or that they were deposited during the biblical Flood. One theory suggested that fossils are God's way of testing out different forms of life, while another alleged that fossils were the works of either devils or gods who placed them where they were found to confuse people.[23]

During the European Middle Ages, early scholars studied the various strata of the Earth and fossils deposited in the different layers of rocks. Even during this somewhat unenlightened time, they noticed that there were significant breaks in any continuous relationship between the strata and types of fossils. From the fourteenth through the sixteenth centuries several naturalists suggested that fossils were actually preserved remains of living organisms. Nicolaus Steno (1638–1686), an early Dutch geologist, correctly connected the existence of fossils, as once-living organisms, to the different layers of rocks that represented different periods of time. Once humans realized that the layers of exposed rock, even on mountains, represented different ages of the Earth's history, they were able to establish the ages of the plant and animal fossils found in these different layers, since it followed that both the organisms and rock layers must have existed during the same geological periods. These layered outcroppings of surface features and their related fossils were among the rationales used to develop the science of organic evolution.

Georges Cuvier (1769–1832), known as the father of paleontology, first classified fossils he found in 1811 or 1812 in Tertiary rocks near Paris, France, as fossil mammals and reptiles. He also developed the technique of comparative anatomy that allows the reconstruction of entire animals of an extinct species using only a few of their bones, and he was also the first to demonstrate that fossils of terrestrial vertebrates had no living counterparts. When he compared this information to different strata of rocks where these particular fossils were found, he concluded that biological succession was punctuated by numerous extinctions (catastrophic interruptions), followed by renewal of new, more advanced species. This theory is the opposite of Darwin's gradual evolution. Cuvier's concept of **catastrophism**, catastrophic events driving evolution, was partly revived in 1972 as "punctuated equilibrium" by the paleontologists Niles Eldredge and Stephen Jay Gould.[24] The debate concerning whether evolution is gradual or punctuated continues due to the imperfect fossil/strata records, but there is no serious scientific dispute as to the role of genetic mutation and environmentally forced natural selection in organic evolution. There are questions concerning the gaps in the geological/fossil history. There is also a question of the fossil evidence related to the evolution of separate (new) species that are capable of interbreeding (macroevolution) and the evolution of the morphology related to changes in body forms and structures within species (microevolution). Regardless of the incomplete geological/fossil record, there exists a rather complete geochronology of Earth and evidence for organic evolution.[25–28]

Summary

Geochronology is the science that uses time-based relationships to determine the age of rocks, strata, and periods of the Earth, as well as the types of organisms that existed during different times in Earth's past. By reading the radiometric records of past geological events and the characteristics of the preserved fossils related to those events, geologists have created a reasonable time scale for both the formation of rocks and the ages of extinct species of plants and animals. Originally, the geological time scale (see Figure 1.1) was based on the relative ages of outcroppings of sedimentary rock strata and the fossil content found in different geographic locations. Today, improved relative and absolute dating techniques continue to be developed and used to fine tune the geological time scale. Radiometric dating is one of the most useful techniques used to determine the ages of both the rocks and the fossils in which they are found. The fields of both geology and evolutionary biology, as well as other fields, benefit from advances in geochronology that provide a more accurate time scale for the evolution of our Earth and life upon it. Geologists and biologists no longer need to rely on just a relative geological time scale based on superpositional evidence with its many gaps and inaccuracies. Today, they use radiometric methods that provide a numerically quantified absolute temporal scale.

Issues

Earth's Age

Many people are confused when they regard the meaning of the scientific concept of the relative or relativity when addressing geological concepts for the age of the Earth. Scientists use both the theory of scientific space-time relativity, which is based on perspectives of the observers, and the statistical concept of probability for correlating attributes of two or more events or objects. In an ideological/philosophical sense or in social/political context the term "relative" is sometimes used to justify a so-called deconstructive approach to science and knowledge, including knowledge about the evolution of the Earth and its life forms. Since every human has his or her own perspective on things, everyone has his or her own truths. Some people tend to use the space-time concept of relativity to justify their beliefs of moral, ethical, and behavioral relativism. This bastardization of the use of the word "relativity" in Einstein's theories of special and/or general relativity indicates a lack of understanding of modern science. Likewise, the current use of the term "quantum leap" to mean a giant step or shift has nothing to do with **quantum theory**, which deals with extremely small quantitative mathematical interpretations of natural phenomena.

No doubt, early humans wondered at what they saw when viewing the heavens and nature around them. Their observations of both the space and the time in which they lived were limited geographically and intellectually. But they did observe that there were patterns and sequences for certain events, such as night and day, floods and drought, earthquakes and volcanoes, and so forth. Regardless of their fears or wishes, these events occurred with regularity for no apparent reason, so it was quite reasonable to believe that some unknown superpower, god, or spirit was responsible for these natural phenomena. It is also only natural to try to invent explanations for unexplained or nonunderstandable events. The Greeks, Romans, and people from other earlier civilizations believed in a great many gods, all of whom exerted or controlled the power of the good or bad events that occurred in their lives. Many of these ancient myths were incorporated into modern religions and have become part of written and observed doctrines. Some of these doctrines are based on the age of the Earth in a form of mythological dogma (e.g., Genesis) or hypothetical geological regions of heaven and hell. For instance, one method used by some religious groups to determine the Earth's age is to count backward the generations of humans described in the Judeo-Christian bibles. This is referred to as the "begat" method. By counting backward the generations listed in biblical texts (so-and-so begat so-and-so, etc.), and assigning a numerical figure to the length of a single generation, it is possible to arrive at a date for the beginning of the

Earth at about 8,000 to 10,000 years ago. This date also seems to coincide with the concept of Genesis and the creation of Adam and Eve as the first humans.

Even with the development of the processes and procedures used for scientific investigations and with the knowledge accumulated over the past several hundred years, many continue to reject the results of geological, archaeological, and anthropological research. The geological time scale is based on a rather complete and accurate age record of the Earth and is not dependent on ancient or modern myths.

Change

The ancient Greek philosopher Heraclitus of Ephesus (c.500 B.C.E.) claimed that everything is in a state of continued and constant change and that nothing remains the same. His famous saying that expresses this philosophy, "A man cannot step into the same river twice," means that both the river and the man have changed after stepping into the water the first time. His contemporaries referred to him as the "Weeping Philosopher" because they considered his philosophy of constant change as a pessimistic view of the world.[29] For centuries, people realized that plants and animals were born, lived a limited period of time, and died. But they did not consider this as change or evolution because the new offspring were basically unchanged from the parents. Many years later and after countless observations of diverse living species and fossils in various geographic locations, the systematic evolution of living organisms became a viable theory. Today, mainstream scientists accept the theory of the evolution of living organisms and natural selection of those best able to survive to reproduce in response to changing geological/biological environments. A similar history applies to the physical changes of the Earth over eons of time.

The concept of creating something out of nothing may have originated with prehistoric people, but it was not rationalized until the ancient Greeks and Eastern civilizations expressed the concept of the Earth and life being created out of some sort of chaotic process. However, this is not what the modern creationism/evolution controversy is about. Today, the debate on the origin, age, and evolution of the Earth revolves around nonscientific fundamental theological explanations of creation as revealed by the literal interpretation of Christian scripture versus the empirical information regarding evolution attained by the application of scientific processes.

There is some controversy over the differences between *micro*evolution and *macro*evolution, both for geology and biology. Most people can accept microevolution, which is defined as a slow genetic change occurring within a population of a species that resulted from a progressive (slow) change in the environment. In other words, they believe evolution to be species specific. It is also possible to observe geological microevolution as volcanoes, earthquakes, storms, and erosion that change surface features of the Earth. For many people, problems arise when they cannot accept that macroevolution involves large-scale changes of both the structure of the Earth and its living organisms over long periods of geological time. This is particularly true for plant and animal speciation that involves the formation of entirely new species through processes of organic evolution. The concept of "convergent evolution" involves species evolving with similar forms, generally due to their way of life in diverse environments.[30] Darwin's concept of "transitional forms" theorizes that some plants and animals contain features from two or more different types of species, thus making it difficult to classify them. Darwin explained the lack of evidence for transitional forms on yet-to-be-discovered species or on fossils that were destroyed many years ago. An example of a recently discovered transitional fossil form is *Archaeopteryx* ("ancient wing") that lived in Bavaria about 163 to 144 mya in the late Jurassic period. It appears to be a cross between a dinosaur and a bird because it possesses a dinosaur's bony tail and mouth with teeth, as well as birdlike features. Other transitional-form fossils have since been found, mainly species with characteristics of both reptiles and mammals. Still, there are opponents to these special theories of evolution. It is important to recognize that Darwin and most current scientists agree that both fossil and geological records are, and always will be, incomplete.[31]

There is overwhelming evidence from the study of fossils, geology, archaeology, anthropology, comparative anatomy, and physiology of living organisms—past and present—that evolution is a natural process of change in both the geological and biological sense. Evidence also exists that more than 99% of all species of plants and animals that ever lived on Earth are now extinct. Over the ages new species have evolved and, in turn, become extinct as new land areas were formed. According to this theory, all species presently existing on Earth, including *H. sapiens,* will at some time in the future become extinct. There is also recent DNA evidence that new species are constantly being formed from existing species. Humans share about 98% of the same DNA as do some other primates, and all living animals (and plants) have some DNA in common. Over the past 4.6 billion years forces have created new geological structures and land forms, just as over approximately the past 3.5 billion years forces have created new life structures and forms. A variety of forces and processes are responsible for the evolution of the physical Earth as well as living organisms that either adapt to change or die out as Earth's natural geological systems and physical environments change. There is also evidence that natural selection in Earth's past geological/biological history was not always steady and that, at times, evolution was either slowed or accelerated by catastrophic astrological or geological events. Thus, microevolution and macroevolution for both the physical and biological aspects of Earth are based on change as natural selection results from geological evolution. Scientific empirical/factual evidence firmly establishes evolution as a viable theory.[32–35]

Notes

1. Graham R. Thompson and Jonathan Turk, *Earth Science and the Environment* (New York: Harcourt Brace, 1999), G.18.

2. Thompson and Turk, *Earth Science,* 314.

3. Isaac Asimov, *Beginnings: The Story of Origins of Mankind, Life, the Earth, the Universe* (New York: Berkeley Books, 1987), 157–66.

4. Isaac Asimov, *Isaac Asimov's Guide to Earth and Space* (New York: Fawcett Crest, 1991), 210.

5. John Brockman and Katinka Matson, *How Things Are: A Science Tool-Kit for the Mind* (New York: Quill, 1995), 234–35.

6. Isaac Asimov, *Asimov's Chronology of Science and Discovery* (New York: Harper and Row, 1989), 26–27.

7. David Bodanis, $E = mc^2$: *A Biography of the World's Most Famous Equation* (New York: Walker Publishing, 2000), 83–84.

8. Steven Weinberg, *Facing Up: Science and Its Cultural Adversaries* (Cambridge, Mass.: Harvard University Press, 2001), 163.

9. Steven Weinberg, *The First Three Minutes* (New York: Basic Books, 1998), 101–21.

10. *Microsoft Encarta,* CD-ROM 1994, s.v. "Chronology."

11. Robert E. Krebs, *Scientific Laws, Principles, and Theories: A Reference Guide* (Westport, Conn.: Greenwood Press, 2001), 216.

12. *Geologic Time Scale,* Department of Geology and Geophysics, University of Alaska, accessed 2001, http://www.uaf.edu/geology/geo_time.html.

13. *Clickable Geologic Time Scale,* accessed 2001, http://www.geol.ucsb.edu/Outreach/TimeScale.

14. *UCMP Web Time Machine,* accessed 2001, http://www.ucmp.berkeley.edu/help/timeform.

15. U.S. Geological Survey, *The Geologic Time Scale,* accessed 2001, http://www.geology.er.usgs.gov/paleo.geotime.html.

16. U.S. Geological Survey, *Major Divisions of Geologic Time,* accessed 2001, http://www.pubs.usgs.gov/gip/geotime/divisions.html.

17. Kenneth W. Hamblin and Eric H. Christiansen, *Earth's Dynamic Systems* (Upper Saddle River, N.J.: Prentice Hall, 1998), 172–86.

18. Hamblin and Christiansen, *Earth's Dynamic Systems,* 186–96.

19. U.S. Geological Survey, *Radiometric Time Scale,* accessed 2001, http://www.pubs.usgs.gov/gip/geotime/radiometric.

20. Thompson and Turk, *Earth Science,* 67–81.

21. U.S. Geological Survey, *Relative Time Scale,* accessed 2001, http://www.pubs.usgs.gov/gip/geotime/relative.

22. *Microsoft Encarta,* CD-ROM 1994, s.v. "Dating Methods."

23. Robert E. Krebs, *Scientific Development and Misconceptions through the Ages* (Westport, Conn.: Greenwood Press, 1999), 91.

24. Niles Eldredge and Stephen J. Gould, "Punctuated Equilibria: An Alternative to Phyletic Gradualism," in *Models in Paleobiology* (San Francisco: Freeman Cooper and Co., 1973), 82–215.

25. David Millar et al., *The Cambridge Dictionary of Scientists* (New York: Cambridge University Press, 1996), 75–76.

26. *Encyclopedia Britannica,* CD-ROM 2001, s.v. "Evolution, Gradual and Punctuational Evolution."

27. Krebs, *Scientific Laws, Principles, and Theories,* 146–47.

28. U.S. Geological Survey, *Fossils, Rocks, and Time,* No. 1999-775-727 (Washington, D.C.: U.S. Government Printing Office, 1999).

29. Krebs, *Scientific Development and Misconceptions,* 79.

30. "Chapter 22: Classification and Evolutionary History," *Online Learning Center,* accessed 2001, http://www.mhhe.com/biosci/genbio/guttman/student.html.

31. *The Fossil Record: Help or Hindrance?*, Center for Learning, College of DuPage, accessed 26 October 1999, http://www.cod.edu/people/faculty/fancher/Fossil.htm.

32. Bruce Vawter, "Ancient Accounts of Creation," in *Microsoft Encarta,* CD-ROM 1994, s.v. "Creation."

33. David L. Marcus, "Charles Darwin Gets Thrown Out of School," *U.S. News and World Report,* 30 August 1999.

34. *Creation/Flood Myths of the World,* accessed 2001, http://www.templar.bess.net/Comp_names/bookofgods/creation3.html.

35. Robert Wright, "The Accidental Creationist," *New Yorker,* 13 December 1999.

2

Geological Forces, Processes, and Systems

Over the period of a single human's lifetime the Earth appears to change very little, but from the time of the Earth's birth ~4.5 billion years ago there have been many geological changes. From the time of recorded history some regions experienced devastating forest fires (caused by both nature and humans) that converted the forested areas into grassland, or landslides caused by earthquakes that altered the courses of rivers, or the creation of new lands and/or islands through volcanic activity. Although to most of us the Earth seems to remain much as it was a few decades ago or even a century ago, this time frame is deceiving because the Earth is constantly changing, albeit over relatively long periods of time.

Occasionally, you may hear the Earth referred to as the "living" Earth, but that confuses the inanimate with the animate, or the inorganic/nonliving (physical) with the organic/living (biological) changes on the Earth. The analogy is based on the fact that both biological life processes and the geology of the Earth are a complex of dynamic systems driven by a variety of forces and processes. A major distinction is that although over periods of billions, millions, thousands, and even hundreds of years a number of physical systems keep the Earth in a constant state of flux and a steady state of dynamic equilibrium, these physical systems do not involve the two main systems related to life, that is, reproduction and metabolism. Physical forces are the *causal* factors resulting in a series of *effects* (processes), which interact to contribute to complex events known as systems. Some forces, processes, and systems responsible for the dynamics of and between the spheres that make up the Earth have only recently been understood. These forces, although seemingly chaotic and random, follow universal, basic scientific laws, such as the law of **thermodynamics**, which is the relationship between heat and energy. All the forces, processes, and systems follow universal physical laws, even though the exact *causes* for some geological *effects* are not always clear. Natural phenomena are what interest earth scientists and drive the search for knowledge and understanding of how our Earth originated, how its structure relates to various forces, and how its environments have evolved.

One of the main sources of Earth's energy is the sun, which affects mostly the outer spheres. In addition, heat is produced by the fission decay of internal radioactive elements, the **kinetic energy** of molecular motion (heat), and stresses mostly near surface spheres (crust). Without the sun's energy the Earth would sooner or later become an ice planet, with surface temperatures approaching absolute zero (−273.15°C or −459.67°F), the point at which molecular motion, and thus heat, ceases. The heat energy produced by Earth's internal forces, along with gravity, is responsible for great pressures that alter the structure of both internal and surface spheres. Some degree of stability in the composition of elements, minerals, rocks, and their structures is achieved over millennia, but since the Earth and its systems are dynamic, complete stability can never be permanently achieved as long as there is some heat inside the Earth and the universe. Heat and other forms of energy create forces that drive processes responsible for the dynamic nature of the Earth. The force of gravity creates stresses and continuously mixes and separates rocks and other matter into layers according to their densities. The rotation or spinning of the Earth eastward on its axis creates **angular momentum**, with **centripetal forces** being counteracted by the somewhat artificial **centrifugal force**.

Many forces are responsible for the way the Earth is structured and its dynamic nature. This not only includes internal structures but surface features and our environment. Chapter 2 addresses the causes and effects of these forces and processes on the spheres of the Earth.

Physical Forces

Force, as defined by Sir Isaac Newton's second law of motion, is based on mass as a fundamental unit in motion within an absolute system consisting of two **vector** units. The equation for force is $F = ma$, where F is force, m is mass, and a is acceleration (speeding up or slowing down of motion). "F" and "a" are both vectors. A force is the *push* or *pull* that gives acceleration to mass—in other words, a force is the applied energy that makes things move. In this book the term "force" is also used in a broader sense as the energy that drives processes, systems, and cycles. There are many physical push and pull forces related to the structure, geology, biology, and chemical and physical systems of the Earth.

The Earth is a dynamic structure that was formed by a number of physical forces, and it is still being slowly altered by those same forces. The sections that follow provide a few examples of these physical forces.

Gravity and Gravitation

Gravity and gravitation are not exactly the same things. The term "gravitation" refers to a *universal* phenomenon, whereas gravity refers to a *local* manifestation of gravitation, such as the attraction of a mass to the surface of a planet, which is also considered to be the weight of a mass. In both cases the phenomenon is a force that acts on all matter at a distance. On Earth we are mostly concerned with gravity where its force is greatest at the surface and decreases toward the center of the mass. At one time in history it was considered impossible for a force to act over a distance, which meant that the force had to actually touch the object in order to make it move, such as moving a rock by pushing it. Galileo (1564–1642) was one of the first scientists to conduct experiments with falling bodies to arrive at a concept of how a force affects the acceleration of falling bodies. He developed a concept of gravity but did not develop it into a theory. Sir Isaac Newton (1642–1727), who developed several laws of motion, derived the formula to explain gravity as the proportionality of the products of two masses attracting each other over a distance, but this force of attraction becomes weaker as related to the square of the distance between the masses. Although we do not know exactly what gravity *is,* we can describe it as a real force that attracts everything with mass regardless of where that mass is located in the universe, that is, gravitation. Current experiments are being conducted to determine if there are so-called gravity waves that travel through space somewhat similar to electromagnetic (light) waves. Also, there is the concept of "gravitons," which, at the level of quantum gravity, may be analogous to the wave/particle duality of light that exhibits properties of both waves and particles; or possibly gravity is related to the quanta and string theories. (Quantum theory describes subatomic particles as point sources, whereas string theory is related more to a single dimensional curve with zero thickness; both are mathematical concepts of theoretical particle physics.) These current experiments may provide evidence for either or both the gravity wave and graviton concepts or some new ideas about what gravity really is.

Gravity is a unique force in that it always pulls, never repels or pushes. Actually, gravity is one of the weakest forces in nature, even though it is exerted over great distances. Gravity exists throughout the universe and is partly responsible for the accumulation of leftover particles from the Big Bang that formed stars, nebulas, planets, and other celestial bodies. After the sun was formed, the force of gravity resulted in the attraction of small bits of matter to larger pieces of matter as they revolved around this massive star. In time, through the process of accretion, the small nucleus of a proto-Earth grew; as more and more bits of matter agglomerated onto the more massive new Earth its gravity also increased, attracting larger meteors and asteroids that were incorporated into the Earth as we know it today. (This process is still going on, but most of the hundreds of meteors striking the Earth every day are dust to sand-grain sized and vaporize in the upper atmosphere before striking the planet's surface.)

The force of gravity is responsible for the three-dimensional (3-D) global spherical shapes of planets. This is due to a natural attribute of gravity that acts as if it originates from the center of a planet and pulls all of the planet's mass toward its center in a spherical pattern. The greater the mass of the planet, the greater the force of gravity (e.g., the moon's gravity is approximately 1/6 the Earth's gravity). Planets, as with most large objects, form into nearly spherical globular structures because, geometrically, it is the only way that the greatest possible amount of matter can be contained within the least surface area of a 3-D shape. The result is that the mass of a large planet is pulled toward the planet's center of gravity. Derek Sears, professor of cosmochemistry at the University of Arkansas, describes the process as follows: "With large bodies and internal heating from radioactive elements, a planet behaves like a fluid, and over long periods it succumbs to the gravitational pull from its center of gravity. The only way to get all the mass as close to a planet's center of gravity as possible is to form a sphere."[1] The collection of the maximum amount of mass around a planet's center is known as "isostatic adjustment." This process does not exist for smaller celestial objects, such as asteroids, because their gravitational force (pull) is much too weak for them to form into spheres. Thus, they maintain their irregular shapes. Gravitational force is also partly responsible for the collection of the densest matter toward the center of planets. Thus, the mass of the outer spheres near the planet's surface (e.g., crusts) is composed of less dense matter. This process forms the Earth's internal concentric layers, or spheres, of matter. The innermost spheres consist of the densest matter, such as the iron core, while less dense magma, rocks, and minerals form the lithosphere and the continental and oceanic crusts.

Scientists have used low-orbiting Earth satellites to measure the fluctuation of gravity for the surface of the Earth. NASA has launched a pair of Gravity Recovery and Climate Experiment satellites (GRACE 1 and GRACE 2) designed to make measurements of Earth's gravity field that are 100 times better than what previous instruments accomplished. These satellites record slight gravitational fluctuations of the Earth's physical features, such as polar ice caps, ocean levels, continental water storage, and weather conditions. These and other studies have revealed several things about the structure of the Earth. By using the concept of spherical harmonics to describe large-scale deviations in the strength of gravity, scientists determined that the Earth is not fluid but instead is composed of a series of more solid spheres. Gravity instrument data indicate that the mass anomalies in the crust are not due to differences between the thickness of the continental and oceanic crusts. Instead, these differences are due to some deeper, more massive spheres, such as the mantle and core. Correlation of these data also indicates that mantle densities under the continental crust are different from the mantle densities under oceanic crust—but they are in equilibrium (see the discussion of isostasy in this chapter.) These gravity data were confirmed by **seismic** investigations.[2]

Coastlines are the most noticeable features between the oceans and the continents. A number of forces and processes are responsible for relatively more rapid erosion and transport of sediment along the coastlines than on any other areas on Earth. Gravitational tides, wave action, changing sea level, melting of glaciers, wind, and currents all affect the formation and destruction of reefs and beaches, and the shapes of coastlines. The highest tides on Earth reach ~14 meters (m) (~46 feet [ft]), whereas the average tides are about ~0.8 m (~2.6 ft). Not only do some coastlines have higher tides than others, but also some continental areas have more tides per day than others. The average time between high tides, worldwide, is 12 hours and 25.5 minutes. The alignment of the sun and moon with the Earth not only produces tides on bodies of water but their gravity also produces a tidal effect on the more solid Earth. This distortion, due to gravitational tidal effects, causes a ~4.5 to ~14 inch bulge of the Earth at the equator.[3] Tidal action is responsible for a breaking effect that slightly reduces the rate of spin of the Earth, but very slightly, to the extent that the day is lengthening only about one second approximately every 62,500 years.[4]

An ocean tectonic plate subducts beneath a continental plate at the boundary where they meet. Once these oceanic slabs of the lithosphere dip to begin their journey to the deeper mantle, gravity moves them along. This process produces what is known as "gravity heat," which is a form of friction and is partly responsible for the heating of magma between the boundaries where the plates meet above the subducting slab. This gravitational heat raises the temperature of the magma, thus reducing its **density**, which can promote eruptions near plate boundaries to produce volcanic activity and flowing **lava**.

Convection

Convection is a force driven by heat flow in either liquid or semisolid matter. The internal heat of the Earth is produced by the decay of radioactive elements and possibly some residual, leftover heat generated when all the meteorites slammed into the Earth as it was being formed. The Earth's crust is composed of granite rocks that contain high levels of radioactive elements, whereas basalt rocks make up the oceanic crust, which exhibits less radioactivity. It was at one time assumed that more heat would flow up through the continents than the oceans, but this proved not to be the case. After numerous measurements scientists determined that the heat flow from both crusts was about equal, which indicated there must be differences in the chemical compositions of the two crusts.[5]

As heat drives molten magma, it expands and becomes less dense, forming convection cells deep in the Earth. Convection recycles hotter, less dense material upward, where it cools, becomes denser, and is driven downward to be reheated again to continue this slow convectional movement. This circulating movement becomes lateral near the subsurface, which assists in the horizontal movements of the continental plates. This process is slow, but sure, over long periods of time and, along with some other processes, is now known as the science of plate tectonics.

Compression and Extension

Compression and extension are the two stressful forces that are most responsible for rock formations found on or near the surface of the Earth. They are created by heat, the convection of magma, gravity, and several other forces. Compression occurs when pressure forces material into a smaller volume; extension is the opposite, that is, expansion. These forces are both created by convection, as well as by surface movements that squeeze and expand Earth's features, resulting in mountain building, the mid-oceanic ridge, lateral rifts, continental drift, and plate tectonics. These will be discussed later.

Scientific Laws

Some of the scientific laws and principles related to forces that drive the physical, chemical, and biological processes and systems of the Earth are thermodynamic principles, conservation of mass, threshold, and feedback.

1. *Thermodynamic* principles include several laws related to heat. The first law of thermodynamics relates to the conservation of energy, which states that energy can neither be created nor destroyed, but in isolated systems, it can be changed from one form to another. (Of course, the exception to this law is expressed as $E = mc^2$, where, in some nuclear reactions, a small bit of mass may be converted to energy.) Some of the forms of energy are mechanical (frictional), chemical, electrical, magnetic, solar, and nuclear. Energy can be defined as the ability of a system to perform work on another system, the result of which is a specified change in that system. The second law of thermodynamics, **entropy**, is crucial to both closed and open systems. The difference is that in an open system, if no new energy enters the system, heat (molecular motion) is lost to its surroundings and the system becomes less organized. A closed system, however, in a sense, does not require an external source of heat (energy) as it approaches a steady state. Many cosmologists assume that the universe is a closed system, whereas the Earth is an open system. As long as the sun is constantly supplying energy, life on Earth will most likely be maintained. In other words, mass and energy become more randomly distributed throughout the universe, and energy (heat/molecular motion) is dissipated (entropy or disorganization increase). As an example, a hot cup of coffee always gets cold, whereas while the reverse—a cold cup of coffee becoming hot—can never occur unless energy (heat) is applied to it. The coffee's heat (molecular kinetic energy) is being dissipated in its surroundings. In time, as the sun ages several billions of

years from now, it will become large enough to obliterate the Earth and the entire solar system before it finally becomes a so-called dead star. Of more immediate consequence, in a much shorter time period, is the possibility that the level of energy received on Earth from the sun may be altered by both natural and human induced determinants, that is, Earth will become too cold or too warm for life to exist. Earth, as an open system, requires a certain quantity and quality of solar radiation to maintain life. The sun's energy output over the past 4 or 5 billion years has varied somewhat, inducing cycles of very warm and very cold periods. Without the sun's energy, Earth would be a dead, ice-ball planet.

2. *Conservation of mass* is a law similar to that of energy in the sense that mass cannot be created nor destroyed but can be altered in form (again, with the $E = mc^2$ exception). For instance, in combustion the material burned changes into gases and ash plus heat and light without the loss of any appreciable mass. Theoretically, the end components of combustion could be reassembled. Chemical equations depict this concept of conservation of mass by using the symbol □ , which signifies an equal mass of atoms and molecules on each side of the equation (e.g., $2H_2 + O_2$ □ $2H_2O$). Similar to the law for conservation of energy, there are exceptions to the law for the conservation of matter, but these exceptions occur at the subatomic level.

3. *Threshold* and *feedback* are two concepts important to the physical forces affecting the evolution of the Earth. They are not exactly forces but instead points at which pressure from forces builds to a certain level where things begin to happen. They do not always occur sequentially: for example, a threshold, such as a stick breaking after being bent to a certain point, is not followed by a feedback reaction. But these effects do occur both inside and on the surface of the Earth, in both the physical and biological environments of Earth. For example, as the result of plate tectonics, two sections of continental plates may move in opposite directions alongside each other, increasing pressure at their junction until a certain threshold of this pressure is reached. It only takes a small change in the relationship of the opposing plates to each other where at a certain point (the threshold), they slip past each other, equalizing the built-up pressure, resulting in an earthquake. Also for example, small climate changes can alter the size of glaciers over a period of time by melting the ice. At some point, a threshold is reached, causing a feedback that results in still greater melting and changes. At the same time, however, at different latitudes, where more sunlight is reflected from snow-covered land areas, other threshold and feedback systems result in colder temperatures and additional growth of glaciers. Historical periodic global ice ages and warming periods are examples of slight perturbations reaching a threshold where feedback results in climate changes. Threshold and feedback systems are normal phenomena that involve many complexities, as do most chaotic systems. One of the goals of earth science is to understand how these physical forces, along with a multitude of causal factors, guide and control complex systems. These causal factors include but are not limited to the inclination of the Earth to the sun, the wobble of the Earth on its axis (precession), the changing levels of energy that Earth receives, the alterations of Earth's orbit, and human factors.

Forces in the Standard Model

Three fundamental forces included in the **Standard Model**, in the mathematical quantum theory of modern physics, are used to describe nature. The Standard Model deals with *fields of force* related to quanta of waves/particles. Particles are quantized manifestations of fields, and quanta are what scientists recognize as particles within these fields.[6] The energy and momentum of fields exist as tiny bundles, or quanta, observable as waves or particles. According to quantum theory, all subatomic or subnuclear particles are really quanta in a force field of energy. A wave or field that is mathematically quantized also has particle-like properties (e.g., energy, mass, momentum, and angular momentum), as manifested by the photon.

One example of the fields of force of the Standard Model is the magnetic field that can be viewed by placing a sheet of paper over an iron bar magnet that is resting on a flat surface and then sprinkling fine iron filings over the paper. Gently tapping the paper that is resting on the magnet shifts the iron filings to trace the bar's magnetic force field. Another example of the fields of force occurs when a wire carrying an electric current produces a force field around the wire.

The three basic forces of the Standard Model are (1) electromagnetism, (2) the weak nuclear force, and (3) the strong nuclear force.

1. The field of electromagnetism consists of quanta of energy that also may act as particle quanta. The quantum theory of electromagnetism (radiation) combines waves and particles, as well as describing the interactions of radiation with electrically charged atoms and electrons. Electromagnetism might be thought of as the quantum theory of light. The electromagnetic spectrum covers a wide range of radiation, ranging from the very long radio-type waves to the very short cosmic/gamma-type radiation.

   - *Photons* are massless particles (bundles known as quanta) of the electromagnetic field associated with visible light that carry the energy, momentum, and angular momentum of the field. Visible light photons exhibit specific wavelengths within the electromagnetic spectrum and exhibit characteristics of either waves in a quantum field or particles within a gravitational field. Photons have the duality of both waves and particles depending upon how they are observed and measured.
   - *Electrons* carry a negative charge, whereas their positive counterparts, *anti-electrons,* also known as "positrons," carry a positive charge. Both are elementary particles within the electromagnetic field. Electrons orbit the nuclei of atoms (at great distances relative to the diameter of the nuclei) as they exhibit different energy levels related to the particular orbit in which they reside. Electron force fields are responsible for the weak interaction involved between chemical elements during chemical reactions. The electromagnetic forces between interacting atoms of elements are much weaker than are the force fields that hold positively charged nuclei **protons** together.
   - *Leptons* are a class of small, elementary quanta of energy exhibiting wave/particle characteristics. As part of weak electromagnetic interactions, they only interact with electromagnetism and gravity. Their counterparts are *antileptons.* Electrons are also considered leptons, as are several other subnuclear entities.
   - The *muon,* also known as the "mu meson," is a type of lepton. It may exhibit a positive or negative charge. Another lepton is the relatively heavy *tau* particle, which is produced by the collision between an electron and positron. Still another lepton is the *neutrino,* which originally was thought to be massless with no charge. Recent research suggests that the neutrino may have some barely detectable mass after all.

2. The weak nuclear force consists of 6 types of **quarks**. Quarks are fundamental quanta particles that compose the protons and neutrons within nuclei of atoms. Weak nuclear interactions are responsible for radioactivity when nuclei of some heavy, unstable ele-

ments decay. A typical nuclear decay reaction is the Fermi beta decay that takes place when a neutron decays, resulting in the production of a $^{+}$proton, an $^{-}$electron, and a neutrino.[7] Radioactivity also produces other elementary forms of radiation as fields of force. The three more common forms of radiation arising from weak nuclear forces are alpha (α: $^{+}$nuclei of helium), beta (β: high energy $^{-}$electrons), and gamma (γ: extremely short wavelength penetrating radiation, such as cosmic radiation).

3. The strong nuclear force consists of massless gluons that are exchanged between the quarks of nuclei and are responsible for holding the particles within nuclei together, that is, gluons overcome the natural repulsion force of the positive-charged protons within nuclei.[8] Quantum chromodynamics is the mathematical theory that describes the strong interactions among quarks. Quarks are basic particles with fractional electrical charges that form elementary particles in nuclei, for example, protons and neutrons. Although quarks and gluons are hypothetical, mathematically they serve to help explain the strong nuclear force of the Standard Model.

Other factors included in the Standard Model, for example, gauge theory, Higgs particles, and bosons, explain the forces that interact with and compose fields of force in nature. The one force of the Standard Model that does not seem to fit this scheme is gravity. Einstein's theory of general relativity states that gravity is the effect of the curvature of space and time. On the other hand, Newton's reductionist theories of motion and gravity describe how gravity performs on smaller scales, such as the solar system and on Earth. Newton's concept of gravity is one of the major forces that formed the spherical structures of the Earth and continually affect the evolution of Earth's features. When an apple falls from a tree it still proceeds, as far as possible, to the Earth's center of gravity. Some physicists still think that the Standard Model might lead to a Final Theory, a Theory of Everything (TOE), or to Einstein's long sought Grand Unification Theory (GUT) that will include gravity. They envisage that gravitational force will fit some future model so that one, rather simple equation will explain all of nature's energy, forces, fields, quanta, and so forth. Leading in this direction is the assignment of a hypothetical particle called the "graviton," which is the quantum of the gravitation field and is related to string theory.[9] If gravity is beyond the Standard Model, then physicists will continue seeking a new model to explain the nature of nature.

Other Forces and Processes

Basic forces and elementary particles affect all matter, not only on the Earth but also in the rest of the solar system and universe. An understanding of how various forces, even at the quantum level, are responsible for how Earth was formed, what its structures are, and how it is evolving is basic to the study of earth science. Several other concepts are related to forces and processes that affect the changing Earth:

- *Pressure:* Atmospheric and water pressure are the weight of the air or water due to the force of Earth's gravity on them. *Weight* is the amount of force exerted on a body or object by gravity, while *mass* is determined by the composition of an object. Weight is determined by the force exerted on an object by gravity. Thus, weight varies according to the mass of the object as well as the mass of the planet involved (and the distance between the two). While weight of an object is determined by gravity, the mass of an object is constant throughout the universe.

- *Density* is the amount or quantity of mass (matter) in a given unit of volume. It is expressed as Density = mass divided by its volume, usually in terms of grams (or kilograms) per cubic centimeter (or cubic meter), but units for any measuring system can be used to determine density. We usually think of density as something closely packed, heavy for its size, or compact. Dense objects almost always sink or displace less dense objects. This is an important concept in earth science and is related to gravity, thermodynamics (heat), compression and extension, and convection, as these and other forces affect and alter the density of Earth's matter.
- The *magnetic force* of the Earth generated in the core extends to form the *magnetosphere* located beyond the **thermosphere** of the atmosphere. The magnetosphere confines the Earth's magnetic field and is modified somewhat as it diverts the supersonic flow of **plasma** (solar wind of ionized gas) around the Earth.

Seismology

Seismology is the science that uses both natural and artificial internal vibrations to indirectly determine the origin, composition, and structure of the Earth, as well as to detect and measure the strengths and locations of earthquakes and nuclear explosions. The instruments used to detect these internal vibrations, called "seismometers" (also referred to as seismographs, seismic detectors, or geophones), receive and record seismic vibrations and convert them to electrical impulses that are recorded and analyzed. The concept is not new. Ancient China and other regions always experienced earthquakes. Thus, the Chinese astronomer-royal and polymath Chang Heng (fl. 2d century C.E.) developed the first earthquake-detecting device in 132 C.E. This first seismograph, made entirely of bronze, was fashioned in the form of a round urn from which the heads of eight dragons protruded, each of which held a bronze ball in its mouth. Each ball was held slightly less tight than the others so that they would drop at different degrees of disturbance. Around the base of the urn, eight open-mouthed frogs sat directly under the dragons' heads. Chang Heng's seismometer consisted of an ingenious internal device based on inertia that converted seismic vibrations to motions that would cause one or more balls to fall out of the dragons' mouths. On one occasion, one of the balls dropped into a frog's mouth without any evidence of a local earthquake. Several days later a message arrived that reported an earthquake about 400 miles from the location of the instrument.[10]

Even today some Chinese scientists rely on the observations of abnormal behavior in animals prior to an earthquake. This ancient concept is based on the belief that animals can sense underground vibrations even before those vibrations can be felt or measured by humans. It was reported in China that cattle, sheep, and horses refused to enter their corrals, rats left their hiding places, shrimp crawled up on dry land, ants picked up their eggs and migrated, fish jumped out of the water, rabbits just kept hopping around, and during the middle of winter snakes came out of hibernation and froze to death. All of these animal activities were observed several months before a severe quake occurred. In late 1974, Chinese scientists used these natural signs, along with the fact that well water spouted out of some wells while other wells dried up or had water become cloudy, to predict the Haicheng earthquake. Heeding the warnings, the local government evacuated schools, factories, and businesses; thus, many lives were saved when the earthquake finally struck on February 14, 1975.[11]

Two basic modern types of seismology are used not only to study the internal structure of the Earth, but also to measure the location and strength of earthquakes.

- *Refraction seismology* locates the approximate depths of horizontal interfaces of Earth's internal spheres by measuring the traveltime of seismic waves that travel at different rates. It can also determine the seismic velocities of the types of rocks existing between interfaces of different internal spheres. Refraction occurs when a signal (wave or vibration) traveling through a medium of one density enters a medium of a different density. The direction of propagation of the seismic signal is bent (refracted) from its normal path, and its velocity changes as it passes through the boundary of the two densities. *Refraction shooting* is a type of seismic survey that measures the degree of refraction, thus presenting a seismic picture of the boundary between internal structures of different densities.
- *Reflection seismology* is a more useful tool that produces detailed, 3-D images of the spheres of the Earth. Reflection occurs when a wave or vibration approaches a surface at a specific angle and is reflected back from the surface at the same angle. Reflection seismology is used to examine internal folding of layers, faults, and other structural features. More recently, it has successfully identified and located petroleum and natural gas fields for exploitation. It is a form of radar, or echo-sounding, that augments other geological methods of exploring and understanding the internal processes and systems that make the Earth a dynamic planet.[12]

Seismology indirectly measures some of Earth's processes and systems, for example, magma plumes, plate tectonics, and earthquakes, that result from a variety of forces within the Earth. Data collected and analyzed by seismologists provide information needed to develop accurate theories and understanding of the structure and changes of the Earth, as well as determining the location and strength of earthquakes. Seismic waves of the same frequency sent into the Earth were found to travel at different speeds as they proceeded through layers of different densities. Seismic waves travel slower in hot rock (magma) and faster in cold or cooler, denser rock material. A common analogy is attempting to determine the state of ripeness of a honeydew melon by tapping its outside skin. A dull sound indicates it is too ripe; a higher pitched, clean sound indicates it may be almost ripe. The energy used in tapping sends a sound vibration through the melon that enables you to indirectly determine its internal structure. Similarly, geologists send seismic waves through the layers of different densities that make up the Earth, thus indirectly determining the structure of the concentric spheres.[13]

Two basic types of seismic waves, body waves and surface waves, are produced by earthquakes or man-made explosives.

- Body waves are the waves that travel into and through the Earth as seismic vibrations, which are different from the waves that travel along the surface of the Earth. The two types of body waves are P waves and S waves. The P waves (*primary* waves) are the major type of seismic waves that are sent through the Earth. They are compression vibrations that alternate between compression and expansion. The process is very much similar to a coiled spring, such as a toy Slinky that is stretched, which has stored energy. When released, the stored energy in the Slinky creates primary waves in the coil as it moves *parallel* to the direction of its propagation. The P waves are longitudinal waves similar to sound waves moving through air, a liquid, or a solid where the particles (molecules) of the medium move back and forth (compression and expansion) in the direction in which the wave is traveling. The P waves travel at different speeds according to the densities of the materials through which they pass. At the Earth's crust, P waves travel at about 4 to 7 kilometers (km) per second (sec) (2.5 to 4.3 miles [mi] per sec). When reaching the mantle, P waves travel at about 8 km/sec (5 mi/sec), while sound in air travels at only about 0.34 km/sec (0.21 mi/sec). As mentioned previously, the denser the medium, the faster sound will travel through it. The S waves (*secondary* waves or shear waves) arrive after the P waves have passed. The S waves, being slower, are received *after* the P waves by an observation instrument that has detected the waves. In S waves the particles (molecules) vibrate back and forth at right angles to the direction in which the wave is moving. This process is very much similar to rapidly moving a garden hose or rope up and down, with the S waves being produced as a hump or wave that transverses the length of the hose or rope. An S wave is sometimes referred to as a "standing wave" because the wave moves *perpendicular* to the direction of the wave motion. The S waves can only be generated and passed through solid material because this type of seismic waves is related to the closeness or compactness of molecules.

- *Surface Waves* are seismic waves similar to ocean waves because of their rolling motion. Surface waves move much more slowly across and through the layers of the Earth than do P or S waves. The two types of surface waves, that is, up-and-down and side-to-side, produce a rolling or heaving of the surface during an earthquake. When combined with the movement caused by body waves, they produce the violent disruption on the surface.[14–16]

The P and S waves are generated some distance beneath the surface at the focus of an earthquake from which they radiate in all directions. The epicenter, just above the focus, receives the waves first, thus they are the strongest. If the Earth were a **homogenous** solid mass, all waves would travel in straight lines out from the focus. But since the Earth's internal structure is not homogenous, the waves are bent to form a differentiated pattern due to the increased pressure with depth and densities of the internal spheres. Seismic waves travel at different velocities in rocks of different densities as well as in material of different rigidity. The S waves travel more slowly than the P waves and, along with P waves, are diffracted off the surface of the liquid outer core, whereas some P waves continue through the core to reach the opposite side of the Earth from the epicenter. The S waves do not travel through the core to the opposite side of the Earth. From about 105° to 140° from the epicenter is a surface area referred to as the "P wave shadow zone" where no waves are recorded. The patterns and intensities produced by waves from both earthquakes and man-made explosions are used both to detect the location and severity of earthquakes and to identify the internal structure of the Earth. The boundaries and nature (composition, density, and rigidity) of the Earth's internal spheres are identified by analyzing seismic data. This is how geologists have determined that the outer sphere of the core is liquid and the inner core is composed of solid iron and nickel. These techniques also identified the core-mantle boundary and the Mohorovicic discontinuity boundary between the upper mantle and the crust/lithosphere.[17–18]

The seismogram is a picture of recorded seismic waves made by a seismometer (seismograph). The seismic data are automatically recorded on a paper chart, photographed, or displayed on a computer screen as a seismogram. The seismometer works on the principle of inertia, as might be stated in Newton's first law of motion, that is, a body at rest tends to remain at rest until a force acts upon it. For instance, if you are trying to read a map while riding in an automobile over a bumpy road, it is difficult to hold it still to see the fine print because the car bounces around and moves your hands, thus it is difficult to focus on the print. In a similar situation, it is even more difficult to write. A simple seismograph can be constructed by applying the principle of inertia by attaching a heavy weight to a spring. In other words, attach a pen or marker to a weight so that it can write on a piece of graph paper and record any movement of the Earth, even as the inertia of the weight tends to keep it immobile. (See the Appendix for a suggested project on how to build a seismograph.) Today, seismologists use more sensitive modern electronic devices and computers that produce 3-D images. Since they are a more accurate means of recording seismic waves, they can determine both the strengths of earthquakes or nuclear detonations and their locations by analyzing seismic data and using geographic triangulation. The seismograph has also assisted in determining the internal structure and composition of the Earth.

Seismology and other sciences provide data that indicate that the structure of Earth's internal spheres are not homogeneously uniform regions. They have identified relatively thin transition boundary layers at depths where various processes and structural changes are taking place. Heat and gravity are the major sources of the energy driving the forces that determine the existence of the thin separation layers and dynamic processes described in the following sections. Also, the boundaries between different internal layers of the Earth are not smooth, distinct divisions. The margins between spheres can be quite uneven.

Asthenosphere

The Oxford English Dictionary (OED) relates the derivation of the word "asthenosphere" to the combination of the Greek words *asthenes* (weak) + the letter *o* + *sphere* (globe), which is interpreted as "sphere of weakness." The OED also attributes the introduction of this term to Joseph Barrell (1869–1919), who published in volume 23 of the *Journal of Geology* (1914) his theory of **isostasy**, which shows that below the lithosphere there is a thick earth-shell that yields to long-enduring strains of limited magnitude—therefore, called the sphere of weakness—the *asthenosphere.* A year later, the same journal mentioned that "The weakest part of the asthenosphere is of the order of one-hundredth of the maximum strength of the lithosphere."[19]

Usually, the asthenosphere is referred to as a separate, but not very distinct, layer just below the lithosphere and above the upper mantle. (The nature of the spheres that make up the core and mantle will be addressed in chapter 5.) At times it is referred to as either the unique upper portion of the mantle or the lower portion of the lithosphere. A major factor determined by seismologists and geologists is that the lithosphere, including the continental and oceanic crusts, floats on the asthenosphere. This attribute is important to the theory of continental drift, tectonic plates, and boundaries (which are discussed in subsequent sections). Also, the layer on which the lithosphere floats is not hard rock but instead is a weak, plastic-like region with a consistency somewhat like Silly Putty or hot road tar.[20] At a depth of about 75 km (46.5 mi) to more than 125 km (77.7 mi), the hard rock of the lithosphere becomes the plastic-like asthenosphere, which extends to a depth of about 350 km (217 mi) to the top of the upper mantle. As previously mentioned, both the temperature and pressure increase with the depth of the Earth, and when it becomes hot enough, hard rock melts. This creates a layer of hot, semisolid, plastic-like material that softens and flows after being subjected to great pressures and temperatures over geological time. This melting process starts in the outer core and mantle and continues in the asthenosphere, which makes it an unstable but dynamic system.

The upward movement of the hot, less dense mantle material creates a pressure that leads to melt production, creating a systematic feedback between the forces of **percolation** and **convection** flow within the asthenosphere and mantle. The lithosphere is thinner under the oceanic crust than under the continents, and thus the pressure created in the asthenosphere is more likely to force hot lava through weak spots on the ocean floor. As the hot magma/lava exits the crust, its fluidity increases, allowing it to flow with less resistance. This percolation not only mixes the Earth's chemicals, minerals, and rocks but also deposits them on the surface, as well as submerges them by **subduction** back into deeper areas to be recycled again over long periods of geological time. It is hypothesized that convection provides the forces in the core that cause the Earth to act as a self-regulating dynamo, resulting in its magnetic field. Large-scale convection in the mantle acts as a heat engine that forces a separation between the lower mantle and the upper, resulting in the material in the central mantle area rising up while flowing down in the outer area due to heat variations.[21]

An example of percolation is the once popular coffee percolator pot. Convection occurs in the pot of water as it is heated. This causes the boiling water to rise in the center stem of the pot and soak the coffee grounds that are in the basket atop the stem. As the coffee water cools, it flows downward at the edges of the pot to be reheated and again rise. Convection force also provides a lateral movement of subcrustal/lithospheric mantle material, resulting in an unstable base for the solid lithosphere to drift over the less rigid asthenosphere, that is, continental drift and plate tectonics. This drift of rigid lithospheric plates tends to move slowly, at a rate of about 2 inches or less each year. Geologists learn about the nature of the internal structure and age of the Earth by examining the chemicals in the minerals and rocks formed from molten material (magma) that is forced up from the asthenosphere through weak spots in the crust/lithosphere (volcanic activity). They also indirectly determine the structure and nature of the internal spheres by tracking the seismic vibrations sent through the Earth by earthquakes and man-made explosions.

"Isostasy" is the concept that the lithosphere is balanced as it floats on the asthenosphere.[22] There are two theories of isostasy that deal with how the rocks of the crust/lithosphere are in floating equilibrium on the denser, semi-fluid, upper region of the mantle known as the asthenosphere. Both theories are based on the fact that the continental and the oceanic crust/lithosphere are of different densities than the deeper mantle. These two types of crust/lithosphere exist side-by-side in alternating blocks. Some blocks are the less dense granite that makes up the continents; others are the more dense, relatively thin basalt that forms the ocean floors. John Henry Pratt (1809–1871) based his hypothesis on the fact that the density of rocks of the continental mountains is 2.5 grams per cubic centimeter (g/cm^3). Thus, they float *higher* than do more dense (3.2 g/cm^3) blocks of material. However, both the continental and oceanic crusts float on the mantle (asthenosphere), which is 3.4 g/cm^3. Pratt's hypothesis accounts for the rise of mountains as blocks of continental mass. The second hypothesis, proposed by Sir George Biddell Airy (1801–1892), suggests that all rocks are of the same general 2.7 g/cm^3 density, and thus *thick* objects float higher than *thin* objects in the same material. This is his explanation of how mountains appear where the crust is thicker and ocean basins evolve where it is thinner.[23] The important aspect of these two hypotheses is that both are based on the concept of isostasy, where blocks of the crust/lithosphere reach equilibrium by floating above the asthenosphere/mantle and thus drift somewhat independent of each other based on their densities.

Mohorovicic Discontinuity

As mentioned, the word "seismic" refers to Earth vibrations, including those produced by earthquakes or artificially induced by explosions.[24] Seismic waves are somewhat like light waves because they move in straight lines if the medium through which they travel is always of the same density. However, if they meet a substance of a different density, they either will be reflected back like a mirror or refracted (bent) at an angle or speed up or slow down. The images produced by modern seismographs and other instruments can identify outlines of even relatively thin internal layers.

In 1909 the Croatian geologist Andrija Mohorovicic (1857–1936), using seismic waves resulting from earthquakes, discovered that these waves produce an anomaly when they arrive near the Earth's epicenter. One type of wave arrived farther away than the other wave type. Mohorovicic hypothesized that a boundary between rocks of different densities caused the waves to refract (bend), thus causing this difference in distance. This boundary is now known as the *Mohorovicic discontinuity*, commonly referred to as the "Moho."[25] From his discovery that waves penetrating deeper into the Earth arrived sooner than waves traveling along its surface, Mohorovicic deduced that the crust/lithosphere lay over a more dense mantle in which the seismic waves traveled faster due to the increased density below the crust.[26] He surmised that this distinct separation of layers was located at about 32 km (20 mi) below the surface. This was the first discovery of a global discontinuity between layers (spheres) of different densities within the Earth's interior. It was later determined that the crust was only 5 to 10 km (3 to 6 mi) thick under the oceans, about 25 km (15.5 mi) thick under continental flatlands, and about 75 km (46.6 mi) thick under continental mountains.[27]

In 1960 the National Science Foundation (NSF) sponsored a project to drill into the Moho.[28] At that time the deepest well drilled was about 10 km (6.2 mi). Therefore, to reach the Moho, it would be necessary to drill about 3 times that depth on land. The project was abandoned due to a lack of resources and the realization that the great heat at that depth would destroy most drilling equipment. Temperatures increase about 15° to 75°C (59° to 167°F) for every kilometer drilled into the Earth; therefore, at just a bit deeper than 25 km the temperature would be almost 2000°C. Iron melts at 1536°C.[29]

Theories Explaining Earth's Geology

Contraction Theory

One of the early nineteenth-century theories that explained the changing geography of the surface topography of the Earth was called the "contraction theory." This theory was based on the concept of a **nascent**, hot, incandescent Earth shrinking as it cooled over time. As all matter cools, it contracts and become smaller as its particles are pressed together. It was assumed that since the Earth also contracted, this shrinking in size resulted in compression of rock material that folded and thrusted into geological formations. In other words, contraction was the causative factor that altered Earth's geological structures. Eduard Suess (1831–1914), an American geologist, made the analogy between the wrinkles on the surface of a dried-out, shrunken apple and the surface of the cooling Earth.

Suess assumed that the original crust was basically in one piece and became unstable as cooling continued, forming oceans and landmasses. He based this idea on the fact that similar types of plants and animals were found on separate landmasses. He also assumed that over eons of time, geological changes would continue.[30] Various versions of the contraction theory were developed, but none withstood the rapidly accumulating evidence that more accurately explained the Earth's geology. Contraction theory was soon superseded, first by the theory of continental drift, and later by the modern theory of plate tectonics.

Continental Drift

The basic concept of continental drift is somewhat like systems of slowly opposing forces creating stresses that, over long periods of geological time, move massive chunks of crust over the Earth's surface. The theory of continental drift asserts that a very old original landmass was broken up by a series of forces that displaced the pieces to form the continents. The concept is rather simple, but collecting and analyzing the evidence for movements of continents was more complicated. Over millions of years these new pieces of landforms drifted to their present locations as Earth's seven continents. The first person to ask questions about and develop a theory of continental formation based on their current positions and structures was Leonardo da Vinci (1452–1519).[31] Puzzled by the existence of marine fossils found on elevated land areas, da Vinci assumed some continents must have been connected. Early sailors may also have recognized the apparent congruent outlines of some coastal areas. In 1620, Sir Francis Bacon (1561–1626) reported the similarity of the coastlines of eastern South America and western Africa and suggested that they were once joined. He noticed similarities in the matching indentations, notches, and capes in the coastlines of the Old World and New World continents. In the eighteenth century, some French and German geologists suggested that the great biblical Flood or a massive tidal wave was responsible for the separation of a massive, single landmass into smaller landforms. By the nineteenth century the idea that such a flood or wave would be strong enough to result in the present formations of rocks and land structures was discounted.[32]

Evidence that continents are not now located on the surface of Earth where they once were thousands of years ago is based on several factors. The most obvious are the similarities of shorelines and geological structures, for example, mountains and present continents, with distant-past land configurations. The movements of the continental and oceanic crusts and underlying lithosphere are the result of convection and the spreading of the ocean floor. Manifestations of continental drift include the following:

1. Fossils of similar past plant and animal life on presently dispersed landmasses
2. Similarities of past climates as evidenced by, for example, fossils
3. Similar ground till and grooves in rocks caused by concomitant glaciation found on separate continents

4. Similar minerals and rocks, as well as ancient magnetic fields locked in crystals, located on different continents

The concept of continental drift was not formalized until Eduard Suess theorized that the southern continents were once combined as one large landmass. He called this great southern supercontinent "Gondwanaland," after the Gonds of ancient India. His theory was based on similarities of geological and plant fossils found on the continents of Africa, South America, Australia, and Antarctica and the subcontinent of India. He observed similar geological structures, including mountain ranges, regions of volcanoes and earthquakes, and coastlines for these landmasses.[33] Suess also gave the name "Laurasia" to a comparable northern landmass that included eastern Canada, North America, Europe, and Asia.[34]

In 1912, Alfred Lothar Wegener (1880–1930), building on Eduard Suess's concept of a southern Gondwanaland, theorized that all land and earth was once connected with the configuration of a supercontinent and that ~225,000 mya, during the Permian period, this landmass began to separate and drift apart to form the present configuration of continents. He called his supercontinent "Pangea," from the Greek word for "all lands" or "Earth." The northern part of Pangea was commonly called "Laurasia," whereas the southern part was called by Suess's name, "Gondwanaland" (see Figure 2.1).[35]

Wegener worked on his theory of continental drift from 1910 until his death in 1930. He based his theory on four important observations.

1. There is a more accurate fit of the edges of the continental shelves of the current continents than there is of their coastlines

Figure 2.1
Over hundreds of thousands of years, physical forces acted on the tectonic plates of the supercontinent Pangea, causing the slow progress of continental drift that resulted in the present geographic locations of the continents.

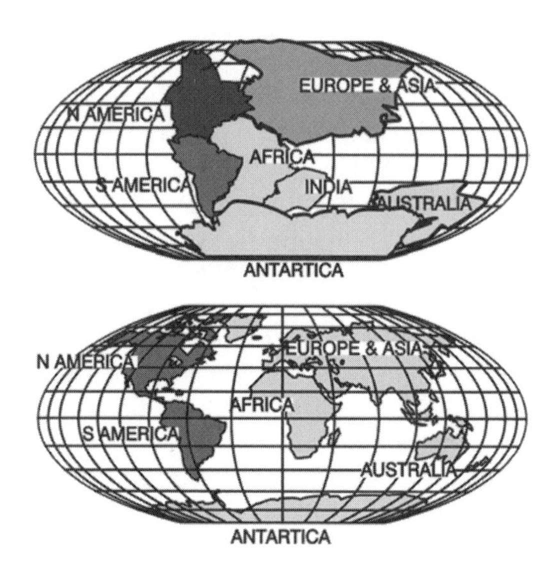

2. Current measurements indicate Greenland is moving westward from the European continent
3. The Earth's crust is composed of two types of rocks: continental crust is composed of a lighter, granite-type rock material, which floats on the heavier but thinner layer of basalt material that forms the oceanic crust
4. Although there are significant differences in plant and animal species found on various continents, there are also great similarities of species found on the now-separated continents, indicating these continents were once connected

Wegener also determined that belts of rocks located in Africa and South America are of the same type. Later, it was determined that about 300 mya the glacier ice sheets that covered South America, Africa, India, and Australia gouged similar parallel striations on rocks found on the different continents. This apparent match of rock markings indicates that these landmasses were once joined.

At first, many scientists, particularly in the United States, disagreed with Wegener's theory of continental drift. One objection to Wegener's theory was that this type of activity could not occur in such recent geological history (225,000 mya) as he had proposed. Another objection was that the softer, lighter rock material that forms the crust would be unable to penetrate the more dense rocks of the ocean floor or to plow their way through oceanic crust (basalt). Another objection questioned whether the continental crust slides over the oceanic crust. In fact, these parts of his theory did prove to be incorrect. Physicists later established that the oceanic crust is too strong for continents to plow through them like a ship through water and that frictional resistance is too great for continents to slide over oceanic crust as he proposed.

Today, however, parts of Wegener's theory are accepted in an updated version to conform to the new science of plate tectonics.[36]Although Wegener's theories of how continents drifted were not exactly correct, his evidence indicating that many million years ago the continents were actually once joined is seldom disputed. When looking at a map of the world it is possible to envisage that at one time Africa and South America fit together, as well as how Southern Australia can be linked to Antarctica. The concept of continental drift is more evidence that the Earth is not a static, immovable mass but instead is dynamic and constantly changing. Wegener's ideas of continental drift have been revived and revised and have led to the modern plate tectonics theory, including plate boundaries.[37] Little was known about the nature of the oceanic and continental crusts until after World War II, when several rapid advances were made by a number of European and United States scientists. One such discovery was the major topological and physical differences between the ocean bed and continental systems. It was during this post-war period that elongated mountain ridges with a height of 2 to 3 km (1.3 to 2 mi) were discovered near the middle of the Earth's ocean beds. Some of these ridges were more than 1,000 km (620 mi) long. Rift valleys associated with these ridges were also discovered. The upward flow of magma over the crests created these ridges as the lava spread outward from the ridge onto the ocean beds. Scientists also discovered that undersea earthquakes were associated with the formation of these mid-oceanic ridges. It has since been determined that the mid-oceanic ridges, as well as some ocean bed trenches, also are related to fracture zones (broken ocean crust/lithosphere), which extend over long distances in the basins of oceans but end where the continental plates meet.[38] (See the next section for more on the theory of plate tectonics.)

Recently, Christopher Scotese, a geologist at the University of Texas at Arlington, predicted that the present configuration of continents will recombine in the future to form a new megacontinent, which he calls "Pangea Ultima." His theory is based on both Wegener's idea that Pangea broke apart in the Permian period to form the present continents and the newer theory of plate tectonics. Scotese added that in about 250,000 million years, as tectonic plates continue to move continents, forces will continue to slide the seven continents into one another, thus forming a new grouping of continents. He predicts that Pangea Ultima will form around a large inland sea, similar to how the Mediterranean Sea originally formed. Using geology of the past, Scotese also predicts that Africa will move north and smash into Europe, Australia will migrate into Asia, and North America's east coast will mesh with Africa's west coast, while South America will swing around to close in a great, central body of water.[39]

Plate Tectonics

During the nineteenth and early part of the twentieth centuries there was no viable global theory to explain the major geological features on the Earth. Not until the 1960s was a new theory explaining these features developed. This theory was, to some extent, based on the old continental drift concept, but it was also drastically enhanced by a variety of new evidence to explain the forces, processes, and systems involved in the movements and interactions of great blocks of the Earth's crust/lithosphere, the Earth's surface features, earthquakes, volcanoes, mountains, and valleys. This new theory soon became known as "tectonic plates," which is the first global theory not only to explain the history of geology but also to be generally accepted by the majority of scientists. Plate tectonics is the unifying theory of geology that explains how the forces of heat energy, gravity, and magnetism drive the systems of interrelated, interacting processes that make the Earth a dynamic planet. Plate tectonics explains a number of geological processes as a series of actions resulting in historic geological changes. This theory also accounts for continued, and possible future, geological changes in the Earth's structure. Some results of these physical changes include the formation of the following:

- Concentric spheres of differing temperatures, pressures, and densities
- Magma and rock formation and flow (convection cells)
- Continental drift/plate tectonics (landmass movements)
- Seafloor ridges, rifts, and valleys
- Earthquakes and volcanoes
- Seafloor spreading as the result of volcanic activity at the ridges
- Formation of continental mountains and valleys
- Paramagnetic and heat anomalies, as well as other geological features

The theory of plate tectonics incorporates research data from special studies that were conducted during and following World War II by scientists from a variety of institutions the world over. These data, when combined, explained many puzzling geological observations related to the phenomena of continental drift. In the late 1960s, Daniel P. McKenzie and Robert L. Parker, who were located at the Scripps Institution of Oceanography (SIO), and Jason Morgan at Princeton University independently established the first model for plate tectonics based on the theory that rigid bodies of crust move relative to each other on the surface of a sphere that has a common axis (pole).[40]

Plate tectonics is an excellent example of how a new theory becomes accepted by scientists when it is based on critical observations and rational explanations of prior and confusing empirical evidence. The new theory of plate tectonics was based on new observations, data from new instruments and technologies, and laboratory research data. Wegener's original concept that continents were composed of granite-type rock drifting over a subsurface layer of basalt was altered to conform to these new data.

The surface of the Earth consists of a relatively thin layer composed of the continental crust and the oceanic crust, which is an even thinner layer. This sphere of crust composes the uppermost layer of the lithosphere, which consists of relatively cool rock ~5 to ~50 miles thick. These two outer spheres (layers) rest on top of the hot, semiplastic, tarlike asthenosphere that itself is sometimes considered the upper part of the mantle. Since the mountain granite rock of the continental crust is thicker and heavier than the ocean crust, it sinks toward the mantle and is partly responsible for the circulation of the material in the asthenosphere. This flow of material is heat driven and is only partly responsible for very slow movement of huge sections (plates) of the Earth's surface.[41] At the same time, the motion of the plates contributes to the continuing process, particularly as ocean plates are submerged (subduction) under the coastal rims of continental plates (convergent boundaries). Since the science of plate tectonics is rather new, it has altered the once-accepted concept of the Earth's structure and forces that affect this structure. The word "plate" originally referred to a large, paving stone–like, solid, firm, slab of rock found in the crust/lithosphere. The word "tectonic" is from the Greek word that means "to build," also as in the word "carpenter." Tectonics is now considered the study of the deformation and motion of the rocks that make up the Earth's lithosphere and crust and the forces, processes, and systems responsible for such deformation. In essence, "plate tectonics" means the building and moving of the fragmented plates of the outermost layer of Earth over the top of a hotter, more fluidlike layer of material. There are basically two kinds of plates—oceanic plates are mostly thinner and move outward from mid-oceanic ridges, whereas continental plates are thicker and move to a lesser degree. Note, however, that although the concepts of continental drift and plate tectonics are somewhat related, they are not exactly the same. The concept of continental drift was based on the idea that large blocks of the crust, driven by unseen forces, plowed through the oceanic crust similar to a ship going through water (see previous discussion of Airy's theory). What really happens is that continents ride passively on mantle material as the oceans spread out as new magma material is intruded through the rift (opening) at the tops of the oceanic ridges. This **viscous**, magmatic basalt moves laterally away from the axes of the ridges, thus causing seafloors to spread away from both sides of the ridges. According to magnetometer and heat flow data, this spreading of the seafloor results in the outward movement of the older seafloor as new material is added at the crest of the ridges and continues to spread horizontally from the axes of the ridges. This results in the expansion of the ocean floor that increases the distance between continents (e.g., compare the fit of the contours of the east coast of South America with the west coast of Africa).[42]

Magnetometry

The original concept that continents (and later plates) move slowly across the surface of the Earth was assumed to be true by empirically matching continental coastlines. There was not much hard research data to support the theory until the middle of the twentieth century. During World War II, the U.S. Navy was interested in research that examined variations in magnetic fields found on the floors of various oceans. It was hoped that **magnetometry** would provide knowledge of how magnetic anomalies might be used to both hide and locate submarines. After the war, the Office of Naval Research (ONR), U.S. Geological Survey (USGS), and the Naval Ordnance Laboratory (NOL), using an airborne magnetometer originally designed to detect submarines, surveyed the magnetic anomalies related to volcanic activity off the Aleutians and other Pacific Northwest islands. In the early 1950s, geologists from the SIO surveyed the Pacific Northwest ocean floor by using a proton magnetometer. A new, adapted version of the old airborne magnetometer was trailed behind an oceanographic research vessel that used a grid map to gather data on the ocean floor off the northwest coast of the United States and southwest coast of Canada. This ship-towed instrument detected and measured undersea magnetic anomalies connected to features on the ocean bed, such as atolls (ring-shaped, coral reef-islets), guyots (flat-top underwater seamounts), and scarps (short for "escarpments," which are cliffs created by faulting or erosion). These findings were encouraging enough to continue the study on the research vessel *Pioneer,* which was in the process of conducting a **bathysphere** survey off the west coast of the United States. After several months of dragging a magnetometer behind the *Pioneer,* scientists had results that were significant and covered a large area of the northwest Pacific Ocean. The data indicated rock magnetism off the west coast arranged in a distinct north-south linear pattern. Other scientists followed up on these data, using improved instruments to discover an even more distinctive magnetic pattern of seafloor rocks with horizontal offset breaks of the ridge by a number of underwater fault lines. These underwater fault lines were later identified as "transform faults" by Canadian geologist John Tuzo Wilson (1908–1993). The magnetic striations became known as "zebra stripes" and were believed to be related to seafloor spreading. It was now clear that, at the mid-oceanic ridges, magma was extruded upward through the crest of the ridge. It then flowed down and outward, which expanded the seafloor away from the ridge (see Figure 2.2).

In the early 1960s, Frederick Vine (1939–1988), a graduate research student working under his supervisor, Drummond Matthews (1931–), compared the magnetic orientation of two undersea volcanoes. One was magnetized in a specific north-south direction, whereas the other was magnetized in the east-west direction. This led to their Vine-Matthews hypothesis, which associates seafloor spreading with reversals in the Earth's magnetic field. Vine and Matthews also speculated that they could determine the age of ancient magnetic rock material on

Figure 2.2
The zebralike magnetic stripes recorded on the ocean floor indicate that, over many millions of years, the seafloor has been spreading and the Earth's magnetic field has reversed itself many times. This is indicated in the images recorded by magnetometers by the striped nature of magnetic rock and mineral sediments on the seafloor.

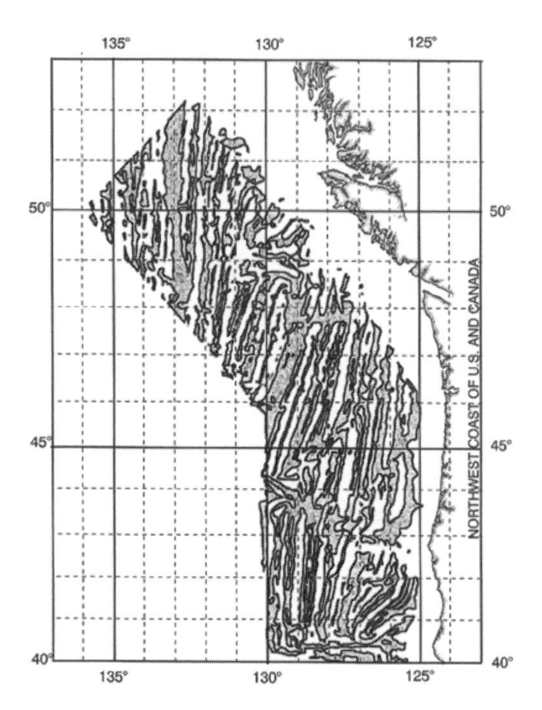

the ocean floor by comparing the patterns of stripes with the reversals of the Earth's magnetic field. Of course, this assumed that the spreading was consistent over time, as the reversal of the magnetic field was irregular. Vine checked this concept by using radiometric dating methods to determine the age of the cooled magma rocks in successive linear zebra stripes. He found a correlation between the ages of the rocks in the stripes to the periods of magnetic field reversals.[43–44]

The polarity and orientation of the Earth's magnetic field has changed many times in the geological past. As magnetic magma material was extruded and intruded on the ocean floor, it cooled to form solid minerals and rocks, and their magnetic particles were aligned according to the Earth's magnetic field. Over time, new magma that extruded from the crest of mid-oceanic ridges spread outward to form a linear pattern of north-south zebra stripes. As the orientation of the Earth's magnetic field changed and more magma was extruded at mid-oceanic ridges over longer periods of time, the new magnetic material acquired a different magnetic orientation than did the preceding stripe. Also over time, this spreading at the base of the ridges extended the width of the ocean floor, resulting in shifting of continental plates.

Summary

Plate tectonics is a global concept that addresses the forces, processes, and systems responsible for mountain building, folding and faults in the crust, the gradual movements of large-scale landmasses, sudden horizontal displacement of land along faults, and igneous and metamorphic processes of rock formation. These processes result not only in building mountains but also in creating the oceanic trenches, ridges, and rifts and in earthquakes and volcanic activity.

One theory proposed that the force related to the spinning effect of Earth's rotation around an axis (two poles) resulted in plate movement. This theory uses geometry proposed by Leonhard Euler (1707–1783), a Swiss mathematician who stated that two rigid bodies can move with respect to each other over the surface of a sphere if they are rotating around a common pole (axis) of the sphere. This concept explains how two separate massive plates near the surface can move with respect to each other when affected by the common spinning axis of the Earth's globe, but only as long as they retain the internal consistency that, basically, continental plates maintain. As a means of measuring this independent motion of plates, that is, not in relation to each other but instead related to separate frames of reference, a series of hot-spot volcanoes located on a chain of islands was used as a common point of reference to measure both the direction and speed of plate movements.[45]

The plate tectonic plate theory also explains the geological history observed on the margins of the continents. These forces created passive-to-active breaking (separating) and buildup (reassembly) of plates over the past 200 to 300 million years. Along with his concept of transform faults (plates sliding past each other), which revived the older idea of continental drift, John Tuzo Wilson, in 1965, coined the term "plate" for the large broken pieces of the lithosphere. After observing this passive-to-active phenomenon, Wilson proposed the following stages for explaining plate tectonics:

- Continents heat up and bulge under great forces.
- At the same time the seafloor spreads, creating rifting. Old, inactive rifts represent passive margins of plates, while new volcanic action produces new rifts as seafloor spreading continues.

- Subduction takes place as one plate slides under another as the seafloors continue to spread.
- Continents collide and join together.

In 1967, Jason Morgan proposed that the Earth's surface consisted of 12 rigid plates (six of them major plates) that move relative to each other. Shortly thereafter, Xavier Le Pichon (1937–) identified the location and types of plate boundaries and their direction of motion.[46–47]

Many other scientists from the 1960s through the 1990s contributed additional research data to firmly establish the basis for plate tectonics, which today is accepted by the majority of scientists as the global theory that explains the geological features of Earth. The understanding of this slow but steady process, which will continue to alter the ocean beds and positions of continents in relationship to each other, is partially based on the following points.

- There are six major plates and several minor ones, including several smaller pieces that fit between larger plates (see Figure 2.3). Each plate is a rigid slab of the crust/lithosphere and is about 100 km (62 mi) thick. Thus, they are referred to as "lithospheric plates."
- The plates move slowly as they are driven by energy produced by upsurging forces created by the internal convection cells of

Figure 2.3
The present configuration of the continents is defined, in part, by the current boundaries of the major tectonic plates located under both the oceans and the continents.

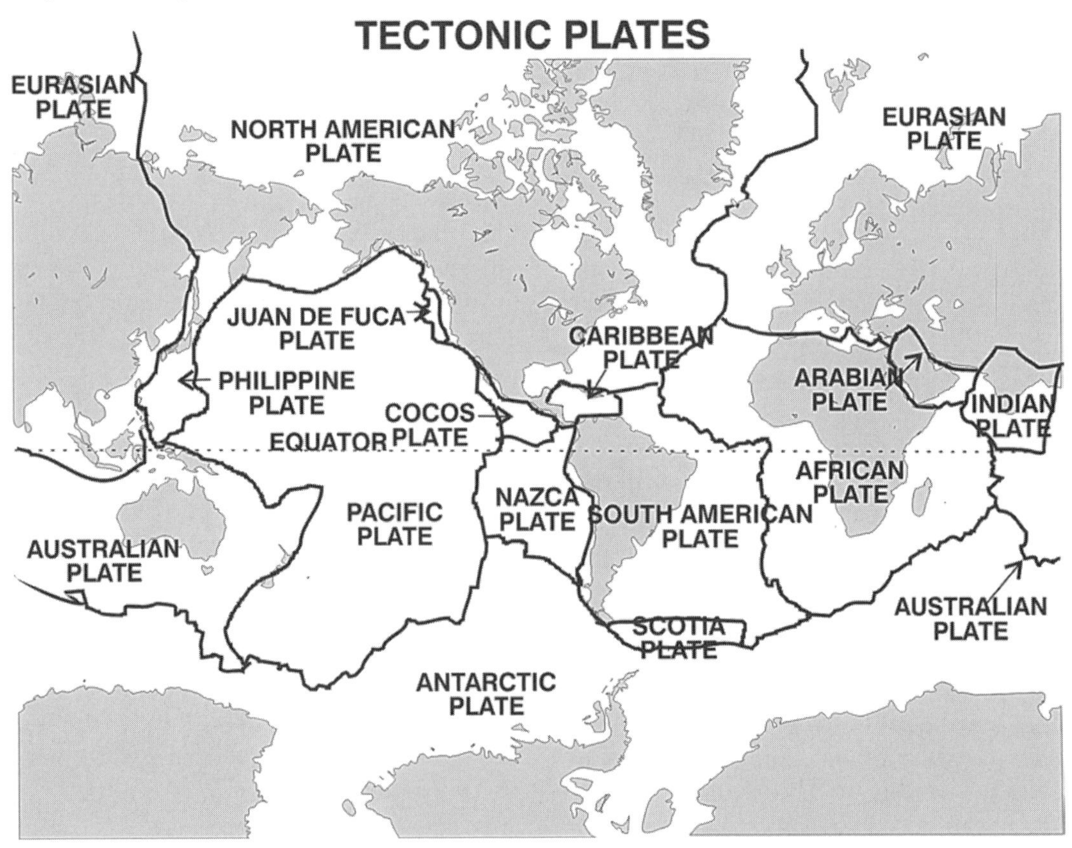

magma that is intruded at the mid-oceanic ridges, which results in seafloor spreading.

- Earthquakes, volcanoes, and other geological activity originate at the boundaries between plates.
- The deep areas of plates exhibit less geological activity and milder earthquakes than occur at the plates' boundaries. The three types of boundaries that separate plates are *convergent, divergent,* and *transform* boundaries (see the section on Plate Boundaries).
- Subduction zones occur when one plate is pushed under another, resulting in the recycling of magma. Subduction can also produce earthquakes and volcanoes.
- The ocean floor, on both sides of mid-oceanic ridges, is slowly but constantly spreading by the addition of new magma as it flows down from the ridges' crests.
- Powerful forces drive the lithospheric plates. The forces and processes responsible for plate movements only recently have become well known. Heat energy supplied by the fission decay of radioactive elements in the mantle and crust/lithosphere plays a major role in the process of magma convection, as does the force of gravity. Convection occurs in a fluidlike medium when a dense upper region displaces a less dense subregion of the medium. This density inversion is reversed as the upward-flowing medium cools, becomes denser, and again sinks toward the center of gravity. Heat-driver convection is the main process responsible for magma intruding at the mid-oceanic ridges (and rifts), spreading the seafloor and resulting in continental drift, mountain building, earthquakes, volcanoes, and crystal growth in minerals.[48]
- As plates move across the asthenosphere of the mantle, and as the seafloor spreads, oceanic plates slide under continental plates (subduction) and trenches and mountains build up above the plates as they collide and overlap. Lava extrudes into weak areas near the surface at the upper edges where plates meet forming chains of volcanoes (island arcs). Volcanoes can also occur near the centers of some of the plates.
- Our knowledge that major tectonic plates move several centimeters per year has been derived from observations made by dating marine magnetic anomalies. At the mid-oceanic ridge the extruding magma cooled, forming magnetic signatures related to the Earth's changing magnetic field. These magnetic signatures were locked into the ocean's crust. As the seafloor spread, it reconstructed the geometry of the oceanic crust (see Figure 2.4). Recent global positioning system (GPS) measurements indicate that the Antarctic, North American, and Eurasian plates have been altered from past positions by the melting of major ice sheets over the past 15,000 years.

Plate Boundaries

Plate boundaries are fracture zones that separate one tectonic plate from another, forming edges that are more or less in conjunction with one plate and another neighboring plate. Two possible analogies are (1) the sutures or seams that join the bone plates that compose the human skull and (2) the seams stitched on a baseball that tie together the edges of the cover. As with all analogies, these are not exactly the same as the boundaries for tectonic plates. The plates are very slowly in motion in relation to each other, and thus their boundaries are disturbed and distorted by a number of forces in a variety of ways. Another analogy is that these plates are large chunks of crust/lithosphere that form a loose skin that

Figure 2.4

As magma rises beneath the seafloor at the mid-oceanic ridges, rifts are created as new material builds up mountains and spills over the edges forming new seafloor, which spreads laterally to the ridge along with the underlying plate. As the oceanic plates move outward, they meet continental plates forcing the ocean crust and lithosphere to form a trench as they dip at converging boundaries between the two plates. Plate tectonics is the global process responsible for this system of plate movement and spreading seafloors that result in the subduction of crust/lithosphere into the inner regions of the Earth to be recycled.

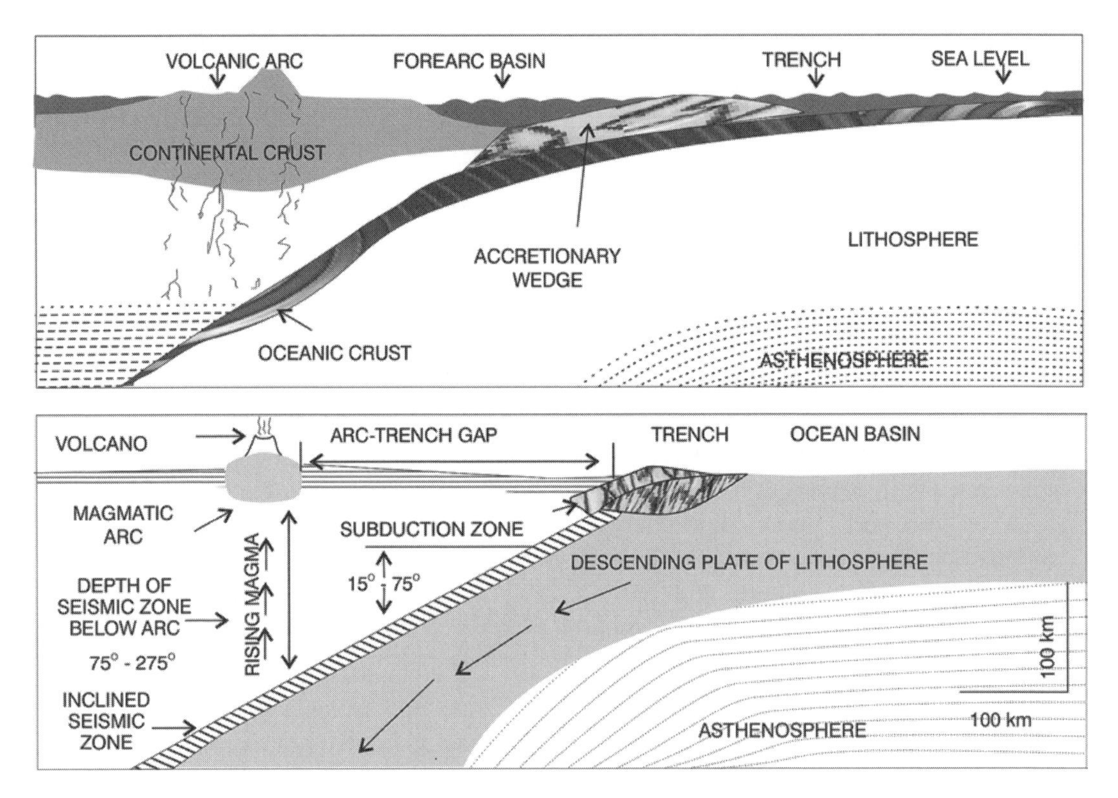

slides over the semisolid asthenosphere as they are driven by a number of forces, including gravity and magma convection cells. In addition, internal heat causes convection currents that form domes of magma that affect plate movement. Although entire plates move, most of the detectable motion is measured at their edges, not at the interior or center portions of the plate. The forces generated by the movements of plates at their boundaries are responsible for the major geological events on or near the surface of the Earth. These forces build mountain ranges, produce volcanoes, and cause earthquakes.

The categories of plate boundaries are described according to the relative movement of one plate to another:

1. *Divergent boundaries* occur when two plates move apart from each other
2. *Convergent boundaries* occur when two plates move toward each other

3. *Transform boundaries* exist where two plates slide horizontally past each other (see Figure 2.5).

Divergent Boundaries

Some edges or boundaries of plates are also known as *constructive plate margins,* because they were constructed by magma upsurges at mid-oceanic ridges; thus constructive margins can only occur between two oceanic plates. Two constructive processes create oceanic ridges. One involves the areas where convection currents raise hot magma and fractured rock upward to add to the crest of oceanic ridges. This volcanic activity is evident along the axes of the ridges, which means that the flowing magma proceeds to spread the seafloor outward, perpendicular to these axes, resulting in what are known as *spreading ridges.* In the second process, as new material pushes the older ocean floor crust aside, undersea earthquakes occur as ridges continue to expand along their axes. Other types of ocean crust ridges, not associated with earthquakes, are referred to as *aseismic ridges.* As the seafloor spreads it moves the continental plates further from the ridges.

Figure 2.5
Three basic types of boundaries are formed by the meeting of separate tectonic plates. (1) Divergent plates create a spreading center resulting in a rift zone that allows magma to rise to the ocean floor to form the mid-oceanic ridge as an undersea mountain chain. (2) Converging plates create a subduction zone where the edge of one plate with ocean crust is forced under another plate creating a trench on the ocean floor. This type of boundary formation results in deep earthquakes and surface volcanoes near the continental boundary. (3) Transform boundaries are formed where two plates slowly move past each other in opposite directions, causing faults in either the continental or oceanic crust/lithosphere that result in shallow earthquakes.

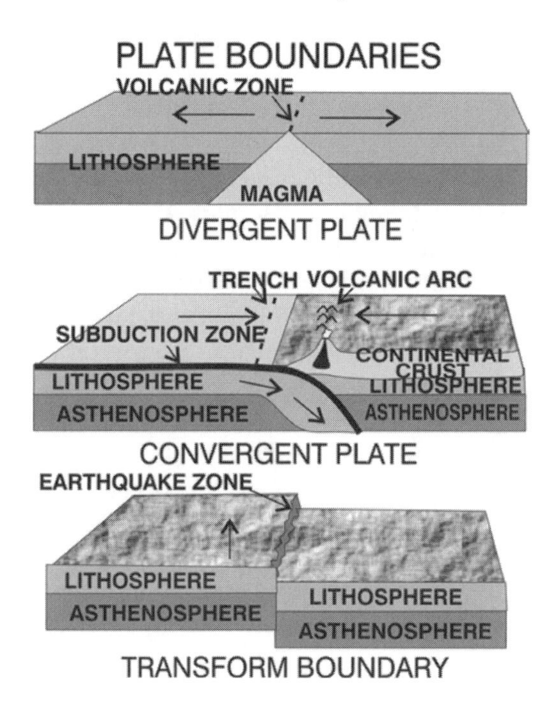

Another type of divergence sometimes takes place near the centers of plates, causing rifting (splitting) that results in two plates forming from a single plate. One such example is the Red Sea and the Gulf of Aden, whose plate rifted about 20 million years ago when it was separated to form the main African plate. In addition to rifting over extensive areas of the Earth's surface, some plates experience *extensional* stresses that result in the separation of boundaries between two crust/lithospheric plates. These divergent boundaries may be located under continents or, more likely, under ocean beds where plastic-like magma oozes up through this weak area to form underwater geological features, such as the submerged global mountain ranges known as Oceanic Ridges. Most of the divergent boundaries occur in ocean crusts because that is where the Earth's crust is at its thinnest.

To the north, the Mid-Atlantic Ridge includes Iceland, which is an example of an islandlike ridge that has risen above sea

Table 2.1
Examples of Divergent Plate Boundaries

Names of Diverging Plates	Orientation
North American plate ⟨ ⟩ Eurasian plate	North-South
African plate ⟨ ⟩ South American plate	North-South
African plate ⟨ ⟩ Antarctic plate	East-West
South Pacific plate ⟨ ⟩ Antarctic plate	West-East
Indian-Australian plate ⟨ ⟩ Antarctic plate	West-East
Indian Ocean plate ⟨ ⟩ East African plate	Southeast-Northwest

level. The Mid-Oceanic Ridge is part of the globe-circling, continuous chain of submarine mountains that rise about 2 or 3 km (1.2 to 1.9 mi) above the surrounding ocean floor but well below the ocean's surface. The Mid-Atlantic Ridge progresses in a longitudinal direction from the Arctic Ocean in the north, southward through the middle of the Atlantic Ocean where it joins with two latitudinal ridges, one progressing eastward between the southern tip of Africa and Antarctica and the other heading westward around the tip of South America. The Mid-Atlantic Ridge is only one segment of the global-oceanic ridge that encircles the Earth to form the globe's largest mountain chain, albeit under water and seldom seen. Divergent boundaries are responsible for a continuous spreading of the ocean floor along the Mid-Atlantic Ridge at the average rate of ~2.5 centimeters (~1 inch) per year or ~25 km (15.5 mi) in a million years. Geologically speaking this is a very rapid motion over a period of millions of years. In the time of the dinosaurs (~65–200 mya) the Atlantic Ocean was a tiny inlet of water, but it has become the vast ocean of today as the continents of North America and Europe drifted further and further apart.[49]

Divergent boundaries, also called rifting, form long narrow continental troughs and valleys, as well as underwater mountains and islands, to the extent that continents are split apart as the land, in a sense, stretches. When divergent boundaries occur on continents, rift valleys are formed as the crust is torn apart, creating earthquakes and allowing magma to rise to the surface. As mentioned, the Red Sea is an example of an active rift that is an extension of the failed East African rift. By the process known as *seafloor spreading,* this rift will grow into a new larger sea sometime in the distant future. In time, this rift, with the Suez Canal at its northern end, will separate Arabia from Africa. The rift is currently ~3,000 km (~1,864 mi) long and ~100 to ~300 km (~62 mi to ~186 mi) wide, and extends through east Africa.[50–51]

In summary, diverging plates move apart from each other at the average rate of 1 to 10 cm (0.5 to 4 inches) per year, commonly resulting in weak areas on the ocean floor where the intrusion of magma into the lithosphere forms new igneous rock. Thus, much of Earth's new ocean crust forms at divergent plate boundaries (see Table 2.1).[52–54]

Convergent Boundaries

Approximately 50% of the ocean floor was formed in the past 65 million years. This rate is equivalent to replacing the entire Earth's surface in only 225 million years, which means that the entire Earth's surface could have been renewed in a relatively short period of geological time. It is obvious that this replacement of the Earth's surface has not increased its overall size in the last 225 million years, so some other system must be responsible for this gradual replacement of the Earth's crust. A number of forces, processes, and systems are responsible for recycling deep material to continually replace Earth's existing surface material.[55]

Tectonic plates that converge (come together) form *destructive* margins (edges), places where the slow but sure movement of massive blocks of crust/lithosphere exert destructive forces. Convergent boundaries are produced by *compressional* stresses and occur when tectonic plates (large blocks of lithosphere) collide and/or move over or under one another. These movements are responsible for important geological activities, including the deformation and renewal of the crust, mountain building, and the metamorphic transformation of igneous rocks. As an ocean crust plate meets a continental crust plate, the thrust of the converging boundary produces a *subduction zone,* in which the descending ocean plate is inclined downward into the mantle and produces folding and faults that result in seismic activity. Subduction of one plate under another is the process by which the Earth's shell recycles itself. The destructive margin of the lithospheric plate that is subducted is changed to the more ductile (fluidlike), denser rock of the asthenosphere due to increases in temperature and pressure at a depth of ~700 km (435 mi).[56] Subduction also creates strong compression at the plate boundaries, piling up unconsolidated sediments and pieces of rock slabs into a wedge similar to the accumulation of snow in front of a snowplow. Unlike the constructive plate margins that are generally found between divergent oceanic plates, destructive margins occur between plates of differing densities, resulting in three different types of converging plate boundaries.

1. *Ocean-to-ocean* convergent plate boundaries can result in several formations.
   (a) Island arcs as curved lines of volcanic islands (Western Pacific)
   (b) Ocean trenches (the Pacific Marianas trench is 11,000 m [36,000 mi] deep, deeper than Mount Everest is high)
   (c) Rising magma, volcanic activity, earthquakes, and rock deformation (Western Aleutian chain off the coast of Alaska and the Pacific Coast Rim of islands)
2. *Ocean-to-continent* convergent boundaries result in mountain building and the opening of trenches in the oceanic crust, as well as deep earthquakes, volcanoes, and rock deformations. One example is the Andes mountain range in South America.
3. *Continent-to-continent* convergent boundaries result in mountains and deep earthquakes, as well as changes is some rock structures. Since there is little or no difference in the densities of the different continental plates, they do not sink into the mantle. In addition, one continental plate pushing into another landmass may result in crumpling and bulging of the crust/lithosphere mass to form mountains as well as other rock formations. The Alps in western Europe, the Appalachians in the eastern United States, and the Himalaya mountains that separate northern India from southern China are all examples of convergent boundaries.[57–58]

More recently a prominent seismic discontinuity was discovered at a depth of 660 km (410 mi) in the southwest Pacific Ocean near Fiji. Seismic evidence indicated that a subduction zone in this region is the subhorizontal remnant of a subducted slab of ocean crust located between the depths of 400 to 660 km. This remnant slab from an ancient subduction zone was detached from the current slab and provides a barrier to the current slab, preventing it from proceeding downward into the lower mantle. It also alters the thermal and other properties of the upper mantle in this region.[59]

Transform Boundaries

Transform boundaries produce *frictional* stresses that occur where two plates move horizontally as they attempt to slide past one another in opposite directions. Transform boundaries are sometimes referred to as *conservative plate boundaries* because mass is conserved. W. G. Ernst, in his book *Earth Systems,* describes them as the type of transform boundaries that create mountain belts: "[conservative plate boundaries] . . . as continental transform boundaries (strike-slip faults) . . . are characterized by straight, mass-conservative horizontal-slip segments, and by locally deformed zones."[60] Subduction does not play a major role in transform margins because neither plate is submerged into the mantle. Transform boundaries occur mostly, but not always, in ocean basins, and they are found perpendicular to mid-oceanic ridges. Although relatively short, they can be wide and represent fracture zones. There are two major classes of transform boundaries.

1. *Ocean-to-ocean* transform boundaries are located at the axis of an oceanic ridge. These are the types of transform margins related to the horizontal breaks that cross the linear zebra stripes of magnetic rock located on spreading seafloors. A good example is found between the East and South Pacific Ocean beds.
2. *Continent-to-continent* transform boundaries form small mountain ranges along fault lines. An example is the earthquake-prone San Andreas fault that extends from southern to central California. Most of California's earthquakes occur at less than 15 km (9.3 mi) in depth and most frequently in the southern part of this fault, where it makes a large bend. The expectation is that a future large quake is building up in the central region of the San Andreas fault.[61–63]

Earthquakes

Most earthquakes occur in the lithospheric boundary zones between oceans and continents at a depth of ~650 km (400 mi).[64] Most are detected at lesser depths along coastal margins, as this is the layer of the Earth where plates slide past each other and the lithosphere is rigid enough to be brittle, similar to the shell of a nut that can be fractured by great force. A variety of forces creates thermal stress due to buildup of heat production and heat loss near the surface as the result of tectonic movement and circulation of magma from the deeper mantle. This near-surface stress builds up potential energy that, when released, suddenly forces blocks of the lithosphere to move rapidly past each other horizontally. These forces produce a stored-up elastic strain that, when released, results in major earthquakes. Although this release of energy and movement is limited to a localized region of rock, it can produce a great movement of blocks of the lithosphere and crust and thus cause serious destruction to surface features. This slippage can produce three types of faults: (1) a *strike-slip* fault, where one plate slips past another without vertical displacement; (2) a *normal fault,* where one slab slips downward from another sloping slab; and (3) a *thrust fault,* where one slab slides up another sloping slab. Over time, a particular fault zone can produce repeated earthquakes. Earthquakes caused by these types of faults are called *tectonic quakes* because they involve the process of shifting tectonic boundaries.[65] *Volcanic quakes* are another type of earthquake that is produced by different processes. The elastic strain resulting from volcanic activity is the result of hydrodynamic motion of the magma reservoir beneath a volcano that releases gas under pressure. This release of energy can also cause a shift in local plate boundaries and create a crack or break in rocks. An analogy for elastic strain is bending a stick, about 2 feet long and 1/4 inch in diameter, until it finally breaks. Under this process, the stick undergoes elastic deformation until the continued bending reaches the point of the stick's greatest weakness, and when it breaks.[66] Another analogy for the buildup of stress is the stretching of a rubber band. The rubber band exhibits elastic deformation as the energy used to stretch it is stored in the elongated band. When the stress is removed, it returns to its original shape as it releases the stored energy. Solids and elastic material have a finite strength that has a threshold limit, before which point they are deformed when a stress is applied to the material. Stress might be thought of as a "force applied over an area." On the other hand, fluids have no finite threshold limit and are not deformed as they continue to flow regardless of the stress applied.[67] Rocks in earthquakes may form a new plastic-like shape when the stress is released, or the elastic deformation may rupture, creating a brittle fracture that results in a large crack called a fault. The San Andreas fault zone (SAFZ) in southern California is an example of the energy of the built-up stress in a transform boundary of two tectonic plates overcoming their mutual frictional resistance. When the stress reaches the breaking point, the accumulated elastic energy is released. The result is a slipping of one plate past the other.[68] The San Andreas fault is ~1,050 km (~650 mi) long and moves ~1 cm (~0.4 in) per year over long periods of geological time. However, during the 1906 San Francisco earthquake the fault line moved as much as ~6.4 m (~21 ft).[69] A January 1, 2001 Associated Press report by Robert Jablon stated that a new 5 km (3 mi) deep basin of unconsolidated underground sediment was located under the San Gabriel Valley just east of downtown Los Angeles. This poses a danger because it is located near fault lines in southern California. Gary Fuis, a seismologist with the USGS, stated in Jablon's AP report that such basins can shake like big bowls of Jell-O due to the relatively fluidlike underlying sediments that have not yet consolidated into more solid rock material. An earthquake occurring near this type of unstable geological structure increases the risk of extensive damage to life and property in the area. This discovery of deep sediment is important in the understanding of the nature of earthquake hazards.

Every year more than a million earthquakes are recorded by a worldwide system of earthquake detecting stations that collect and analyze seismic vibrations within the Earth that are caused when the energy in deformed rock is suddenly released.[70] Both earthquakes and man-made ground-level explosions create seismic waves. The point below the surface, where the movement that caused the earthquake is located, is referred to as the *focus,* while the point on the surface just above the focus is called the *epicenter.* Seismic waves are generated from the focus area. Two major and different types of vibrations (S waves and P waves) result from these seismic waves.

Measuring Earthquakes

As previously mentioned, seismology is the science of detecting and analyzing the waves and vibrations that earthquakes send through Earth's internal spheres (see the section Seismology). Thus far, the data collected have provided scientists with the information to develop accurate theories regarding the structure of the earth, as well as the location and strength of earthquakes. The recently developed global positioning system (GPS) of orbiting Earth satellites was designed to assist in accurately locating specific positions on the surface of the Earth. Geologists have found new ways to use this system to study both volcanic and earthquake activity. As a supplement to using seismometers, tiltmeters, and strainometers to detect Earth movements, geologists now are using grids of GPS instruments to detect very slight shudders and heaving of the Earth that are preludes to more destructive earthquakes. Japan's Geographical Survey Institute now gathers data from a grid of more than 1,000 GPS sensor stations located about 15 miles apart near known fault zones. These stations detect and record very slight oscillations that were previously undetectable. These data are then sent every 15 seconds to a central computer that can provide adequate warning time for possible disastrous earthquakes.[71] This new system is expected to improve the rather inexact data systems now used to provide early warnings for possible earthquakes.

The two main scales used over past years to measure the intensity of an earthquake are the Richter scale and the Mercalli scale. The Richter scale, developed in 1935 by Charles Francis Richter (1900–1985), is an absolute scale based on the **amplitude** of the waves. Richter used $\log^{10}$ (logarithm base 10), or a tenfold increase in power for each numerical increase in the scale, and magnitude 1 on the scale represents an earthquake just barely measurable. Thus, an earthquake measuring 2 on the Richter scale is ten times stronger than one measured at 1, and one reading 3 is ten times greater than one recorded as a 2 magnitude, and so on. This means that an earthquake measuring 9 has a magnitude 100,000,000 times greater than an earthquake measuring 1 on the scale. One measured at 8 is the equivalent to 5,643,000 metric tons of TNT explosive.[72]

In addition to measuring the amplitude of seismic waves, the Richter scale can measure the magnitude or extent of earthquake severity. Table 2.2 shows a comparison of Richter magnitude to earthquake severity.

The other earthquake-detection scale was developed by Giuseppe Mercalli (1890–1914) in 1902. (A 10-point scale early version of the Mercalli scale was proposed by Michele Stefano De Rossi [1834–1898] in 1878). Both Mercalli's and De Rossi's scales used indirect and subjective descriptions of the damage to buildings, as well as how the quake affected the behavior of people. These

Table 2.2
Estimates of Severity of Earthquake Damage Related to the Richter Scale

Richter Magnitude	Earthquake Effect
Less than 3.5	Generally not felt, but recorded by instruments.
3.5 to 5.4	Often felt, but rarely causes damage.
Under 6.0	At most, slight damage to well-designed buildings.
	Can cause major damage to poorly constructed buildings over small regions.
6.1 to 6.9	Can be destructive in areas up to about 100 kilometers across in population centers.
7.0 to 7.9	Powerful earthquake. Can cause serious damage over large areas of the Earth.
8 or greater	Great earthquake. Can cause serious damage in areas several hundred kilometers across.

Source: J. Louie, *Richter Earthquake Damage Scale*, accessed 2001, http://www.seismo.unr.edu/ftp/pub/louie/class/100/magnitude.html 10/9/1996.

Table 2.3
The Modern Mercalli Earthquake Intensity Scale

I. People do not feel any Earth movement (a wave with minimum amplitude, but measurable by seismographs).
II. A few people might notice movement if they are at rest and/or on the upper floors of tall buildings.
III. Many people indoors feel movement. Hanging objects swing back and forth. People outdoors might not realize that an earthquake is occurring.
IV. Most people indoors feel movement. Hanging objects swing. Dishes, windows, and doors rattle. The earthquake feels like a heavy truck hitting the walls. A few people outdoors may feel movement. Parked cars rock.
V. Almost everyone feels movement. Sleeping people are awakened. Doors swing open or close. Dishes are broken. Pictures on the wall move. Small objects move or are overturned. Trees might shake. Liquids might spill out of open containers.
VI. Everyone feels movement. People have trouble walking. Objects fall from shelves. Pictures fall off walls. Furniture moves. Plaster in walls might crack. Trees and bushes shake. Damage is slight in poorly built buildings. No structural damage.
VII. People have difficulty standing. Drivers feel their cars shaking. Some furniture breaks. Loose bricks fall from buildings. Damage is slight to moderate in well-built buildings and considerable in poorly built buildings.
VIII. Drivers have trouble steering. Houses that are not bolted down might shift on their foundations. Tall structures such as towers and chimneys might twist and fall. Well-built buildings suffer slight damage. Poorly built structures suffer severe damage. Tree branches break. Hillsides might crack if the ground is wet. Water levels in wells might change.
IX. Well-built buildings suffer considerable damage. Houses that are not bolted down move off their foundations. Some underground pipes are broken. The ground cracks. Reservoirs suffer serious damage.
X. Most buildings and their foundations are destroyed, and some bridges are destroyed. Dams are seriously damaged. Large landslides occur. Water is thrown on the banks of canals, rivers, and lakes. The ground cracks in large areas. Railroad tracks are bent slightly. Bridges twist and may fall.
XI. Most buildings collapse. Some bridges are destroyed. Large cracks appear in the ground. Underground pipelines are destroyed. Railroad tracks are badly bent.
XII. Almost everything is destroyed. Objects are thrown into the air. The ground moves in waves or ripples. Large amounts of rock may move. Total damage in some areas.

Sources: J. Louie, *Mercalli Earthquake Damage Scale,* accessed 2001, http://www.seismo.unr.edu/ftp/pub/louie/class/100/mercalli.html 10/10/1996; Patricia Barnes-Svarney, ed., *The New York Public Library Science Desk Reference* (New York: Macmillan, 1995), 390; W. G. Ernst, ed., *Earth Systems: Processes and Issues* (New York: Cambridge University Press, 2000), 405.

scales restrict use of the system to populated areas where these effects could be recorded.[73] Seismologists seeking information on the severity of earthquake effects routinely use a modified modern version of the Mercalli intensity scale. This scale does not require any instruments to record the severity of the earthquake. Instead, it relies on firsthand reports for damage assessments. The intensity ratings are given in Roman numerals, using I as low severity and XII as high or greatest damage. Table 2.3 presents the scale used by the Federal Emergency Management Administration (FEMA) of the U.S. Government.

Up to this point we have been discussing earthquakes that mostly occur on continents, but earthquakes also occur underwater on the oceanic floor, possibly more often than on land. Oceanic earthquakes that occur on the seafloor are usually the result of the collision of tectonic plates that form a rift (valley), ridge (mountain), or a transform boundary (fault) perpendicular to an oceanic rift valley or ridge. When an earthquake occurs just off a coastline or in the oceanic crust at ~5 to ~200 km (~3 to ~125 mi) offshore, it can cause great damage, not because of the vibrations or shaking of the land nearby but because of the giant waves, called tsunamis, that are generated and that cause flooding in distant coastal areas (see the section on tsunamis).

Volcanism

Volcanic activity and earthquakes are intimately associated with each other, and both usually occur near plate boundaries. Volcanoes are phenomena resulting from internal forces that circulate magma in convection cells. Magma that is extruded from weak spots in either the oceanic or the continental crust results in either lava flows or violent volcanic eruptions. Not all volcanic activity takes place at plate boundaries. One exception is the chain of volcanoes of the Hawaiian Islands located in the Pacific Ocean ~3,200 km (~2,000 mi) from the nearest plate boundary. In 1963, John Tuzo Wilson, who advanced the theory of plate tectonics and transform boundaries. also proposed the theory of stationary hot spots in the mantle over which plates could move. His concept, which contradicted the accepted plate tectonic theory was, at first, not well received. Later, seismographic and other studies proved him correct. Hot spots are located throughout central fractured regions of plates, not just at plate boundaries. A well-known example is the sustained, ancient area of volcanic activity under the Hawaiian Islands. The islands of Hawaii are located at the southern end of an ~6,000 km (~3,728 mi) volcanic trail of hot spots known as the Hawaiian Ridge–Emperor Seamounts oceanic chain. This chain reaches north on the floor of the Pacific plate to meet the Aleutian Trench, which extends from Alaska westward on the seafloor. More than 25 hot spots have been identified worldwide, including the volcanic regions of the Galapagos Islands, Iceland, and the Azores, as well as those off the southeast coast of Africa and the Hawaiian Islands. In addition, there are a number of hot spots under the North American plate. One is the region located under Yellowstone National Park. This particular fractured region exhibits thermal energy that fuels more than 10,000 hot pools and springs, including the geyser Old Faithful and numerous other boiling mud holes.[74]

An early increase in earthquake activity heralded the eruption of Mount Saint Helens, in Washington state, as well as many major volcanic eruptions in Alaska and in other countries located around the Pacific "Ring of Fire." These volcanic areas are on the edges of plate margins, either where continental and oceanic plates push against each other or where the oceanic crust is subducted under the continental crust creating both earthquakes and companion volcanoes.

A volcano may be formed by the opening of the crust/lithosphere through which magma erupts and forms mountains and oceanic ridges, or volcanic activity occurs where fracture cracks deeper in the Earth lead to magma chambers. Since the magma is less dense than the surrounding solid rock and is also under pressure, it rises to the surface through either these cracks or vents. If the pressure in the chamber is great enough, magma will exit through a weak place in the crust/lithosphere, either slowly to form a lava flow or explosively in a fiery display of shooting magma and rocks (see Figure 2.6).[75]

Types of Volcanoes

Volcanoes are defined according to several factors, including their physical form, size, type of magma, and activity. Table 2.4 describes several characteristics of volcanoes.

One of the most striking examples of a violent eruption is a *caldera,* which forms a large circular basin that may include many volcanic vents. If it erupts in the ocean, islands will be formed. In 2000, Walter L. Friedrich published the book *Fire at Sea. The Santorini Volcano: Natural History and the Legend of Atlantis* in which he describes Santorini, a group of volcanic Greek islands that formed a caldera basin in a region where tectonic plates collided about 3,500 years ago, during the Bronze Age. The Minoan eruption that took place in 1883 in the Santorini caldera changed the then existing islands as it covered them with a 60 m (197 ft) layer of **pumice**. As recently as 1950 a volcanic eruption occurred, and in 1956 a strong earthquake devastated the town of Oia on one of the islands in the caldera's basin.[76]

Figure 2.6
Volcanic activity occurs when molten magma, which is less dense than rock, rises from hot mantle plumes, through the lower pressure regions of the asthenosphere, to weak spots in the crust/lithosphere where it either oozes out slowly or erupts to form volcanic cones at plate boundaries or weak spots near the central areas of plates.

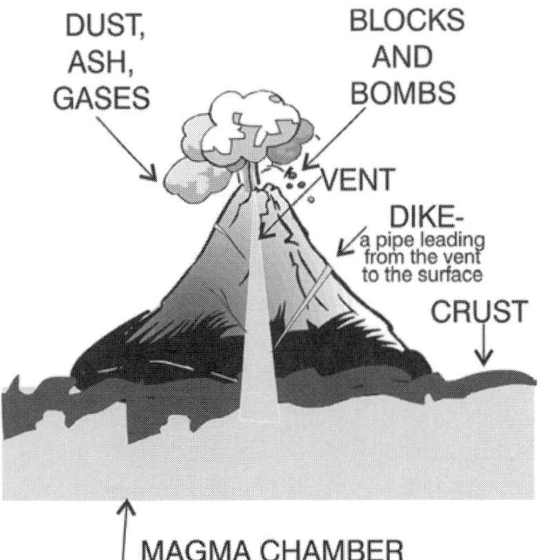

An example of a rather unique volcanic eruption occurred on July 17, 2001. The 500,000-year-old Mount Etna volcano is located in the Mediterranean Sea on the island of Sicily off the coast of the so-called boot of Italy. Mount Etna is not only the tallest volcano in Europe but also the most active. Its last previous eruption was in 1992.[77] Mount Etna's lava flows in 2001 were less viscous than the volcanic lava flows of Mount Saint Helens in 1980. This means that Mount Etna had not built up lava domes that create great pressures within the cone that result in violent explosive eruptions, as it had done during its 1992 eruption. Instead, the lava flowed freely from four fissures (vents) on the sides of Mount Etna's volcano, and the volcano spewed out some gas and ash from a poorly defined cone. The freely flowing lava destroyed some property, but people were able to evacuate without injury. One of the most amazing events that volcanologists witnessed was the formation of hoop rings of steam and aerosol gases blown out of Mount Etna's side vents. These steam rings are similar to the smoke rings blown from the mouths of cigarette or cigar smokers. While scientists do not know exactly how these rings were formed, they believe that they were caused by the high steam content of the gas erupting from the volcano's vents. Also, as compared

Table 2.4
Types of Volcanoes

Type	Form	Size	Magma	Activity	Examples
Basalt plateau	Gentle slope	Large area 1–3 km thick	Basalt	Gentle eruption	Columbia River Plateau
Shield volcano	6°–12° slope	Up to 9000 meters high	Basalt	Gentle, some fire	Hawaii
Cinder volcano	Moderate slope	100–400 meters high	Basalt or andesite	Ejection of pyroclastics	Paricutin, Mexico
Composite volcano	Alt. layers of flow	100–3500 meters high	Several types of magma + ash	Often violent	Vesuvius, Mt. St. Helens
Caldera	Explosive, central depression	Less than 40 km in diameter	Granite	Very violent	Yellowstone; San Juan Mountains

Source: Graham R. Thompson and Jonathan Turk, *Earth Science and the Environment* (New York: Harcourt Brace, 1999), 142.

to other volcanoes, an exceptionally low amount of ash spewed out into the atmosphere from this 2001 eruption of Mount Etna.[78–79]

Mud volcanoes, another type, are not actual magma-type volcanoes. They are formed by the compression of fine-grained, water-rich minerals, such as clay, that are mixed with gases to produce a pressurized slurry that bubbles to the surface as a small mudflow. Even so, since they may accompany lava flows, mudflows can be dangerous and may be considered as a type of volcano. A number of these 100 cm (3 ft) sized dome-shaped mud pies resemble large cow pies and are found on the seafloor of the eastern Mediterranean Sea. These mudflows have been discovered on the edge of the subduction area of the African plate that dips beneath the Eurasian plate.[80] In addition to creating these relatively small but numerous seafloor mud pies, mud can also bury forests and villages as it accompanies molten magma flowing down volcanic cones.

Magma

It has been obvious since the beginning of human history that the stuff ejected by volcanoes is hot enough to glow, melt, and flow down the sides of volcanic cones. This material seems to exit from the bowels of the Earth and, no doubt, led to the concept of Hades. Just what that stuff was and where it came from were questions that had to wait for modern science to answer.

Magma is, in essence, molten rock that forms at various depths in the Earth. But rocks do not melt in the exact way as a chunk of ice does as it warms up. Rocks are composed of a variety of minerals that do not all have the same melting points, because one type of mineral may have a specific **chemical bond** that can be broken at a specific temperature, while another type of mineral will have a different structure that can be broken at a different temperature. For example, the albite oxide minerals (e.g., garnet and quartz) melt at approximately 1118°C (2050°F), whereas the anorthite silicate minerals (e.g., feldspars and andesine) melt at a much higher temperature, ~1553°C (2800°F).[81]

The physical forces, energy, and conditions necessary to form different types of magma are (1) heat and temperature gradients, (2) pressure gradients and gravity, (3) density gradients, (4) water availability, and (5) viscosity.

Heat and Temperature

As previously mentioned, Earth's internal heat is generated by both the fission decay of radioactive elements and some residual heat left over from the formation of the planet 4.5 billion years ago. The difference between these two sources of heat is that heat from radioactivity is constantly renewed as it is dissipated outward toward the surface, whereas internal heat is dissipated and temperatures decline at the more shallow spheres closer to the surface. There is also a small amount of friction heat generated near the surface by the motion of tectonic plates as they slide over and against each other.

When most things become hot enough, they change from a solid to a plastic-like semi-fluid, liquid, plasma, or gas. Rocks are no different than any other solid and will melt when temperatures reach a point that breaks the bonds of the rocks' minerals. The *geothermal gradient* is the *rate* of change of temperature related to the distance when proceeding toward the center of the Earth. The lower crust contains a relatively high percentage of radioactive chemical elements that raise temperatures to the point where some, but not all, minerals begin to melt. Therefore, at the lower region of the crust/lithosphere, ~50 to 250 km (~30 to 150 mi), there is only partial melting of **felsic** type rocks, while at a greater depth, where temperatures range greater than ~1300°C (2400°F), **ultramafic** rocks begin to melt.

Molten magma seeps out of cracks in the Earth's surface or explodes in violent eruptions. Once it reaches the surface, magma is referred to as "lava," which, when cooled, becomes different forms of rocks. The three most common types of rocks resulting from cooled volcanic magma are basalt, andesite, and rhyolite. Their differences depend on their chemistry, which is related to how they were formed. (See chapter 4 for the classification of rocks.)

Pressure and Gravity

Most of us personally have experienced the increase in pressure when either diving or swimming under water. The same phenomenon exists for the atmosphere and the interior of the Earth. The deeper one drills down into the Earth the greater the pressure—as well as temperature. The force of gravity is constantly pulling the outer spheres toward the center that constantly squeezes deeper rocks, which also increases the temperature at which these rocks will melt. At the surface of the crust, where there is normal air pressure (14.7 pounds/square inch), sodium feldspar crystals will melt at about 1100°C (1950°F); at ~100 km (60 mi) depth. Where the pressure is ~35,000 times greater than on the surface, a temperature of about 1400°C (2550°F) is required to melt the same mineral crystal. In other words, the greater the pressure (depth), the greater the average amount of heat (temperature) required for rocks to become magma. Conversely, as semimolten magma is pushed outward toward the crust by convection, the pressure is reduced, as is the temperature required to melt the rocks to form magma. In addition to the decrease in pressure at the outer layer, the rifting of tectonic plates as they diverge (spread apart) also reduces pressure on rocks, thus allowing them to melt at lower temperatures to form magma.[82]

Density

Density is determined by relating the volume of an object to the object's mass (density = mass ÷ volume). It might be thought of as how much stuff (mass) there is in a given space (volume). But, and this is a big but, densities are affected by both the degree of heat (temperature) and the pressure exerted on that particular mass. When heat is added to matter, the motion of its molecules increases due to an increase in its temperature. Therefore, molecular motion also increases as rocks melt, resulting in the rock particles taking up more space. Thus, the same mass of rock material expands to occupy a greater volume, if allowed to do so. These rocks will have a greater volume and thus a lower density than those rocks that did not expand and melt. (Note that even at normal temperatures and pressures different types of rocks [and other matter] have a wide range of densities due to their varying molecular structures.) If magma is kept under extreme pressures, even if the temperatures are hot enough to cause most minerals to melt near the surface, minerals will continue to be solid (or plastic-like semi-liquid) because the increased pressure restricts molecular motion. This prevents the rocks from expanding to become lower density magma. At a depth of ~40–60 km (~25–37 mi), when fracturing in the upper mantle and lower lithosphere results in a decrease in pressure, magma is able to intrude into a higher level of the lithosphere, becoming more fluidlike as it becomes hotter and expands. If there is adequate pressure, this melted magma will be forced up to the surface in the form of lava flows or volcanoes.

Decompression melting results when pressure on a hot solid substance is reduced, allowing for greater molecular motion that results in melting (magma) followed by intrusion and cooling. This concept is also known as "pressure-releasing melting."

Water

Since the water molecule is polar, that is, a molecule with regions exhibiting slight negative and positive charges, it can attract charged particles (ions) of other substances. When the polar molecules of water attract ions of the surface of rock crystals, the bonds of the crystals' structure are weakened, thus requiring a lower temperature to break these bonds. If water was not involved, such rock crystals would require a higher melting temperature. This phenomenon is amplified when a small amount of water is under great pressure—the rock material's melting point is scientifically reduced. In other words, at the same pressure, *dry* rocks have higher melting temperatures than *wet* rocks. Thus, the amount of water affects the melting point and consistency of magma. Since rocks melt at lower temperatures when water is added, asthenosphere rocks that are hot but not melted can become semi-liquids when water is present.

As the lithospheric plate sinks into the mantle at a subduction zone of the oceanic crust, water is driven into the mantle. As the subducting plate forces the oceanic plate downward into the mantle, water-rich oceanic crust is also carried to the upper mantle. As this water is introduced into the hot, dry rock material of the upper mantle, it lowers the temperature required to melt the rocks. This allows the less dense, but still under pressure, molten magma to rise to the surface. Thus, water, as well as pressure, is an important factor in the formation of magma at the levels where oceanic plates drive water-rich crust beneath continental plates. A combination of the water reducing the melting point of rock, the friction of the plate movement, and the pressure release all add to the melting of rock in the upper mantle and the creation of magma that rises through the crust to produce volcanoes. The most important of these three factors is the induction of wet ocean sediment and water that lowers the rock's melting point.[83]

Viscosity

Viscosity is the property of a substance that exhibits resistance within itself that affects its flow rate. For instance, water flows very easily. Therefore, it is considered to have a very low viscosity, whereas molasses, being thicker (denser) than water, has a somewhat higher viscosity. Hot molten magma is also viscous. The viscosity or flow of cold rocks is measured by the rate of their shearing under stress. When molasses is heated, it becomes thinner and flows more readily. Heat generally reduces viscosity of substances. Viscosity can be described as the opposite of the physical property known as fluidity.

The percentage of silicate in magma also affects its physical properties and consistency (see Figure 4.1 in chapter 4 for the tetrahedral structure of the element silica). Silica is similar to carbon in that its tetrahedral links can join together to form long polymerized chains. If the magma contains an abundance of silica, it forms viscous lava that results in a granitic rock that has a longer silica chain structure than basalt magma that contains less silica. These long chains of silica in granite-type magma bind the material tighter, thus producing magma that is more viscous and resistant to flow. This means that it may harden before reaching the surface and may form intrusive rock. In contrast, magma with low concentrations of silica forms basaltic magma that has low viscosity and can flow rapidly down a mountainside or across a valley.[84]

Mantle Plumes

Mantle plumes have yet to be directly observed, but there is adequate evidence to support their existence. A mantle plume is a column of hot, plastic-like rock arising from the mantle, through the asthenosphere, into the lithosphere. Although mantle plumes are not a major geologically important phenomenon, they are being studied because they may lead to new 3-D seismic technologies useful for petroleum exploration, as well as providing a better understanding of smaller areas of rising magma.

It is assumed that plumes originate near the interface of the Earth's outer, hot liquid-iron core and lower mantle. This heat forms less dense magma, resulting in greater buoyancy that allows it to rise into the upper mantle to form domes that enlarge to produce buoyant mantle plumes. Experiments indicate that a new plume starts as a bulbous head with a pipelike tail that extends down to the lower mantle/outer core interface. As new upward-moving hot magma arrives in the dome through the deeper tail, it enlarges like a balloon as it cools and becomes less dense. Some of the new, hot magma from the tail shoots through the balloon to escape through a hot spot in the crust/lithosphere to form a volcano, a barrier reef, a sea atoll, or a seamount. If the plume erupts on the ocean floor, it can form a large oceanic plateau; or, as the lithosphere continues to move, plumes can form a narrow island chain of hot spot volcanoes over volcanic seamounts. They also can form mid-oceanic ridges, for example, the Hawaiian Islands.[85–86]

A recent study of seismic waves from beneath the Pacific Ocean and the southeastern Asian continent by scientists at the University of California at Berkeley, as reported in *Science* magazine, indicates new evidence of superplumes. Scientists located two superplumes of molten rock several thousand kilometers wide (or possibly several smaller ones grouped together) extruding through the boundary between the Earth's upper and lower mantle/outer core region at about 400 miles deep. They theorized that the heat from these plumes may be responsible for the low viscosity of the asthenosphere that lubricates the movement of surface plates that dip beneath each other (subduction), thus feeding volcanoes where these plates meet.[87]

Plutons

Plutons (plutonic rock) are formed in two ways, first when magma moves under the surface of the Earth flowing into fractured rock, and second when magma pushes into layers of sedimentary rock. When the magma cools, it forms tabular slabs of igneous rocks that range from a few centimeters to several hundred meters in thickness. These spread over rather large areas, so they are much broader in area than they are thick. Plutons are classified by their positions in relation to country rocks that have already been formed. Country rocks are preexisting rocks that have been subjected to intrusions of magma or have enclosed mineral deposits. This process takes place as heat convection forces molten magma upward into the fractures of country rocks. Plutons vary greatly in size and shape. Also, if surface erosion has occurred over the area where they formed, plutons may be exposed and easily observed. A variety of intrusive magma tabular pluton formations are named according to their shapes.

1. *Dikes* are flat, slablike magma formations that form a wall-like structure that cuts across layers of rocks. Examples can be viewed in exposed areas in the region known as the four corners, where the states of Colorado, New Mexico, Arizona, and Utah join.
2. *Sills* are similar in shape to dikes, but the magma was intruded between parallel layers of rocks. Two famous examples of sills exist on the Gettysburg Civil War battlefield in Pennsylvania. During the battle known as Pickett's Charge, General George Pickett led the Confederate Army from a lower sill called Seminary Ridge up a higher-level sill, called Cemetery Ridge, behind which the Union Army waited. Since this wall of rock afforded the northern army a natural advantage in their eventual victory over Pickett's forces, it can be said that geology of the battlefield helped determine the outcome of this battle and, ultimately, the Civil War.
3. *Batholiths* (from the Greek *bathos,* which means "depth") are huge massive domes or saucer-shaped, magma-formed rocks much larger than a dike or sill. They can be several kilometers thick and hundreds of kilometers in diameter. There are two basic kinds of batholiths. The first, *laccoliths* (rocklike), are large lenticular structures with flat bottoms and arched or domed tops similar to **plano-convex** lenslike shapes. *Laccoliths* are formed when felsic magma is forced into rock cracks and layers faster than it can spread out, thus pushing up the layer of rock into a dome mountain. Many laccoliths are granitic-type rocks. The second type of batholiths is *lopoliths* (in Greek *lopolith* means "saucer"). They may also be lenticular-shaped, as are laccoliths, but they are sunken in their central part due to the sagging of the underlying country rock. Lopoliths can be rich in minerals and dense crystals. A large, partly sunken lopolith in South Africa is a source of the metal platinum. Another example may be seen on the western shore of Lake Superior in the United States, where the country rock has been eroded from the surface to expose the lopolith.
4. A *Boss* is an intrinsic form of magma that pushes up a small, rounded mound of resistant bedrock. However, the magma does not reach the surface. Instead, it causes a relatively small bulge in the country rock layer above it.
5. *Stocks* are irregularly shaped magma plutons covering relatively small areas of less than 75 square kilometers (km^2) (47 square miles [mi^2]). If pluton stocks are larger than 75 km^2 (45 mi^2), they are called *batholiths.* Both may be easily seen if the overlying rock is eroded away exposing their surface magma.[88–90]

Risks from Nature's Forces and Systems

The U.S. Geological Survey (USGS) is charged with the responsibility for investigating the nature of Earth's forces that cause devastation to life and property. On the other hand, the Federal Emergency Management Administration (FEMA) is the major governmental body in the United States responsible for preventing, surveying, and estimating damages caused by nature's forces. The USGS supports and conducts geological research on the mechanisms responsible for volcanoes, earthquakes, and other natural geological phenomena in order to more accurately predict where and when Earth's dynamic forces will result in dangerous situations that may adversely affect both humans and property. Although all risks from natural geological forces cannot be avoided, it is possible to reduce the risks to human life through the knowledge of factors that provide early warnings of implied disasters.

Risks from Earthquakes

It is estimated that there are approximately 7,000,000 earthquakes (based on a historical yearly average) per year that measure 1 on the Richter scale, about 6,000 measuring 5, 20 at a magnitude of 8, and only one is measured every few years above an 8. None has ever been measured over a 9 (but one or two have been estimated as being greater than a 9).[91] It is also estimated that there are about 8,000 earthquakes every day, of which only about 1,100 are strong enough to be felt. Approximately 40 moderate earthquakes occur every year, and a major quake every two years may be severe enough to release energy equal to about 10,000 times that of the first atomic bomb.[92] A 2001 article by Paul Recer of the Associated Press reports the possibility of an earthquake of magnitude 8.1–8.3 in the Himalayan Mountain region where the Indian tectonic plate slides under the Asian plate at the rate of about 6 inches per century. If this prediction proves to be accurate, thirty million people living in India, Nepal, Pakistan, Bangladesh, and Bhutan will be at great risk.[93]

Seismologists and other scientists long for the day when they can predict with a high degree of probability the exact time that a specific fault will produce an earthquake. However, this ability is not currently possible. The techniques used by seismologists to predict with a relatively high degree of certainty that a quake may occur along a particular fault line are studying the history of large earthquakes in specific areas, and measuring the rate at which strain increases and builds up in the rocks at the fault site. Neither of these techniques is exact. Nonetheless, scientists do know that if a region has experienced four magnitude 7 or larger earthquakes during the past 200 years and if these shocks occurred randomly in time, they can assign a 50% probability that another magnitude 7 or larger quake will occur in that region in the next 50 years. Obviously, this is not very assuring to those living in a fault zone area, since these odds also mean that a major earthquake could occur anytime in the near future. A more accurate method of estimating the chances of an earthquake's occurrence is to measure and keep records of how fast the strain accumulates at the plate boundaries and then attempt to determine the critical level at which the plates will slip into new positions. Both of these methods, as well as other techniques, are now being used and tested along part of the San Andreas fault in California. Another obvious solution is to refrain from building or living near an earthquake fault. The more reasonable compromise is to construct buildings that can withstand a magnitude 7 or 7.5 earthquake. A number of construction techniques are available, including placing tall buildings on rollers so that the Earth can move under them without shaking the building. Another technique that has been used for many years is to build structures that can swing and sway without collapsing as the Earth moves or to use extra bracing in construction.[94]

The earliest recorded earthquake occurred in China in ~1177 B.C.E. The earliest recorded earthquake in Europe happened in ~580 B.C.E., while the earliest recorded quake in the Americas took place in Mexico in the fourteenth century.[95] The deadliest earthquake occurred in the year 1201 in northern Egypt with an estimated death toll of more than 1 million people.[96] In recent history there have been a dozen or so major earthquakes. Some examples are listed in Table 2.5.

Risks from Volcanoes

On any given day there are about 1,500 instances of volcanic activity on the Earth. While most volcanoes are not violently erupting, many are oozing and fuming lava and gases, and some even threaten populations that live and farm the surrounding fertile soil.[97] Geologists and seismologists are continually perfecting techniques to measure the buildup of forces that lead to both earthquakes and volcanoes. Despite some evidence that suggests that it may be possible to predict future eruptions of large volcanoes, as yet there is no known process that can determine exactly when an active, or even a dormant, volcano will erupt with great violence. Table 2.6 lists most of the deadliest recorded volcanic eruptions. Interestingly, some of the most violent volcanoes

Table 2.5
Famous Earthquakes

Year	Location	Richter Number
1811	New Madrid, Missouri (a series of quakes)	8.0 to 8.3
1899	Yakutat Bay, Alaska	8.3 to 8.6 (est.)
1906	San Francisco, California	7.7 to 8.25 (est.)
1960	Chile	~9.? (questionable)
1964	Alaska	8.5
1971	San Francisco, California	6.5
1976	Tangshan, Chile	8.2
1976	Hebei Province, China	7.7*
1985	Mexico City, Mexico	8.1
1988	Armenia	6.9
1989	Loma Prieta, Califorina	7.1
1990	Northwest Iran	7.7
1994	Bolivia, South America	8.3 (deepest)
1994	Northridge, California	6.8
1995	Kobe, Japan	6.8
2001	San Salvador, Central America	7.6
2001	Ahmedabad, India	7.9
2001	Seattle, Washington	6.8

Sources: Patricia Barnes-Svarney, ed., *The New York Public Library Science Desk Reference* (New York: Macmillan, 1995), 391; *This Dynamic Earth, Plate Tectonics and People,* accessed 2001, http://www.pubs.usgs.gov/publications.text/tectonics.html.
*This 7.7 earthquake in China killed as many as 800,000 people.

Table 2.6
Volcanic Eruptions with 500 or More Human Fatalities

Year	Volcano Location	No. of Deaths	Major Cause of Death
c.147 B.C.E.	Greece	16,000	Ash and lava flows
79	Vesuvius, Italy	10,000	Ash flow and rock falls
1631	Vesuvius, Italy	3,500	Mudflows, lava flows
1586	Java, Indonesia	20,000	Ash, lava, tsunami
1640	Komagatake, Japan	700	Tsunami
1741	Oshima, Japan	1,475	Tsunami
1772	Papandayan, Indonesia	2,957	Ash flow
1783	Asama, Japan	1,377	Ash and mudflows
1783	Laki, Iceland	9,350	Starvation
1792	Unzen, Japan	14,300	Volcano collapse, tsunami
1814	Mayon, Philippines	1,200	Mudflow
1815	Tambora, Indonesia	92,000	Starvation
1845	Nevado de Ruiz, Colombia	700	Mudflow
1877	Cotopaxi, Ecuador	1,000	Mudflow
1882	Galunggung, Indonesia	4,011	Mudflows
1883	Krakatoa, Indonesia	36,417	Tsunami
1885	Nevado de Ruiz, Colombia	25,000	Mudflows
1902	Mt. Pelee, Martinique	29,025	Ash flows
1902	Soufriere, St. Vincent	1,680	Ash flow
1911	Taal, Philippines	1,335	Ash flow
1919	Kelut, Indonesia	5,110	Mudflows
1951	Hibok-Hibok, Philippines	500	Ash flow
1951	Lamington, Papua, New Guinea	2,942	Ash flow
1963	Agung, Indonesia	1,184	Ash flow
1982	El Chichon, Mexico	2,000	Ash flow
1991	Pinatubo, Philippines	800	Roof collapses
2000	Popocatepeti, Mexico	unknown	Area evacuated
2001	Mount Etna, Italy	unknown	Advanced warning
2002	Afghanistan	not determined	Collapsed houses

Sources: "The Deadliest Eruptions," adapted from Russell J. Blong, *Volcanic Hazards: A Sourcebook on the Effects of Eruptions* (Orlando, Fla.: Academic Press, 1984), 424; *Scientific American Science Desk Reference* (New York: John Wiley and Sons, 1999), 132–33.

caused few deaths because they occurred in very remote areas or they erupted far in past history before deaths were recorded. The numbers of deaths listed in Table 2.6 are estimates.

By far, the most deadly volcanic eruptions have occurred in Indonesia, followed by areas in the Caribbean, Japan, Central America, and the Philippines. The 1980 eruption of Mount Saint Helens in Washington state, after 123 years of quiescence, does not even come close to making the list because there were only 61 deaths reported. However, great numbers of wildlife were killed and much of the surrounding forest was devastated. An amazing fact is how fast vegetation reestablished itself on the slopes of Mount Saint Helens in just a few years, including the beginnings of new forests. One ancient volcano not included in Table 2.6, because there is no actual record of it, is the eruption on Santorini Island in the Mediterranean Sea, estimated to have occurred in ~1650 B.C.E., that presumably resulted in many thousands of deaths. This eruption also led to the myth of the legendary lost city of Atlantis.

It was reported that the Popocatepetl volcano, located about 40 miles southeast of Mexico City, created problems with trying to protect people living near this potential disaster area. The 17,886-foot volcano had been shooting out vapor, ash, and rock on-and-off since December 1994, but with no major eruption. This may have given the residents of villages located near the volcano's base a sense of security. One citizen claimed that it was burning for six years and people are destined to live with 'Popo' the volcano. During the end of the year 2000, many reporters kept the local villagers informed of the volcano's progress while government officials kept emergency facilities opened. Yet few people wished to be evacuated or paid much attention to warnings, even after Popo became more active. This volcano continues to be potentially dangerous, and a major eruption could result in tremendous damage and loss of life.

To minimize the loss of life from these deadly, but natural, events, governments must formulate more effective systems to warn populations of the dangers of imminent—or even potential—volcanic eruptions. The problems associated with adequate warning systems are compounded by two facts. First, many people already live near these areas, and second, despite the known risks, more people continue to settle in areas near volcanoes. The reason for this migration may be the fertility of volcanic soil that aids in food production. Other possible explanations are cultural reasons, tradition, lack of options, complacence, or just human nature to ignore possible future risks. Whatever the reasons, the job of the government agencies was complicated, which raised the probability that lives will be lost in the future.

At one time it was thought that there was little relationship between earthquakes and volcanoes. In late 1990, a magnitude 7.8 earthquake occurred along the Philippine fault and was followed just a few hours later by the awakening of the 500-year-old nearby sleeping giant, Mount Pinatubo, which proceeded to erupt over the next several months. It is estimated that from 200 to 1,000 megatons of supercritical volatiles arose from 5 to 10 km beneath the Earth's surface. In addition to basaltic and dacitic magma, this volcano spewed out ~20 megatons of sulfur dioxide (SO_2). Pinatubo continued to erupt and flow through the next several months, producing the largest stratospheric volcanic aerosol cloud of the twentieth century. The climatic effect was substantial. The air temperatures over North America, Europe, and Siberia were warmer than normal, whereas temperatures in Alaska, Greenland, and the Middle East were significantly lower than normal. There was a rare snowstorm in Jerusalem, and the low temperatures destroyed the coral at the bottom of the Red Sea. Research since this time indicates that similar climatic changes occur following every erupting tropical volcano—past and present. As a result, global climates will continue to be affected by future explosive volcanoes. Following Mount Pinatubo's eruption, it was determined that the stratospheric SO_2 acted as a major contributor to ozone destruction, causing the ozone hole over Antarctica in October 1992.[98–99]

Risks from Tsunamis

The word *"tsunami"* is derived from the Japanese words *tsu* (harbor) and *nami* (wave). Tsunamis are sometimes referred to as "harbor waves," or "seismic sea waves." At times, they are also called "tidal waves," but this is inaccurate because they are not influenced by tidal effects except to the extent that they may be more massive during high tides. In other words, during high tides they can be more destructive.

Eighty percent of tsunamis are giant waves generated by underwater earthquakes. About 20% result from underwater landslides, and just a few are caused by volcanoes. These events are triggered at subduction zones located near shorelines where one tectonic plate slides under another. In about 1600 B.C.E. a giant tsunami that was caused by the collapse of a volcano on the Greek island of Thera (now Santorini) was credited with wiping out the Bronze Age Minoan civilization on Crete and other islands in the Aegean Sea.[100–101]

About 80% of the tsunamis occur in the Pacific Ocean and are caused by either underwater earthquakes, landslides that occur underwater, or volcanic eruptions either on land or on the ocean floor. They usually range from about 1 m (3 ft) to more than 61 m (200 ft) in height, but they can be as high as 150 m (500 ft). A tsunami's wavelength can range from 200 m (656 ft) to 965 m (1968 ft) from crest-to-crest, and with a period (time between crests) ranging from 5 minutes to more than a few hours. An earthquake on the floor of the ocean can produce a tremendous push as it displaces a large area of seawater, thus creating a destructive tsunami wave. As waves reach a shallow, sloping, offshore coastline, their amplitudes (half the height of a wave's crest above the troughs adjacent to the crest) increase due to the bottom becoming shallower. This causes the waves to stack up in height, because the same amount of water now occupies a reduced depth at the shoreline. Much of the energy stored in the approaching massive wave is expended in increasing the wave's magnitude and, thus, its destructive force. In the open ocean tsunamis can reach speeds up to 950 km per hour (590 mi/hour), about the speed of a passenger jet airplane. However, their speed is reduced (as their height is increased) by friction with the shallow ocean bottom as they approach land.[102–103] Tsunamis are not destructive in the open, deep sea because their amplitude (height) is only about 1 m (3 ft) and their crests are spread far apart over great distances. Large ocean-going vessels seldom notice tsunamis in open water. As mentioned, a tsunami becomes a destructive killer as it reaches shallow coastal waters. Two relatively recent earthquake-caused tsunamis resulted in considerable damage to the Hawaiian Islands. An earthquake originating in Valdez, Alaska, in 1964 created a giant tsunami wave that hit Hawaii about 6.5 hours later. Another tsunami, originating in 1960 in Concepción, Chile, traveled over a thousand miles in ~14 hours to reach Hawaii, where it killed 61 people and caused about $24 million in damage.[104] (See Figure 2.7 for an artist's conception of a tsunami.)

One of the best eyewitness accounts of a very destructive tsunami was reported in the May 1999 issue of *Scientific American* magazine. This report concerned retired Colonel John Sanawe, who lived near the end of a sandbar on the northern coast of Papua New Guinea and survived the major tsunami:

> July 17th, 1998 was just another tranquil tropical day when just after the first earthquake shock struck only 20 kilometers offshore he saw the sea rise above the horizon and then spray vertically perhaps 30 meters. First, Colonel Sanawe heard a sound like distant thunder, and then a sound like a nearby helicopter that faded as the sea slowly receded below the normal low-water mark. After several minutes of silence he heard a rumble like a low-flying jet plane as the first 3 or 4 meter (10 or 13 feet) high tsunami wave

Figure 2.7
Tsunamis are generated at the site of earthquakes on the seafloor that result in the displacement of huge volumes of water, causing a surface wave that spreads in all directions. In the open ocean the wave is only a few meters high, but as it reaches the shallow shoreline, the water rises up to form destructive waves that can exceed 150 m in height.

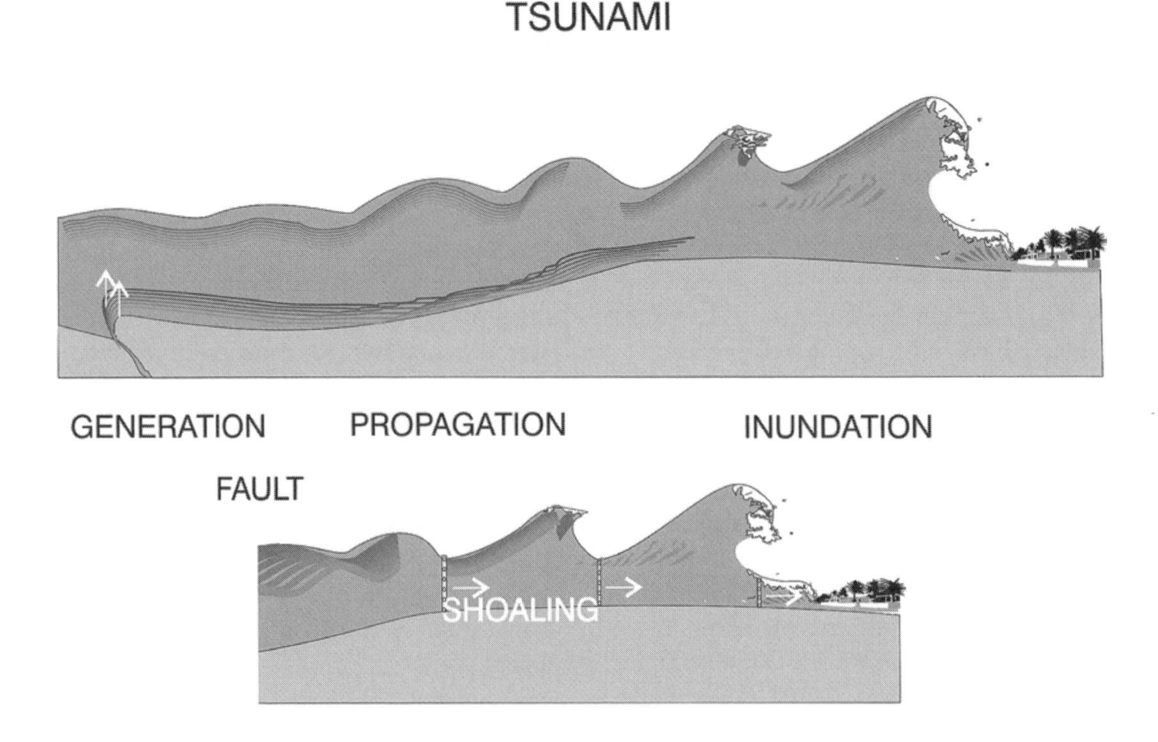

came ashore. As he tried to outrun the wave, it overtook him as the second larger wave flattened the village and swept him a kilometer (almost a mile) into an inland mangrove forest. Colonel Sanawe survived, but more than 2,200 villagers did not.

It was determined that this 7.1 magnitude earthquake in Papua New Guinea produced waves up to 15 m (50 ft) high that struck within 15 minutes. Since this is a relatively small, low-lying island, there were few shelters or places of safety to which the islanders could flee. Thus, the number of casualties was quite high. Even small tsunamis are among the most powerful waves on Earth. In 1792 a tsunami caused by a volcanic eruption was reported to have killed more than 15,000 people in Unzen, Japan. And one of the most destructive tsunamis on record was caused by the eruption of the Krakatoa volcano in the Sunda Straits between the islands of Java and Sumatra in Indonesia in 1893. This tsunami wave washed away 165 coastal villages, killed more than 36,000 people, and reportedly was as high as a twelve-story building.[105] During the decade 1990–2000 there were more than 80 tsunamis reported, but most caused little damage because they struck undeveloped coastlines. Eleven were major disasters that struck coastlines from Java to Chile, killing more than 4,000 people. Today, scientists are placing tsunami sensors along crucial subduction zones to detect, measure, and give early warnings to coastlines threatened by potential tsunamis.[106]

Summary

Physical forces that follow scientific laws are responsible for the processes that drive the complex internal systems of the Earth. These forces affect the cyclic nature of the interchange of magma, minerals, and rocks between the various internal spheres of the Earth. One result of this cyclic system is plate tectonics, which is a global geological concept that explains the constant but slow movement of large chunks of the lithosphere across the asthenosphere and upper mantle. Subduction occurs when one tectonic plate slides under another as their boundaries meet, sending surface material deeper into the Earth's spheres. This process constantly recycles oceanic crust and lithospheric rock down into the asthenosphere, the upper mantle, and even to the edge of the outer core where the subducted material is assimilated into the lower mantle. The deep magma is then recirculated to the surface by convection currents that are created by heat and internal pressure that force hot, new, less dense, viscous magma to rise to the surface. This recycled magma intruding into the mid-oceanic ridges, along with undersea volcanic activity, result in the spreading of the seafloor. As magma flows up and over the crests of mid-oceanic ridges, the ocean floor on both sides of the ridge diverges (spreads), resulting in the expansion of the ocean bed to the point where it submerges under the margins of continental crusts into deeper regions of the Earth. The convergence (meeting) of continental crust/lithospheric plates at their boundaries creates mountains, earthquake fault lines, volcanoes, and other geological formations. The rate of continental plate movement is less than an inch per year, whereas ocean plate movement is a bit faster. Although plate movement is relatively slow, some geological events can be very rapid and violent, as with earthquakes caused by violent slippage at plate boundaries or volcanoes at weak boundary areas and hot spots formed near the Earth's surface.

How information on plate tectonics was discovered and verified over the years involves many people and the use of various techniques and instruments for observing, gathering, recording, and interpreting data about the Earth. Historically, the evidence for continental drift was based on limited observations, namely, (a) the apparent congruity of the coastlines of continents separated by oceans, which suggested that they were once joined (for example, the east coast of South America seems to match the outline of the west coast of Africa.); (b) the similarity of rock types and geological formations on different continents noticed by early geologists; and (c) that plant and animal fossils seemed to be similar on different continents, which could mean that they were once connected.

More recently, improved instrumentation and data analysis methods have been developed that greatly increased the geographic knowledge that led to the acceptance of plate tectonics as being mainly responsible for the dynamic nature of the Earth. These techniques include geological land surveys, geological ocean surveys by ship, sensitive air and ocean magnetometers, improved seismology (including 3-D techniques) and sonars, and global surveys by space satellites. Some of the information gleaned from using these instruments and techniques are (a) the identification of mid-oceanic ridges, with symmetrical deepening about the ridges, a central rift at the axes of the ridges, and undersea volcanic action at the ridges; (b) discovery of numerous transform faults that cut across the ridges at approximately right angles; (c) spaced magnetic stripes extending outward and parallel to the ridges and the periodic reversal of the Earth's magnetic poles over long periods of history that concentrated the magnetic magma in opposite directions to produce these stripes on the spreading ocean floor; (d) the subduction of ocean slabs beneath continental coastal areas, producing coastal trenches and volcanic activity; (e) deep earthquakes indicating that these ocean slabs dip deep into the Earth's mantle; (f) shallow earthquakes that outline the boundaries where plates meet, and mountain building at the boundaries of meeting continental plates; and (g) radar and space satellite measurements that detect actual plate motion.

One of the benefits resulting from Earth's system of plate tectonics is the cyclic processes that formed many of our metallic ores and minerals. These and other natural resources are concentrated near past or present boundaries of continental plates. Obviously, volcanoes can cause short-term damage, but over the long term they benefit mankind as volcanic rock weathers and breaks down to form fertile soils. Most of the metallic minerals (copper, lead, zinc, silver, gold, etc.) are found deep in the roots of extinct volcanoes associated with geologically ancient subduction zones. Fossil fuels such as petroleum and natural gas are the products of deeply buried and decomposed organic matter that accumulated in subterranean and submarine sediments. Geothermal energy in the form of hot water or steam is harnessed from natural heat sources produced at shallow fractures in the Earth's surface. The use of these natural resources has sustained human civilization, both in the past and surely in the future. These and other forces, processes, and systems were active on the Earth soon after its formation ~4.5 billion years ago and will continue their dynamic activity for many more billions of years.

Notes

1. "Why are Planets Round?," *Scientific American* Web site, Ask the Expert, Derek Sears, accessed 20 February 2002, http://www.sciam.com.
2. Gordon J. MacDonald, "How Mobile is the Earth?", in *Plate Tectonics: An Insider's History of the Modern Theory of the Earth,* ed. Naomi Oreskes (Boulder, Colo.: Westview Press, 2001), 119.
3. Patricia Barnes-Svarney, ed., *The New York Public Library Science Desk Reference* (New York: Macmillan, 1995), 400.
4. Isaac Asimov, *Isaac Asimov's Guide to Earth and Space* (New York: Fawcett Crest, 1991), 94.
5. MacDonald, "How Mobile is the Earth?," 199–201.
6. Steven Weinberg, *Facing Up: Science and Its Cultural Adversaries* (Cambridge, Mass.: Harvard University Press, 2001), 97.
7. Robert E. Krebs, *Scientific Development and Misconceptions through the Ages* (Westport, Conn.: Greenwood Press, 1999), 157.
8. Weinberg, *Facing Up,* 61.
9. Weinberg, *Facing Up,* 65.
10. Robert Temple, *The Genius of China* (London: Prion Books, 1999), 162–66.
11. Kenneth W. Hamblin and Eric H. Christiansen, *Earth's Dynamic Systems* (Upper Saddle River, N.J.: Prentice Hall, 1998), 481.
12. Alan E. Mussett and M. Aftab Khan, *Looking Into the Earth: An Introduction to Geological Geophysics* (New York: Cambridge University Press, 2000), 65, 84.
13. Graham R. Thompson and Jonathan Turk, *Earth Science and the Environment* (New York: Harcourt Brace, 1999), 109.
14. Thomas McGuire, *Reviewing Earth Science: The Physical Setting* (New York: Amsco School Publications, 2000), 42.
15. Thompson and Turk, *Earth Science,* 110–11.
16. Hamblin and Christiansen, *Earth's Dynamic Systems,* 472–73.
17. Hamblin and Christiansen, *Earth's Dynamic Systems,* 486–92.
18. Thompson and Turk, *Earth Science,* 123–27.
19. *The Oxford English Dictionary,* 2d ed., CD-ROM, s.v. "asthenosphere."

20. Thompson and Turk, *Earth Science,* 89.

21. Robert L. Bates and Julia A. Jackson, *Dictionary of Geological Terms* (New York: Anchor Books/Doubleday, 1984), 110.

22. Thompson and Turk, *Earth Science,* 101.

23. Edward J. Denecke Jr., *Let's Review: Earth Science* (Hauppauge, N.Y.: Barron's Educational Series, 1995), 129.

24. Bates and Jackson, *Dictionary of Geological Terms,* 456.

25. Edwin Simons Robinson, *Basic Physical Geology* (New York: John Wiley and Sons, 1982), 102.

26. Denecke, *Let's Review,* 125.

27. Hamblin and Christiansen, *Earth's Dynamic Systems,* 486–88.

28. John Daintith, Sarah Mitchell, Elizabeth Tootill, and Derek Gjertsen, *Biographical Encyclopedia of Scientists,* vol. 2 (Bristol, England, and Philadelphia, Pa.: Institute of Physics Publishing, 1994), 627.

29. James E. Bobick and Margery Peffer, *Science and Technology Desk Reference,* 2d ed. (New York: Gale, 1996), 235.

30. Naomi Oreskes, ed., *Plate Tectonics: An Insider's History of the Modern Theory of the Earth* (Boulder, Colo.: Westview Press, 2001), 4.

31. *Microsoft Encarta,* CD-ROM 1994, s.v. "Leonardo da Vinci."

32. Jon Erickson, *Plate Tectonics: Unraveling the Mysteries of the Earth* (New York: Checkmark Books, 2001), 9.

33. Robert E. Krebs, *Scientific Laws, Principles, and Theories: A Reference Guide* (Westport, Conn.: Greenwood Press, 2001), 310.

34. Erickson, *Plate Tectonics,* 11.

35. Thompson and Turk, *Earth Science,* 102.

36. Krebs, *Scientific Laws, Principles, and Theories,* 338.

37. Thompson and Turk, *Earth Science,* 103.

38. *Encyclopedia Britannica,* CD-ROM 2000, s.v. "Plate Tectonics: Gestation and Birth of Plate Tectonics."

39. "Geology: One World," Science and Technology, *Popular Science,* January 2001, 26.

40. Oreskes, ed., *Plate Tectonics*, ix–xxiv, 3–27.

41. Mayde Creek Junior High School, *Earth's Layers and Plate Tectonics,* MCJH-Earth Science-PlateTectonics, accessed 2001, http://www.katy.isd.tenet.edu/mcjh/mcjhhsc/PlateTectonics/platetec.

42. John G. Sclater, "Heat Flow Under the Oceans," in *Plate Tectonics: An Insider's History of the Modern Theory of the Earth,* ed. Naomi Oreskes (Boulder, Colo.: Westview Press, 2001), 133–36.

43. Ron Mason, "Stripes on the Sea Floor," in *Plate Tectonics: An Insider's History of the Modern Theory of the Earth,* ed. Naomi Oreskes (Boulder, Colo.: Westview Press, 2001), 31–45.

44. Frederick J. Vine, "Reversal of Fortune," in *Plate Tectonics: An Insider's History of the Modern Theory of the Earth,* ed. Naomi Oreskes (Boulder, Colo.: Westview Press, 2001), 46–66.

45. *Encyclopedia Britannica,* CD-2000, s.v. "Plate Tectonics: Plate Motion."

46. "The Birth of Plate Tectonics," *Introduction to Plate Tectonics,* accessed 2001, http://www.volcano.und.nodak.edu/vwdocs/vwlessions/plate_tectonics/part11.

47. McGuire, *Reviewing Earth Science,* 51–55.

48. W. G. Ernst, ed., *Earth Systems: Processes and Issues* (New York: Cambridge University Press, 2000), 98.

49. "Understanding Plate Motions. Divergent Boundaries," accessed 2001, http://www.pubs.usgs.gov/publications/text/understanding.html.

50. Thompson and Turk, *Earth Science,* 91–93.

51. Hamblin and Christiansen, *Earth's Dynamic Systems,* 460, 521–28.

52. *Plate Tectonics: The Action is at the Edge,* accessed 2001, http://www.aqd.nps.gov/grd/usgsnps/pltec/pltec3.

53. Stanley Chernicoff, *Geology: An Introduction to Physical Geology* (New York: Houghton Mifflin, 1999), 18–20.

54. Mussett and Khan, *Looking Into the Earth,* 313–20.

55. Mussett and Khan, *Looking Into the Earth,* 320.

56. Robinson, *Basic Physical Geology,* 147.

57. Thompson and Turk, *Earth Science,* 92–95.

58. Hamblin and Christiansen, *Earth's Dynamic Systems,* 496–528.

59. Harry W. Green, "A Graveyard for Buoyant Slabs?", *Science,* 29 June 2001, 2445–46.

60. Ernst, *Earth Systems,* 95.

61. Thompson and Turk, *Earth Science,* 92–95.

62. Hamblin and Christiansen, *Earth's Dynamic Systems,* 534–51.

63. Denecke, *Let's Review,* 107, 132–34.

64. Macdonald, "How Mobile is the Earth?," 121.

65. *Encyclopedia Britannica,* CD-ROM 2001, s.v. "Earthquakes: Causes of Earthquakes."

66. McGuire, *Reviewing Earth Science,* 41.

67. MacDonald, "How Mobile is the Earth?," 114–15.

68. Thompson and Turk, *Earth Science,* 108.

69. *Encyclopedia Britannica,* CD-ROM 2001, s.v. "San Andreas Fault."

70. Hamblin and Christiansen, *Earth's Dynamic Systems,* 470–71.
71. Kathy A. Svitil, "Science's Favorite New Technology: GPS," *Discovery* 21, no. 3 (March, 2002), 73.
72. Barnes-Svarney, *The New York Public Library Science Desk Reference,* 391.
73. Daintith et al., *Biographical Encyclopedia of Scientists,* vol. 2, 754–55.
74. Jacquelyne W. Kious and Robert I. Tilling, *This Dynamic Earth: The Story of Plate Tectonics* (Denver, Colo.: U.S. Geological Survey; Washington, D.C.: U.S. Government Printing Office, 1999), 47–52.
75. Denecke, *Let's Review,* 99–100.
76. Walter L. Friedrich, *Fire in the Sea. The Santorini Volcano: Natural History and the Legend of Atlantis* (Cambridge, UK: Cambridge University Press, 2000).
77. "New Eruptions Mark a Summer of Volcanoes," *Valley Morning Star* (Harlingen, Tex.), 31 July 2001.
78. Richard Stone, "Volcanology: Etna Eruption," *Science,* 3 August 2001, 774–75.
79. Jurg Alean and Marco Fulle, "Etna Hoops it Up: BBC News," accessed 2002, http://news.bbc.co.uk/hi/english/sci/tec/newsid_696000/696953.stm.
80. Linda Rowan, "Geology: Mediterranean Mud Pies," *Science,* 13 July 2001, accessed 2001, http://www.sciencemag.org/content/vol293/issue5528/twil.shtml.
81. Chernicoff, *Geology,* 72.
82. Chernicoff, *Geology,* 72.
83. Chernicoff, *Geology,* 74.
84. Thompson and Turk, *Earth Science,* 132–36.
85. Hamblin and Christiansen, *Earth's Dynamic Systems,* 586–610.
86. Thompson and Turk, *Earth Science,* 133–34.
87. Barbara Romanowicz and Yuancheng Gung, "Superplumes from the Core-Mantle Boundary to the Lithosphere: Implications for Heat Flux," *Science,* 19 April, 2002, 513–16.
88. Denecke, *Let's Review,* 101.
89. Bates and Jackson, *Dictionary of Geological Terms,* various entries.
90. Chernicoff, *Geology,* 78.
91. Hamblin and Christiansen, *Earth's Dynamic Systems,* 474.
92. "The Terrible Power of Earthquakes," accessed 2001, http://www.aolsvc.worldbook.aol.com.
93. Paul Recer, "Researchers: Massive Earthquake Could Rock Himalayan Mountains," *Valley Morning Star* (Harlingen, Tex.), 5 August 2001.
94. Kaye M. Shedlock and Louis C. Parker, *Earthquakes—General Interest Publications,* U.S. Geological Survey, http://www.pubs.usgs.gov/gip/earthq1/earthqkgip.html.
95. Shedlock and Parker, *Earthquakes.*
96. "Most Deadly Earthquakes in History," accessed 2001, http://www.aolsvc.worldbook.aol.com.
97. "NewWatch. Resources. Ready to Rumble," *Science,* 22 June 2001.
98. C. G. Newhall et al., "To Make Grow: Perspectives: Pinatubo Eruption," *Science,* 15 February 2002, 1241–42.
99. Alan Robock, "The Climatic Aftereffect: Perspectives: Pinatubo Eruption," *Science,* 15 February 2002, 1242–44.
100. Robert Koenig, "Researchers Target Deadly Tsunamis," *Science,* 17 August 2001, 1251–53.
101. "How Earthquakes Cause Damage: *Tsunamis,*" accessed 2002, http://www.aolsvc.worldbook.aol.com.
102. Barnes-Svarney, *The New York Public Library Science Desk Reference,* 389–90.
103. Bates and Jackson, *Dictionary of Geological Terms,* 539.
104. Kious and Tilling, *This Dynamic Earth,* 71.
105. "Tsunamis. Natural Hazards. Plate Tectonics and People," accessed 2002, http://www.pubs.usgs.gov/publications/text/tectonics.html.
106. Koenig, "Researchers Target Deadly Tsunamis," 1253.

Earth's Structure

The Geosphere

Humans always have been curious about their surroundings. It is part of our unique nature. The earliest ancestors of modern humans surely speculated on how the heavens, Earth, and its life forms came into being, often attributing the existence of things animate and inanimate, as well as the forces that altered the land such as the weather and the seasons, to various gods or unknown powers. More than two thousand years ago, Eratosthenes (c.276–c.94 B.C.E.) and Aristotle (384–322 B.C.E.) were among the first to contemplate the Earth on a more rational and empirical basis. But even they did so more as a philosophical exercise than a scientific endeavor. It was not until the Renaissance and the Enlightenment that humans began a systematic examination of the Earth and the physical processes that affected the Earth, as well as its environment. Up until then, much of the study of the Earth was based on speculation and assumptions rather than gathering and analyzing empirical data. The science of geology began in the seventeenth century, when the study of the Earth relied less on philosophy and more on the development of scientific principles. Rene du Perron Descartes (1596–1650), Pierre Perrault (1611–1680), and John Ray (1627–1705) contributed to the nascent science of geology. In the eighteenth century, Abraham G. Werner (1750–1817), a German mineralogist, developed a classification of rocks based on their external or physical characteristics, from which he formulated his theory of Neptunism (after the Greek god for the oceans), which stated that rocks were the result of the **crystallization** of mineral deposits in the oceans.[1] Another early geologist, James Hutton (1726–1797), claimed that the evidence for the formation of rocks indicated that they were not the result of water but instead of heat. Therefore, he referred to his theory as "Plutonism," from the Greek god Pluto of Hades. Hutton's concept, based on his observations of rock formations, indicated that the solidification of a molten mass formed the Earth.[2] Hutton further proposed that the physical processes responsible for the geological past were forces that continue in the present. This concept of rocks forming over a long period of history is known as the principle of **uniformitarianism**. It does not mean that change occurs at a uniform rate, however, but instead that "the present is the key to the past." A related concept is the law of **superposition**, where layers of sediment formed sedimentary rocks with older layers found beneath more recent sediments. Hutton theorized that the heat of the Earth caused sedimentary rocks to fuse into the granites and flints since he claimed the **igneous** rocks could not have been formed without heat.[3] Hutton's principle of uniformitarianism is somewhat different from the doctrine of **catastrophism**, proposed by Georges Cuvier (1769–1832), where sudden, violent worldwide events resulted in the modifications of the Earth's crust. These ideas encouraged Sir Charles Lyell (1797–1875), Cuvier, and others to explain the law of faunal succession, which states that fossils found in layers of older rocks represent a succession of life forms signifying changes in time as well as changes in living organisms. Thus, early geologists provided much of the rationale for Charles Darwin's theory of organic evolution. In 1972, Niles Eldredge and Stephen Jay Gould updated the old theory of catastrophism by restating it as their hypothesis of punctuated equilibrium. Punctuated equilibrium asserts that evolution does not occur slowly and continuously over extended periods of time, as in Darwinian evolution, but instead evolutionary changes are evident within relatively short periods and occur episodically, not continuously.[4] Possibly these evolutionary events result from catastrophic geological events, such as a large asteroid crashing into the Earth or major cyclic changes in the Earth's climate.

Modern geology consists of a number of related sciences that study the Earth's history, composition, and the dynamic processes that affect its structure. It is not always recognized that the Earth, as with life, involves dynamic systems. For the most part, physical and chemical forces that cause changes in and on the Earth do so over long periods of time. Thus, humans during a single lifetime do not easily recognize the dynamics of geological changes. On the other hand, there are forces that result in earthquakes and volcanic activity, which do cause relatively rapid geological changes over short time periods.

The term "geosphere" has now become synonymous with what humans, since the beginning of time, have referred to as the "Earth." Early scientists recognized boundaries between layers of soil and rocks and folds in outcropping of rocks. As they studied the makeup of their world, they continued to use the ancient concepts of concentric spheres when describing and identifying layers or regions of the Earth. Thus, we have several main spheres of the Earth, each with several concentric subspheres, not always sharply defined. Early geologists referred to the Earth as being divided into two major classes of structures: (1) "concenters," or large concentric regions (core, silica rocks, and crust); and (2) subdivisions of these concenters called "envelopes," or "geospheres."

These different spheres are defined not only by their depths but by their different chemical and physical properties. A brief discussion of these spheres is presented in the following list; they are addressed in more detail in subsequent chapters.

1. Earth's *crust* is a thin, outer surface layer composed of minerals, solid rock, soil, sediments, and water. There are two distinct types of surface crust.

   A. *Continental crust* is the layer that forms the continental landmasses. It varies in thickness from ~20 kilometers (km) (~12.5 miles [mi]) to a maximum of ~70 km (~44 mi) where mountains have formed. The United States's crust varies in thickness from ~30 km (~19 mi) to ~50 km (~31 mi). The surface crust consists of minerals, rocks, and soils of many types that are constantly being altered by a variety of forces. The continental crust, although thicker, is less dense than is the oceanic crust.

B. The *oceanic crust* lies below ~4 km (~2.5 mi) of water and consists of several sublayers, that is, marine sediments, submarine basalt-type rocks (a dark-colored, fine-grained, extrusive igneous rock, and a layer of gabbro (a dark-colored, coarse-grained, intrusive igneous rock). The combination of these rocks results in the oceanic crust being denser than the continental crust.

2. The *lithosphere* is the rock area that covers the Earth to a depth of approximately ten kilometers. Most geologists consider the rock portion of the crust and the upper mantle to compose the lithosphere. Because it is less dense than the region just below it, there is a tendency for lithosphere to float over the semifluid, pitchlike asthenosphere. The mountains located in crust/lithosphere are only slightly denser than the rest of the rocks, so they dip a little deeper into the next layer.
3. The *asthenosphere,* a relatively weak layer, or envelope, located just below the lithosphere, is known as the "zone of movement." It extends from ~100 to ~350 km (~62 to ~218 mi) in depth. It is a relatively narrow layer, consisting of a hot plastic, tarlike substance. Because it is less firm than solid rock, it is thus less dense and therefore causes seismic waves to weaken when they are sent through it. The asthenosphere is sometimes also considered part of the upper mantle.
4. The *Mohorovicic discontinuity* (Moho) is a transition zone located between the bottom of the crust and top of the mantle where the P-wave velocity of seismic signals is sharply changed, indicating that the mantle is more rigid than is the crust.
5. The *mantle* is a large sphere located between the core and the crust that is composed of several subspheres with a total thickness of ~2,900 km (~1,800 mi). The mantle is divided into two large regions and an intervening layer.

A. The *upper mantle* extends from the base of the crust to a depth of ~400–450 km (250–280 mi). It also included the asthenosphere and lithosphere.

B. A *transition zone* separates the upper mantle from the lower mantle. These two mantle spheres do not have a clean separation of physical properties. The upper mantle is less dense than the lower mantle. Minerals are compressed in the transition zone so they do not change their composition. The transition zone between the two mantles is located at a depth from ~400 to ~700 km (~250 to 435 mi) below the Earth's surface. This zone exhibits a discontinuity in densities between the upper and lower mantles.

C. The *lower mantle* is located from a depth of ~700 to ~2,900 km (~435 to ~1,800 mi) and is larger than the upper mantle. It lies just above the Earth's outer core, where there is another transition zone between these two spheres of different densities.

6. The *mantle-core boundary* is the most pronounced transition zone between any two of Earth's spheres. There is a significant discontinuity of densities between these two spheres that has been determined by seismic signals that travel much faster through the dense core than through the less dense mantle. This transition zone between the lower mantle and outer core lies ~2,900 km (~1,800 mi) below the surface of the Earth
7. The *core* or central sphere is composed of metals. Its radius equals ~3,486 km (~2,166 mi) of the total Earth's radius. The core is approximately the size of the planet Mars. It is composed of two distinct spheres—the outer, liquid core and the inner, solid metallic core.

A. The *outer core* has a radius of ~2,200 km (~1,350 mi). Since it is liquid, it creates a shadow zone to seismic waves that cannot penetrate it (similar to an opaque object that creates a shadow from light waves).

B. The *inner core* experiences extraordinary pressure and, thus, has a solid metal composition. Pressures in this region of the Earth are about 3 million times that of surface atmospheric pressure, and temperatures reach ~5000°C (~9000°F). Its boundaries lie ~5,100 km (~3,170 mi) below the surface, with a radius of ~1,300 km (~800 mi). Seismic signals are able to distinguish a velocity discontinuity at the interface between these two spheres of the core[5–8] (see Figure 3.1).

Figure 3.1
The geosphere consists of three main layers: (a) the crust/lithosphere below the soil; (b) the mantle, consisting of semi-liquid, plastic-like magma that comprises 80% of the Earth's mass; and (c) the iron/nickel central cores.

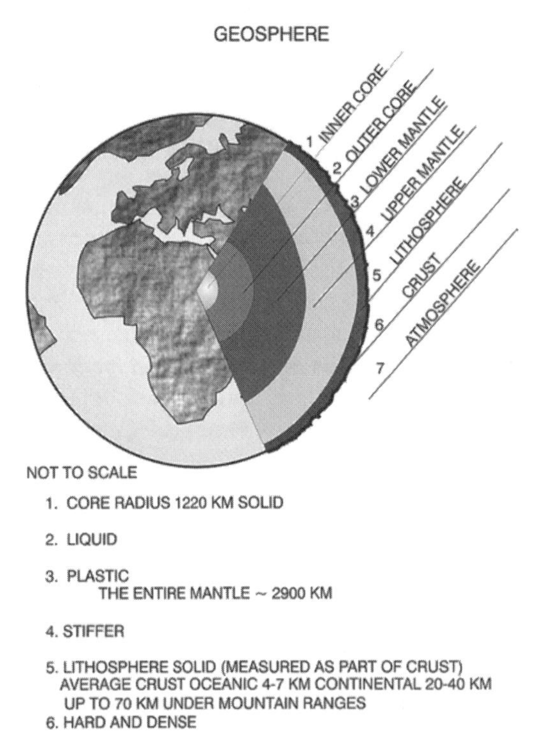

Several other major spheres, including the *hydrosphere* (water) and *atmosphere* (air), also are discussed in subsequent chapters. These two regions, along with the upper rocks, soil, and oceans, are combined to form the *biosphere.* The biosphere also consists of several subspheres, physical and chemical processes, and major systems. The biosphere is where life exists.

Earth's Outer Spheres: Crust and Lithosphere

The geosphere is often defined as the solid portion of the Earth. However, when referring to the entire globe, the definition may include the air (atmosphere) and water (hydrosphere) portions of the Earth. For our purposes, the geosphere is composed of the crust, lithosphere, asthenosphere, mantles, and core (see Figure 3.1).

The crust and lithosphere are the layers of rocks and minerals that make up what lies directly beneath the continental soils and the bottom of the seas. These spheres make up the continental landmasses and the oceanic crust.[9–10] Modern geology defines the lithosphere and the crust as the part of the Earth's surface that is slowly moving relative to well-defined boundaries of belts of seismic, volcanic, and tectonic activity.[11] The uppermost part of the lithosphere contains minerals, metals, rocks, and chemicals that humans, over the years, have learned to use to their advantage to make life easier.

Although some other planets have similar granite and basalt type rocks, plate tectonics (a global system of oceanic and continental plate movement), as far as we know, only exists on Earth. This movement of blocks of continental lithosphere is produced by the subduction (one tectonic plate sliding under another) of hydrated ocean crust consisting of partial melted mantle rocks that are rich in water. Because continental crust and lithosphere are mostly granites, they are less dense than the denser basaltic oceanic crust. Thus, they remain near the Earth's surface. The temperature differences of the magma deep in the Earth are responsible for the less dense magmatic granite rock material rising above the more dense basalt rocks.

Continental Crust

The continental crust is that portion of the Earth where humans and other air-breathing life forms exist. It is distinct from the oceanic crust in a number of ways. This outer layer of the Earth, including soils, experiences much erosion and disruptions due to a variety of internal and surface forces. It is also somewhat unique in its chemical composition. The crust is enriched by the following elements brought near the surface by magmatic activity: Al, Si, Ca, in addition to Na, as well as P, K, which, by weathering, provide nutrients for plants.[12] The Earth's continental crust is thought to be mainly composed of rocks rich in the elements *si*lica and *al*umina, thus the word "sialic," which is related to the granite layers of rock that form the crust of the Earth.[13–14] The crust makes up about 0.6% of the Earth's total volume, but only 0.4% of its mass, which is an indication of its lower density as compared to densities of Earth's internal spheres.[15] While the continental crust varies in thickness, it is about 4 times thicker than the crust under the oceans. Also, the continental crust's upper layer is somewhat less dense than the lower region that extends to the lithosphere, while the lithosphere is also somewhat denser at its lower regions.[16] It is somewhat amazing to realize that the density of the crust is only 2.8 grams per cubic centimeter (g/cm^3), but the *average* density of the entire Earth is 5.518 g/cm^3. Therefore, it follows that the densities of the layers below the crust are much greater than 5.518 g/cm^3. For example, the average rocks of the lithosphere have a density of 3.3 g/cm^3, while iron has a density of 7.9 g/cm^3. Note that 1 cm^3 of water weighs 1 g—thus water has the density of 1 at ~4°C. Therefore, the average density of the entire Earth is 5.518 times the density of water. This difference in density is significant for several reasons. First, as the Earth formed and cooled the heavier elements and compounds were forced toward the globe's center by the force of gravity. In addition, the spinning of the Earth also helped force the less dense material toward the surface. This is similar to the separation of the lighter in weight (less dense) substance from the more dense by using a mechanical centrifuge. Second, since the continental landmasses are less dense, they float on the heavier materials that make up deeper regions of the Earth. Gravity is constantly pulling all the mass of the rocks that make up the outer spheres toward the center of the Earth, thus increasing the pressure on the inner spheres. But rocks can only be compressed so much. To achieve this level of density, which exists for most of the Earth's interior, the inner spheres must be composed of something with a density greater than rock (i.e., iron and nickel). Rocks can only be squeezed to a certain extent before they experience a phase change, in which the atoms and molecules of solids are rearranged to form a liquid, and another phase change from liquid to solid will take place due to even greater pressure and heat.[17]

Land Surface Features

When the Earth is viewed from outer space, the outlines of the continents are quite evident, but the topography of mountains, ridges, plateaus, and deserts are difficult to distinguish. When viewing photographs of the Earth taken by astronauts, its surface appears very smooth, with land areas interspersed within bodies of water, all of which are sporadically covered by clouds. One reason for this is that even Earth's highest mountains are not really very high (relatively) when compared with the total size of the Earth. Consider the size of the Earth in relation to the size of its mountains and compare it to the size of an orange and the height of the bumps on its skin. The bumps that form the rough skin of the orange are greater in height in relation to the orange's total size than are the mountains in relation to the Earth's size. In other words, taking proportions into consideration, the surface of an orange is much more uneven than is the surface of the Earth's continents. Even so, continental surface features affect our everyday lives.

The continental crust has undergone repeated cycles of being built up and worn down over millions of years as a result of a number of forces and stresses. These same processes continue today, albeit at a relatively slow rate. Descending cold, dense slabs of the lithosphere are replaced by hot, lower density magmatic materials from the upper mantle. This process is partially driven by the heat created by the decay of radioactive elements deep in the Earth that creates convection cells that move magma into the upper regions of the mantle and asthenosphere. Also, pressure and friction are created by the movement of large plates of the lithosphere and contribute to the processes. The most outstanding geological surface features of Earth's continents are mountains, which are created by stress exerted by three different types of plate boundaries, whereas the main topographical features of the oceans are basins, trenches, and ridgelike volcanic mountains. The forces and stresses responsible for these geological features are related to plate tectonics (discussed in chapter 2):

1. *Extensional stress* occurs at divergent boundaries of landmasses. This type of stress pulls apart continental boundaries and separates them from each other. Examples are what happens at mid-oceanic ridges that can be accompanied by faults and troughs.
2. *Transform stress* occurs along boundaries. This stress imbalance can create both faulted and uplifted rocks. Transform stress tends to move pieces of crust laterally past one another. An example is the San Andreas fault zone in California.
3. *Compressional stress* is evident at the boundaries where two landmasses are squeezed together, creating geological formations such as large folds and mountain ranges by pushing the crust upward. These formations are the result of deformations in the crust that occur as continental boundaries collide as a result of moving plates. Examples are the Rocky Mountains and the Appalachian Mountains in the United States, and the Himalayas in Asia. The Appalachian Mountains are much older and more eroded than the Rocky Mountains, but at one time they were much higher than the Rockies are today.[18] The Himalayas were formed ~200 million years ago by compressional stress due to collision forces exerted when India and southern Asia/China crashed together. Before the Himalayas were formed, India was an island, like Australia, located south of the equator near present-day Australia and Antarctica. At this time in the Earth's history the place where the Himalayas are now was just a coastline. These mountains formed by compressional stresses created when two landmasses collided are now the Earth's highest point at 29,022 feet (ft) above sea level.[19]

Some of Earth's mountain ranges resulted from the thickening of the crust due to the meeting of deeper plates, which created stress. Less dense molten magma rises, creating stress responsible for folding rocks in the crust and pushing some of the crust into mountain ridges (see chapter 2 for more details.) Also, as the crust thickens and expands it becomes denser, resulting in the condition of equilibrium known as **isostasy**, which floats this slab of denser lithosphere over less dense material below it, similar to an iceberg in seawater. The process of isostatic compensation is an adjustment of different portions of the lithosphere that have different densities. It is based on gravity that adjusts the mass (densities) of topographical features. This process also contributes to the buildup and thickening of the Earth's outer crust, thus forming some mountain ridges and ranges.

Forces That Shape the Earth's Surface

Geomorphology is the science that examines the forces and processes that determine the external geological configurations on the Earth's surface. Geomorphologists classify and describe the nature, origin, and development of landforms and how these forms are related to underlying structures.

Surface Dynamics: Erosion

Both the universal force of gravity and the sun's energy are *external* forces that do their job of leveling out the hills and mountains that are built up by the *internal* Earth forces just described. These forces do their job by weathering and erosion—creating a constant fluctuation between the processes of construction and destruction of geographical features. Erosion, driven by gravity, is accomplished by both **degradation** and **aggradation** and moves rocks and soil on the surface from one place to another, where they are deposited. Three processes are responsible for erosion.

1. The crustal rocks on Earth's surface (minerals and rocks) are weathered, or broken down, into smaller pieces and particles. This happens by mechanical (physical) means such as fracturing, thermal expansion and contraction (including freezing and thawing), abrasion by wind and sand, or when plant roots expand as they grow, fracturing rocks. Weathering can also result when organic and chemical activity occur, such as what happens when plants grow in the cracks in rocks and dissolve the rock material, or by oxidation, by **hydrolysis**, and by the dissolving of minerals. Stalactites and stalagmites are formed when dissolved minerals precipitate in limestone caves. However, the weathered material remains where it has been broken down and accumulates on that site until some force is exerted that moves it to another site.
2. The next step is the movement of the degraded material that results from weathering from one place to another. When weathered material is relocated to another area, often in smaller pieces, it is the processes of erosion that remove and redistribute it. The processes, or agents, of erosion are gravity, rain, running water, wind, and glaciers. A common misconception is that eroded soil is lost. It is not lost. It is just moved from one place on to another—and not always where humans may want it deposited. The final repository for all eroded material is the ocean, where these eroded particles form sediments on the seafloor.
3. Finally the weathered, eroded, and transported material is placed in an area where it accumulates (aggradation). Such areas are usually lower in elevation than where the weathered material originated. Although water does not flow uphill, other forces, such as wind, can erode sand and soil to form dunes, and so forth. In summary, gravity and the sun's energy are the main forces that drive the agents of weathering and erosion that degrade the crust and move it to new locations on Earth. These two main forces are also responsible for water movement that can form **karsts** and other geological features, and gravity can cause mud,

land, and snow slides, as well as other topological features.[20] (See chapter 6 for more on weathering.)

Other Surface Features

Many other continental surface features are the direct consequences of Earth's internal and external forces, gravity, and assistance from the sun's energy. There are four major surface landforms:

1. *Plateaus* are large, elevated areas of relatively flat land. They are usually formed by either long-term erosion of adjacent elevated areas or by ancient floods of basalt rock that formed new surface area that was uplifted and eroded. Plateaus may be classified according to their locations:

    (a) **Intermontane** plateaus are usually large, flat areas formed by long-term erosion of adjacent mountains that fills flat valleys. They may exhibit high pinnacles. Examples are the Colorado Plateau and the Columbia River Plateau.

    (b) Piedmont plateaus are found between mountains and the sea. An example is the Piedmont region of North Carolina.

    (c) Continental plateaus are broad areas of tablelands. An example is the Central Plains in the United States.

2. *Hills,* also referred to as ridges and badlands, are areas elevated about 200 to 700 meters (m) (656 to 2,297 ft) and differ in structure according to their bedrock and drainage patterns. Examples are found in most parts of the world and in the upper Appalachian regions, the Potomac Valley, the Mohawk Valley of New York state, and the Badlands located in the Black Hills of southwest South Dakota and northwest Nebraska.
3. *Plains* are classified according to their roughness, origins (erosion/alluvial—sediments deposited by running water), and location (coastal/ interior). Major plains are found in Europe, Mexico City, northern China, Russia, Canada, South America, and the central United States. Much of the farming is done on the plains of the world, while the great masses of humanity live near coastal areas.
4. *Mountains* are higher in elevation than other landforms and are characterized by steep slopes that do not extend very far into valleys. They have small summits (peaks, crests, horns, pinnacles) as compared to their bases, and they range in height above sea level from about 700 m to more than 8,846 m (2,297 ft to more than 29,000 ft).[21] Mountains are found on every continent.

Other landforms include foothills, ridges, buttes, mesas, canyons, faults, cliffs and landslides (with talus, which is the debris at the base of cliffs), overhanging rock ledges, caves, sinkholes, deserts, beaches, and dunes.

Where high mountains and low hills exist, there will also be valleys. Not surprising, at the bottom of most valleys there is evidence of degraded aggregated material caused by weathering and erosion. Evidence of either ancient streams or rivers may be present; at the same time, water currently flows in these same valleys. Over long periods these flowing highways of water transported the products of the crust's weathering and erosion to the lower coastal areas of the continents and finally into the oceans. These tiny pieces of quartz and other rocks, minerals, and soil were deposited to form beaches and deltas at the end of rivers. Deltas, some covering great distances and depths, extend out from continental shelves into the sea. Alluvial fans are formed as fast-moving streams and rivers deposit soil, sand, and other sediments downstream. Alluvial fans are much more recent than ancient **dendritic** deltas that form over long periods of time where major rivers meet the sea.

Effects of Surface Features on Humans

It might be said that the nature of surface features on the continents (and geology in general) had as much to do with the development of the nature of Homo sapiens and other animals as did social and cultural influences. Land and water features were, and still are, partially responsible for determining how and where people live and conduct their affairs. Archaeologists have discovered ancient settlements in various parts of the world that have much in common. Our ancestors preferred to settle on higher ground, usually near a source of water, and often in areas with overhanging rock formations or large caves where there was an abundant supply of animal and plant food. Obviously, not all animals and plants were found in such ideal places, so some early humans followed food supplies by adapting more nomadic lifestyles, thus ending up in new and faraway places. Geology was a major determinant of where prehumans, and later H. *sapiens,* migrated and settled over the past millennia.

In the article "Who Were the First Americans?" published in *Scientific American* magazine, author Sasha Nemecek summarized recent evidence for the paths early humans followed to arrive in the Americas, presumably to obtain adequate food supplies but also possibly to satisfy their urge to explore. The possible routes that early Americans followed were (a) the Bering land bridge from northeastern Asia to northwestern North America; (b) the Pacific coastal water route from the eastern coastal areas of Asia to what is now Alaska along the western coastal areas of North and South America; (c) the South Pacific crossing from Australia to the southern regions of South America; and (d) the Atlantic crossing from northwestern Europe through Greenland to northeastern North America. Nemecek also reported that these early people probably traveled in small tribes of 15 to 30 individuals, hunted mammoths, fished, and gathered plants for food. One theory is that the ancestors of Native Americans left Siberia for the New World about 10,000 years ago. However, recent genetic studies indicate that humans may have arrived in the Americas about 20,000—and possibly as long as 30,000—years ago, just before the Ice Age abated. During ice ages, tremendous amounts of water from the oceans are tied up as ice on glaciers, thus lowering sea levels, which made passage possible along the northern routes described above. A question still to be answered is exactly when the last major Ice Age retreated, raising the ocean level to the extent that travel over land and ice bridges to the New World was no longer possible. Based on artifacts, Nemecek hypothesized that migrations of these early Americans required only about 100 years by boat along the Pacific Coast to 1,000 years by land routes to spread from North America to the southern parts of South America.[22]

In addition to geological factors that influenced human migration to new lands, geological changes also influenced environmental factors that drove natural selection, greatly influencing human evolution, as well as plant and animal evolution.

At one time it was thought that humans migrated out of Africa to Europe and Asia about 50 to 60 thousand years ago. More recently, Michael Balter and Ann Gibbons, in *Science* magazine, reported research based on skulls and stone tools that indicates humans migrated out of Africa to the Middle East and China about 1.1 to 1.9 million years ago. But humans did not reach Italy until 900,000 years ago, or Western Europe until about 450,000 to 780,00 years ago.[23] The assumption is that geographic features (mountains) and climate (glaciers) impeded migration out of Africa westward but not eastward. In the January 15, 2001 issue of *USA Today,* Dan Vergano reported that evidence from DNA tests and comparison of prehistoric human skulls demonstrated that the Eve theory, which claims humans descended from an archaic human species out of Africa, is incorrect.[24]

The following is an example of how Earth's surface features and climate affect current society. Some years ago a professor asked his economic geology class the following question to be answered in a detailed essay. "How did the last glacier that covered the upper part of the United States affect the price of butter in Chicago?" Before you read the answer, try to figure it out for yourself. The answer goes something like this: As the North American continental glacier moved south about 20,000 to 10,000 years ago, it scooped out depressions in the surface carrying with it great loads of soil and large boulders. About 8,000 to 10,000 years ago, as the Earth became warmer, the glaciers receded, dumping their loads of soil and rocks, becoming unstratified glacial drift and till that were deposited in mounds called moraines. As the soil and rocks were scooped out of the surface, numerous depressions were formed that became lakes and ponds. Part of this area is now the lake state of Wisconsin, known as the Dairy State. The lakes provide water, and the glacial till is now fertile soil supporting grassy pastures, ideal regions to raise milk cows, thus butter. And since Wisconsin is relatively close to Chicago, the price of butter is less than it might be in Chicago if it had to rely on a more distant dairy region.

The shapes of the continents, their juxtapositions to each other, the changes in surface features over long periods of time, alterations in climate, and changes in the sources of foods, both directly and indirectly influenced the arrival and history of early civilization on all continents. It might be said that plate tectonics and other geological changes have and still are altering the small portion of continental surface (30%), as well as the major portions of the Earth's surface that are water (70%), to continue to influence how we live.

Oceanic Crust

Oceanic crust and continental crust are both part of the lithosphere. Volcanic action under the seafloor often results in the oceanic crust rising to form island arcs that become new landmasses. There are several major distinctions between the continental and oceanic crust, namely thickness, density, and chemical/physical composition. Continental crust is about four times thicker than oceanic crust. Oceanic crust averages ~6 km (~4 mi) in thickness and consists mostly of basaltic material composed of a dark-colored intrusive diorite (a type of plutonic, glasslike calcic plagioclase and pyroxene igneous rock). The continental crust is mostly composed of less dense granite (a quartz-bearing felsic plutonic rock). Both are heat formed (plutonic) but contain different chemicals and minerals. Geologists formerly referred to the lower layer of oceanic crust as "sialma" (as opposed to "sialic" for the continental crust). These terms relate to the first letters of their chemical elements (Si, Al, Mn, and Ca). The crust under the ocean floor is composed mainly of *si*lica, *al*uminum, and *ma*gnesium.

Another main distinction between the two types of crust is density. The continents rise high above the sea because, in a sense, they float on the lower, denser mantle layers. This tendency is referred to as "isostasy," which is a condition, determined by both surface and satellite gravity observations, that the total mass of material per unit area in the Earth's upper layer is equal. In other words, the gravity exerted by the thicker, but less dense, continental crust is equalized by the thinner but denser oceanic crust. The two types of crust in the upper sphere maintain gravitational equilibrium as a result of their different densities. Isostasy is an important concept related to the theory of plate tectonics.[25–26]

The edge boundaries of the continental crust form flat platforms where they meet the oceans. At greater distances from the margins of the continents these coastal shelves become steep slopes down to the floor of the oceans. Thus, the differences in both density and thickness of these two crusts have been responsible over millions of years for the formation of saltwater ocean basins. These ocean basins contain more than ~97% (by volume) of the total water existing on Earth. Less than ~3% of the water in Earth's hydrosphere is nonsaline freshwater, and only ~1% of that freshwater is available for human use and consumption. (For more information on the nature of the hydrosphere, saltwater, freshwater, and the hydrological cycle, see chapter 7. For more information on the forces that are responsible for the dynamics that move continents, form oceans, and give birth to islands, see chapter 2.)

The Seafloor

In the past, little was known about the bottoms of the world's oceans because they were under seawater. Information about the structure, shape, and composition of the seafloor was difficult to obtain until both ships and oceanographic instrumentation were developed. With the invention of modern deep-sea vessels and electronic technology capable of collecting data at these tremendous depths, significant information has been added to the store of knowledge concerning the ocean beds.

The following are some of the things we do know about the ocean floor. **Echo sounders** and **seismic profilers** are used to map the topography of the seafloor. The low-frequency sound waves (sonar) that bounce off the bottom of the ocean provide geologists with a profile of its surface topography. Other devices are also used, such as different types of dredges and pipe core collectors that bring up sediments and volcanic rocks for study. Magnetometers towed by ships have been used to map magnetic anomalies located on both sides of mid-oceanic ridges. These data led to the concept of seafloor spreading and movement of continental plates (tectonics). Over the past decades new submersible vessels capable of carrying several scientists cruise the ocean bottom to observe and photograph the ocean beds. And unmanned, robot undersea vessels with video equipment can go even deeper than manned **submersibles**, to ~30,000 ft deep. Other devices employed to study the seafloor and ocean crust are digital seismographs, sensitive magnetometers, instruments to measure gravitational fluctuations, and drilling deep into the oceanic crust. During the last half of the twentieth century undersea surveys learned a great deal about what is at the bottom of the oceans.

Seafloor Features

Some of the structures located in the oceanic crust are described in the following list.

1. *Continental shelves* are where oceans begin. These shelves are almost flat and cover ~8% of the shallow ocean floor. They may extend several hundred miles and increase in depth from just a few meters to ~200 m (~645 ft). The ends of the continental shelves dip steeply to form the *continental slopes* that continue downward until they intersect with the deeper *abyssal plains.* At one time the seafloor was considered to be a large flat plain, but now there is adequate evidence for a great variety of unique topographical

structures important to the understanding of the global nature of plate tectonics.

2. *Ocean sediments* are thicker at the continental margins than they are in the deeper parts of the oceans. This is partly due to the disposition of eroded material carried by rivers into the coastal regions of the oceans. Sediments are thinnest on ridges and mountains in the middle of the oceans because they are usually located some distance from eroded debris deposited closer to the shoreline. There are three types of ocean sediments:

   - *Terrigenous* sediments are composed of materials that originate on the continents and are deposited in the ocean. (Terrigenous sediments move from land □ shelf □ slope □ alluvial plain □ deep seafloor.)
   - *Pelagic* sediments are composed of clay and microskeletons of tiny creatures called plankton that settled to the ocean floor. Since these are sticky and mucky, they are called "oozes."
   - *Hydrogenous* sediments are chemical precipitates that over time gather on the seafloor. An example is the rich deposits of manganese nodules scattered over much of the ocean bottom. The seafloor is rich in several other minerals as well, but it has not proven economically feasible to mine them.

3. O*ceanic trenches* are deep, narrow, elongated depressions that are more-or-less located on the mid-ocean floors. They reach a depth of ~10 km (~6 mi) and have steeply sloping walls. Trenches, which are sites for earthquakes and volcanoes, may be found nearby. The deepest point in the Earth's crust is the Marianas Trench in the Pacific Ocean at ~11,040 m (~36,220 ft).

4. The *mid-oceanic ridge* circles the globe and is found near the center of most oceans. It is often named according to the ocean basin in which it is found (e.g., the Atlantic Mid-Oceanic Ridge). In a few cases, such as near Japan and the Juan de Fuca straight located between Washington state and Vancouver, British Columbia, the ridge is closer to the coastline. The mid-oceanic ridge is where the seafloor starts to spread out from each side of the ridge. There are some volcanoes located near the ridge, but most of the new seafloor is formed by heat-driven convection cells deep in the mantle that bring less dense magma that is intruded at the crests of the ridges. This basaltic magma flows outward from both sides of the axes of the ridges, resulting in the expansion of the seafloor at the ridge. As the spreading of the seafloor continues perpendicular to the ridge, it slowly moves the continental plates away from the crests of the ridges as the older, colder, and thus denser oceanic crust sinks under the continental crust at the leading edge of the plate, to be recycled deep into the mantle. The Earth-encircling ridge forms a continuous mountain chain that is ~75,000 km (~45,000 mi) long. (See chapter 2 for more on the forces that create the ridge and plate tectonics.)

5. *Underwater volcanic action* forms **seamounts**, **guyots**, hills, and islands.

6. *Underwater earthquakes* can cause giant tidal waves, called "tsunamis," that are devastating to people and property located on low-lying coastal areas.[27–28]

Unlike on the continental surface, erosion on the ocean bed is not of great importance. This is mainly because the deep sea currents are very slow and do not create the same conditions for eroding topographical structures found on the ocean beds that wind and water do on landmasses.

The ocean floor also contains hydrothermal vents, similar to geysers, that are mainly located at the crests of ridges. These vents are where pressure from hot water and gasses rupture the seabed and precipitate a variety of materials along the mid-oceanic ridges. These vents were first detected by comparing heat-flow measurements at the crests of ridges with other regions of the seafloor. Some of this material forms an array of minerals and chemicals, including sulfur compounds that are used as a substitute for oxygen to support exotic deep-ocean life.

In chapter 4 the composition of the continental crust, oceanic crust, and the lithosphere are addressed. Why and how the minerals, rocks, and other natural resources are found in these relatively thin spheres, as well as why they are important to humans, also are examined.

Notes

1. Robert L. Bates and Julia A. Jackson, *Dictionary of Geological Terms* (New York: Anchor Books/Doubleday, 1984), 345.
2. Bates and Jackson, *Dictionary of Geological Terms,* 392.
3. John Daintith, Sarah Mitchell, Elizabeth Tootill, and Derek Gjertsen, *Biographical Encyclopedia of Scientists,* vol. 1 (Bristol, England, and Philadelphia, Pa.: Institute of Physics Publishing, 1994), 447–48.
4. Daintith et al., *Biographical Encyclopedia of Scientists,* vol. 1, 364.
5. Anna Clayborne, Gillian Doherty, and Rebecca Treays, *The Usborne Encyclopedia of Planet Earth* (Tulsa, Okla.: EDC Publishing, 1999), 16.
6. Stanley Chernicoff, *Geology: An Introduction to Physical Geology* (New York: Houghton Mifflin, 1999), 303–15.
7. Alan E. Mussett and M. Aftab Khan, *Looking into the Earth: An Introduction to Geological Geophysics* (New York: Cambridge University Press, 2000), 32.
8. W. G. Ernst, *Earth Systems: Processes and Issues* (New York: Cambridge University Press, 2000), 59–63.
9. Mark J. Crawford, *Physical Geology* (Lincoln, Nebr.: Cliff Notes, 1988), 222.
10. Patricia Barnes-Svarney, ed., *The New York Public Library Science Desk Reference* (New York: Macmillan, 1995), 377.
11. *The Oxford English Dictionary,* 2d ed., CD-ROM, s.v. "lithosphere."
12. Jan Kramers, "The Smile of the Cheshire Cat," *Science,* 27 July 2001, 619–20.
13. *The Oxford English Dictionary,* 2d ed., CD-ROM, s.v. "sialic."
14. Bates and Jackson, *Dictionary of Geological Terms,* 466.
15. Barnes-Svarney, ed., *The New York Public Library Science Desk Reference,* 377.
16. John Tomikel, *Basic Earth Science: Earth Processes and Environments* (New York: Allegheny Press, 1981), 108.
17. Isaac Asimov, *Isaac Asimov's Guide to Earth and Space* (New York: Fawcett Crest, 1991), 35–37.
18. Graham R. Thompson and Jonathan Turk, *Earth Science and the Environment* (New York: Harcourt Brace, 1999), 162.
19. Thompson and Turk, *Earth Science,* 167.
20. Tomikel, *Basic Earth Science,* 80.
21. Tomikel, *Basic Earth Science,* 33–38.
22. Sasha Nemecek, "Trends in Archeology: Who Were the First Americans?," *Scientific American,* September 2000, 80–87.
23. Michael Balter and Ann Gibbons, "A Glimpse of Humans' First Journey Out of Africa," *Science,* 12 May 2000, 948–50.
24. Dan Vergano, "Ancestor Theory Falls From Grace," *USA Today,* 15 January 2001.
25. *Grolier Multimedia Encyclopedia,* CD-ROM 1997, s.v. "Earth, Structure and Composition."
26. Bates and Jackson, *Dictionary of Geological Terms,* 275, 468.
27. Crawford, *Physical Geology,* 141–51.
28. Thompson and Turk, *Earth Science,* 93–95.

Earth's Resources

Minerals and Metals (Mineralogy and Metallurgy)

The minerals, metals, and rocks found on the surface or near the surface of the Earth have been of great interest and utility to humans from earliest times. For millennia, people were limited to wooden digging sticks, spears, and clubs, but they eventually learned how to make and use stone axes, flint and obsidian cutting tools, and arrow and spear points, as well as how to use the prettier minerals for ornamentation. They ground up pure copper and iron chunks to form powders to mix with oils for use as body and pottery paints. They also converted somewhat pure chunks of native copper and iron meteorites into useful implements.

The earliest mining of a metal took place ~5000 B.C.E. when the Egyptians learned to smelt copper ores to extract metallic copper. One theory is that malachite copper ore was dropped into a fire, resulting in the formation of copper droplets that early Egyptians made into beads. This discovery was most likely followed by the deliberate smelting of malachite ore.[1] Pure copper was much too soft for practical use; therefore, it was soon alloyed with arsenic and later with tin to form the metal bronze, which could hold a sharper edge. Later zinc was added to copper to form brass. These discoveries ushered in the Bronze Age and the Brass Age.

The Iron Age followed the Bronze and Brass Ages as mining and metallurgical techniques improved when humans discovered how to increase the heat used to extract metals and to shape the resulting metals into tools and weapons. The smelting of iron requires much hotter fires than those needed to extract copper, tin, zinc, and lead from their ores. Thus, smelters were trained to use charcoal and a bellows to increase the flow of air to obtain the higher temperatures required to smelt iron from its ore. Early European miners found it difficult to smelt copper from some of its several different ores. Consequently, they kept getting other metals (later identified as cobalt and nickel) that were also contained in the ores, causing them many problems. German miners named the unknown metal that kept interfering with copper smelting *kupfernickel,* or "Old Nick's copper," which means the "devil's copper." In 1751 Axel Fredrick

Cronstedt (1722–1765) separated this new metal from copper ore and named it "nickel" after "Old Nick's Copper."[2]

Early humans used many chemicals and minerals, as well as plants, for pharmaceutical purposes in attempts to heal the sick. Salt, a common mineral, was always in demand, especially by tribes living in areas where salt deposits were not readily available. By the sixteenth century, humans discovered many new minerals and metals by systematically studying mining and metallurgy techniques. Some even tried alchemy as a way of transmuting less desirable metals into gold. Humans have practiced both underground and surface mining for hundreds of years in response to the demand for the items that could be produced from the mined minerals, metals, and rocks. A number of stages are required for the utilization of mined mineral and metal resources: first is exploration; next is extraction; then it must be processed, followed by refining, leading to manufacturing, and finally marketing of a useful product. Today, worldwide mining is a multibillion dollar industry, and civilization, as we know it, could not have developed without it.

Georgius Agricola (1494–1555), the Latin name for the German metallurgist Georg Bauer (both of the Latin names mean "farmer"), authored several books on geology, mining, and minerals, some of which are still in existence. Agricola was the first to classify minerals by their physical properties, that is, color, weight, transparency, taste, odor, texture, solubility, and combustibility. He also was one of the first to distinguish between types of earth, stones, gems, and so forth. Although his book *De natura fossilium* (1546) described his classification system, many of the terms he used differ from those used by modern geologists. Georgius Agricola is now known as the father of the science of mineralogy.[3–4]

A *mineral* is defined as a substance exhibiting the following four features:

- It must be a solid and found in nature.
- It must be composed of **inorganic** matter (nonliving matter).
- It possesses a specific chemical composition, or range of compositions, no matter where found on Earth.
- Its atoms are arranged in orderly, regularly repeating patterns of atoms in distinctive forms of solid crystals.

Rocks are defined as combinations of one or more minerals, excluding mineral fuels (gas, oil, and coal), which are organic in nature.

Metals are chemical elements that are **opaque**, may be lustrous when polished, and in general are good conductors of heat and electricity.[5–6]

Chemical Nature of Minerals

Atoms are the smallest units of elemental matter. Different types of atoms are responsible for the distinct chemical and physical properties of each natural element. Molecules are combinations of atoms, either of atoms of the same element (e.g., O_2 or H_2) or of two or more different types of atoms that form molecular compounds (e.g., NaCl or H_2SO_4). Minerals are composed of a variety of atoms of different elements as well as molecules of compounds. Understanding minerals requires some understanding of the structure of inorganic matter and how atoms interact when they combine to form different inorganic solids.

An excellent example is the structure of the elements silicon and carbon. Both have a **tetrahedron** structure with four **covalent-bonding** (electron sharing) properties. The carbon atom can bond onto other carbon atoms to form chains or rings, or, as an example, the central carbon atom can bond electronically with four hydrogen atoms to form the methane molecule (CH_4 tetrahedron). This property of carbon is partially responsible for the multitude of organic compounds that make up living substances. The silicon atom is similar in structure to the carbon atom in that it, too, can bond ionically and covalently with other atoms. For example, when a central silicon atom forms bonds with two oxygen atoms, silicon dioxide (silica, SiO_2) is the resulting compound. Silica is the main compound found in many common inorganic minerals and rocks. The characteristic tetrahedron structure for carbon and silicon is responsible for many distinctive inorganic, as well as organic, types of molecular structures (see Figure 4.1).

One might say that since carbon is the main element found in living cells and tissues, it determines the organic world, whereas silicon mainly determines the inorganic (nonliving) world. There are, however, exceptions. For example, diamonds and graphite are pure carbon but are considered inorganic because the forces and processes by which they are formed (heat and pressure) are not biological processes. Cyanide (CN, carbon + nitrogen) is an example of another carbon-containing inorganic compound.

Any example of a mineral has a similar chemical composition no matter where it is found. As the Earth formed, the elements were unevenly distributed throughout its structure. The concentration of specific elements varies markedly between and among different spheres. For instance, the element iron composes only about 5% of the Earth's crust but almost 100% of its core. Also, oxygen is found in greater concentrations bound up in molecular compounds within the crust and mantle than in the core (or even the atmosphere).

The most common chemical elements that comprise the Earth's crust by weight are listed in Table 4.1 by percent of their presence. These 13 elements encompass more than 98.5% by weight of all the Earth's elements. All the other elements combined comprise less than 1.5% of the total. The chemical elements form a large variety of inorganic compounds that are found in the crust. They are referred to as minerals and crystals. In turn, rocks are composed of great varieties of these minerals.[7]

Figure 4.1
The electronic structure of the silicon atom is similar to that of the carbon atom because both have four valence bonding electrons that form vertices of a tetrahedron. Carbon atoms, as well as silicon, form inorganic minerals. While carbon is also an important element in organic (living) matter, silicon is not. The silicon atom strongly bonds with four oxygen atoms to form a *silicate-oxygen tetrahedron.* About 95% of the Earth's crust and rocks are composed of silicon-based minerals.

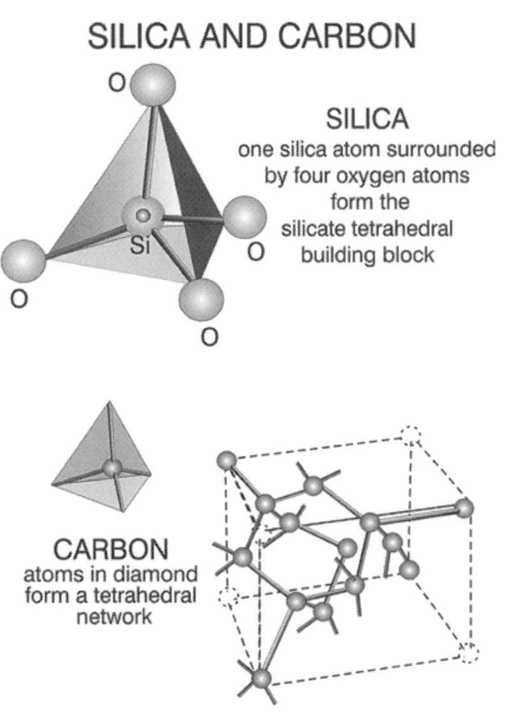

Table 4.1
Earth's Common Elements

Element	Approximate Percentage
Oxygen	46.6
Silicon	27.7
Aluminum	8.1
Iron	5.0
Calcium	3.6
Sodium	2.8
Potassium	2.6
Magnesium	2.1
Titanium	0.4
Hydrogen	0.2
Carbon	0.2
Phosphorous	0.1
Sulfur	0.1

Physical Nature of Minerals

In addition to chemical composition, the physical properties are also important distinguishing features of minerals. A number of physical characteristics determine the nature and classification of minerals:

- All minerals have some form of a crystalline structure.
- Minerals vary by the type and size of the atoms that formed the crystal.
- The type of chemical bonds that holds atoms together (i.e., ionic, covalent, metallic, or van der Waals force) determines the mineral's structure.
- The conditions of temperature and pressure under which the minerals are formed produce different characteristics.[8]

Minerals are identified by using the following major physical characteristics:

- *Crystal structure*—internal arrangements of atoms
- *Cleavage*—breaking of a mineral *along* its crystal plane, revealing its structure
- *Fracture*—a rough, faultlike break, or a conchoidal (curved conch shell-like) break *across* a mineral's crystal plane
- *Colors*—can be distinct for some minerals
- *Luster*—appearance of light reflected off the surface, for example, glassy (shiny) or earthy (dull)
- *Shapes*—crystals with distinctive faces and angles between the faces
- *Hardness*—a mineral's resistance to abrasion
- *Specific gravity*—determination of density (mass/volume)
- *Streaks*—colors of a fine powder of the mineral
- *Striations*—narrow ridges or bands along the mineral's surface
- *Reaction to acids*—extent and speed of chemical reaction to acids that are distinctive for most minerals
- *Magnetism*—some minerals are magnetic, others are not
- ***Fluorescence***—ability of mineral to glow in **ultraviolet** (UV) light[9–10]

Color and shape are poor indicators of minerals, whereas luster and hardness are useful and easy tests. In 1812 Friedrich Mohs (1773–1830) developed the Mohs scale for mineral hardness. The scale ranges from 1 to 10, based on the idea that a mineral with a higher number will scratch all minerals with lower numbers on the scale.[11] (See the Appendix for a simple project to determine the hardness of minerals.) Some examples are listed in Table 4.2.

Since the Mohs scale of 10 was first introduced, several synthetic materials have been produced that are harder than 10 (e.g., silicon carbide). Therefore, the modern Mohs scale has been expanded from 10 to 15.[12]

Table 4.2
Mohs Scale for Hardness of Minerals

Mohs Scale Mineral	Common Characteristic or Use of Mineral
1. Talc (softest mineral)	Soft rock used for talcum powder, dusting
2. Gypsum	Used in building cement, plaster of Paris
3. Calcite	Limestone, building materials, optics
4. Fluorite	Fluorspars, flux for metal refining, emery wheels, dentifrices
5. Apatite	Constituent of bones and teeth, phosphate rock
6. Orthoclase	Rock-forming mineral, e.g., feldspars, igneous rocks, granites
7. Quartz	Piezoelectric devices, oscillators, glass
8. Topaz	Found in tin ore, gemstones, birthstone for November
9. Corundum	Abrasive, emery, rubies and sapphires
10. Diamond	Hardest mineral, brilliant gemstone, abrasive

Source: Graham R. Thompson and Jonathan Turk, *Earth Science and the Environment* (New York: Harcourt Brace, 1999), 30.

There is also a hardness scale for metals. A special test called the Brinell hardness test uses a steel ball dropped under force onto the metal to be tested. The size of the dent in the surface of the metal is related to the scale.

In addition to minerals being categorized by their physical properties (e.g., hardness), they also are categorized by chemical composition. Some examples of minerals classified by chemical composition are listed in Table 4.3.

Crystallography is the science that studies crystal structures by passing x-rays through mineral crystals. This forms a **diffraction** pattern of the arrangement of atoms in the mineral that is either projected onto a photographic film or recorded by a computer program. This procedure became possible shortly after Wilhelm Conrad Roentgen (1845–1923) discovered a form of radiation that could look through objects. Since these rays were of an unknown nature, he called them "x-rays." In 1912, Max Theodor Felix Von Laue (1879–1960) was awarded the Nobel Prize for physics for his discovery that x-rays, passing through crystals, were diffracted according to the interatomic distance of the atoms in the crystals; thus, each crystal material produced a photographic image distinctive for that type of mineral. The only father/son team to win a Nobel Prize was Sir William Bragg (1862–1942) and his son, Sir Lawrence Bragg (1890–1971), when they used diffraction to discover the x-ray pattern of the atomic structure of sodium chloride (salt). This led to the development of the diffractometer, a modern device that identifies minerals according to the arrangement of their atomic structures[13] (see Figure 4.2).[14]

(A) *Isometric* crystals can be cubic in shape. Isometric crystals have three axes of equal length at right angles to each other. (Some examples and their common uses follow each category in this list.)

1. Diamonds (**allotropic** form of carbon, C) crystallize isometrically with carbon atoms covalently bonded by single bonds as octahedral structures rather than cubes. This structure results in Mohs 10 hardness. These types of crystals are also made synthetically and used in abrasive grinding wheels and jewelry.

Table 4.3
Chemical Classification of Minerals

Chemical Group	Examples	Common Name (Uses)
Oxides	Iron oxides, Fe_2O_3, Fe_3O_4	Iron ores, hematite and magnetite (steel)
	Chromite, $FeCr_2O_4$	Chromium ore (chrome on automobiles)
	DiHydrogen oxide, H_2O	Water (not exactly a mineral; ice, drinking, etc.)
Sulfides	Sphalerite, ZnS	Zinc ore (galvanized steel)
	Pyrite, FeS_2	Fool's gold (source of sulfur, sulfuric acid)
	Cinnabar, HgS	Mercury ore (thermometers, electrical switches)
Sulfates	Gypsum, $CaSO_4$	Plaster of Paris (walls, wallboard)
	Barite, $BaSO_4$	Barite (drilling mud, bowling balls)
Native Elements	Gold, Au	Gold (jewelry, plating, electronic circuits)
	Copper, Cu	Copper ores (wiring, electronics)
	Sulfur, S	Sulfur (matches, fireworks, acid)
	Graphite, C	Carbon (lubricants, electrodes)
	Silver, Ag	Silver (jewelry, photographic film)
	Platinum, Pt	Platinum (auto exhaust catalyst, jewelry)
Halides	Halite, NaCl	Salt (common table salt)
	Fluorite, CaF_2	Fluorspar (flux in steel making)
	Sylvite, KCl	Potash ore (fertilizer, source of potassium)
Carbonates	Calcite, $CaCO_3$	Calcite (gravel, Portland cement)
	Dolomite, $CaMg(CO_3)$	Dolomite (paper making, fertilizers, ceramics)
	Aragonite, $CaCO_3$	Aragonite (ocean sediments, pearls, seashells)
Hydroxides	Limonite, $FeO(OH){\bullet}H_2O$	Iron ore (pigments, pig iron)
	Bauxite, $Al(OH){\bullet}H_2O$	Aluminum ore (metal construction, airplanes)
Phosphates	Apatite, $Ca_5(F_3Cl,OH)PO_4)_3$	Apatite (laser crystals, fluoridation, acid)
	Turquoise, $CuAl_6(PO_4)_4(OH)_8$	Turquoise (opaque gem, Dec. birthstone)

Sources: Richard L. Lewis, *Hawley's Condensed Chemical Dictionary,* 12th ed. (New York: Van Nostrand Reinhold, 1993), 32; Database of more than 3,570 minerals and gems, Geology Department, University of Wisconsin, Madison, Wisconsin.

2. Halite (common table salt, rock salt, and sea salt; NaCl) are cubic crystals with many uses—to produce metallic sodium and chlorine; as a food preservative; for supercooling, metallurgy, ceramic glazes, herbicides, fire extinguishers, medicines, nuclear reactors, seasoning, and small amounts required in a healthy diet.
3. Other examples of minerals that exhibit isometric crystals are galena (plumbous sulfide, PbS), lead ore; fluorite or fluorspar (calcium fluoride, CaF_2), used in electronics, lasers, and high temperature lubricants; and pyrite (ferrous/iron sulfide, FeS_2), fool's gold, often mixed with other metals, used to manufacture sulfur, sulfuric acid, and sulfur dioxide.

(B) *Tetragonal* crystal systems are rectangles with three axes at right angles to each other, but unlike isometric crystals, only two or three of the axes are equal to each other.

1. Rutile/ilmenite (titanium dioxide, TiO_2), titanium ore, is found in igneous rocks and beach sand. It is a white paint pigment and used in cosmetics and the manufacture of glass and ceramics, ink, and fabrics.

Figure 4.2
Under varying conditions of temperature and pressure, minerals form six basic three-dimensional crystal systems. A. Cubic (isometric) systems have three axes of equal length at right angles to each other (e.g., pyrite, galena, diamond). B. Tetragonal systems are composed of rectangles with three axes at right angles to each other (e.g., zircon, rutile). C. Orthorhombic (rhombic) crystals have three mutually perpendicular symmetric axes at right angles to each other (e.g., topaz, sulfur). D. Monoclinic crystals exhibit two of the three axes that are not at right angles, while the third axis is at right angles to both, but none are equivalent in length (e.g., gypsum, borax, muscovite). E. Triclinic crystal systems have three unequal axes that are mutually oblique; they lack symmetry as compared to other crystal types (e.g., turquoise). F. Hexagonal and trigonal crystal systems have similar compositions, with two axes aligned to each other at 120°, with the third axis at right angles to the other two (e.g., calcite, corundum, dolomite).

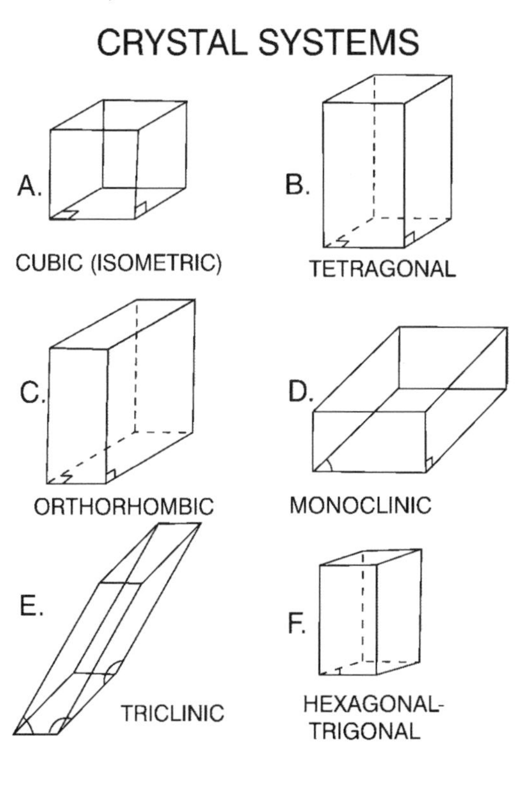

2. Zircon (zircon silicon oxide, $ZrSiO_4$) is found in igneous rocks and sediments. It is a gemstone and is used as an abrasive; in the manufacturing of enamels and porcelains; and as a catalyst.
3. Cassiterite (stannic oxide or tin dioxide, SnO_2), tin ore, is used as a catalyst; in the manufacture of cosmetics; in perfume preparation; as a textile mordant; and as a polishing powder.

(C) *Orthorhombic* crystals have three axes at right angles to each other, but unlike cubes and tetragonal crystals, none of the axes in rhombic crystals are equivalent in length.

1. Topaz [$Al_2SiO_4(F,OH)_2$] is a transparent/translucent crystal found in siliceous igneous rocks, is also known as citrine or topaz quartz, and is the gemstone for November.
2. Chalocite, or chalcosine (cuprous sulfide, Cu_2S), is found with other copper minerals; has a metallic, dull black luster; and is a major source of copper.
3. Other examples are marcasite (iron sulfide, FeS_2, similar to pyrite), chalcopyrite or copper pyrite ($CuFeS_2$), sulfur (S, brimstone), and stibnite (antimony trisulfide, Sb_2S_3).

(D) *Monoclinic* crystals have two of the three axes not at right angles, but the third axis is at right angles to both, while none of the axes are equivalent.

1. Muscovite is a mica-like sedimentary mineral [hydrous potassium aluminosilicate, $KAl_2(AlSi_3O_{10})(OH)_2$], similar to sanidine ($KAlSi_3O_8$), and flaky like some sandstones. Industrial uses include making enamels, ceramic ware, glass, abrasive wheels, fertilizer, roofing material, and soaps.
2. Azurite (both natural and artificial mineral), cupric carbonate [$Cu_2(OH)_2CO_3$], is a major ore of copper and a deep blue/purple ornamental stone. It is related to the mineral malachite [$Cu_2CO_3(OH)_2$], which is a minor ore

of copper and a bright green, solid ornamental stone used for pyrotechnics, pigments, hair pomades, fungicide, and as a feed additive.

3. Other examples are gypsum (hydrous calcium sulfate $CaSO_4 \cdot 2H_2O$) and borax (hydrated sodium borate $Na_2B_4O_7 \cdot 10H_2O$).

(E) *Triclinic* crystals have no axes at right angles to each other, and none of the angles are equal. These types of crystal minerals are rare.

1. Turquoise $[CuAl_6(PO_4)_4(OH)_8 \cdot 5H_2O]$, found as prismatic crystals with pyramidal terminations in aluminum-rich igneous rocks, are 8 on the Mohs scale and are found as red, yellow, and blue-green to sky-blue gem stones.
2. Sanidine (potassium aluminosilicate, $KAlSi_3O_8$), also known as glassy feldspar or ice spar, is a clear, glassy crystal found in volcanic rocks such as trachyte. It is used to make pottery, enamel, ceramic ware, glass, and abrasive wheels.

(F) *Hexagonal* and *trigonal* crystals are similar and are composed of two axes aligned at 120° angles to each other. The third axis forms a right angle with the other two but is a different length. Examples of hexagonal crystals include the following.

1. Graphite (carbon, C) is opaque, soft, greasy to touch, gray to black in color, conducts electricity, and is found in metamorphic rocks. It is used in lead pencils, paint, electrodes, lubricants, and as a moderator in nuclear reactors.
2. Beryl [silicate of beryllium and aluminum, $Be_3Al_2(SiO_3)_6$] is ore of beryllium, a highly toxic carcinogen. It is the lightest construction metal known, is used to make strong alloys, forms emerald and aquamarine gemstones, and is used as a coating in fluorescent lamps.
3. Two other examples are calcite (calcium carbonate $CaCO_3$) and apatite (a form of calcium phosphate, with fluorine $[Ca_5(PO_4,CO_3)_3(F,OH,Cl)]$).

Examples of trigonal crystals include the following.

1. Arsenic (As) is a nonmetallic, yellow, crystalline solid element, used as an alloy agent for other metals, as insecticides, in semiconductors, and in the pharmaceutical industry.
2. **Corundum** (aluminum oxide, Al_2O_3) forms ruby and sapphire gemstones and is very hard (9 on the Mohs scale). As emery it is used as an abrasive.
3. Two other examples are cinnabar (mercuric sulfide HgS, ore of mercury) and dolomite [a carbonate of calcium and magnesium, $CaMg(CO_3)_2$].

The six basic geometric patterns or shapes of mineral crystal structures are dependent on how the particular crystal breaks (either by fracturing or cleaving). These breaking patterns are also used to identify different types of minerals (see Figure 4.3).

A number of common metals are usually obtained by extracting them (by smelting and electrolysis) from their mineral **ores**. Although they are not in themselves minerals, they are important to modern society, as well as being nonrenewable natural resources. Examples of these common metals and their uses are given in the following list.

- *Aluminum* (Al) is the most abundant metal element in the Earth's crust. Bauxite, the ore of aluminum, is known as alumina or aluminum oxide (Al_2O_3). It is mixed with various minerals and clay and may be gray, yellow, or reddish brown. Bauxite ore is imported from Guinea, Jamaica, and Australia. Deposits are also

Figure 4.3
Minerals composed of crystals representing the different structural systems (see Figure 4.2) fracture in different patterns and forms.

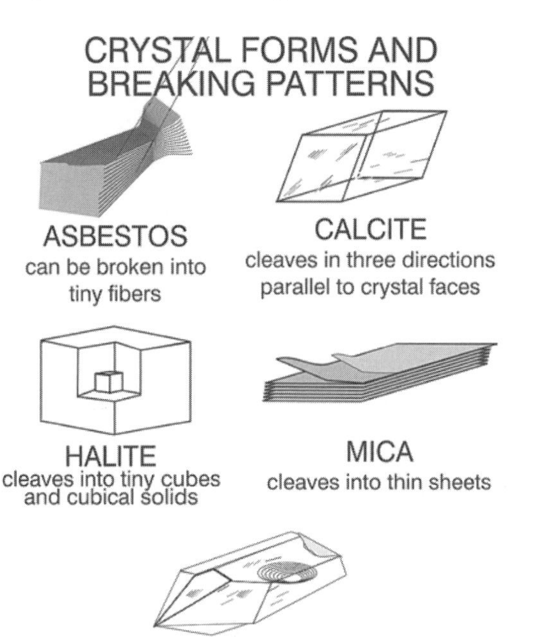

found in Brazil and India. Aluminum has many uses in the construction and transportation industries, including airplanes and automobiles. Popular uses are as wrapping foil and soft drink cans.

- *Chromium* (Cr) is extracted from the mineral ore chromite (Cr_2O_3), 90% of which is imported from South Africa, Russia, Cuba, Turkey, and Zimbabwe. Its main use is protective plating in metallurgical industries, for example, chrome trim on automobiles. It is **carcinogenic**.
- *Cobalt* (Co) is a very hard metal extracted from several different mineral ores imported from Africa, Cuba, Zaire (Congo), Zambia, and Canada. It is mixed with iron and other metals to make specialty alloys for jet engines. It is used also as a paint dryer, an alloy to make cutting tools, and as a chemical catalyst.
- *Copper* (Cu) is extracted from the copper ore chalcopyrite ($CuFeS_2$) and other ores that are mainly imported from Chile, Russia, Canada, Zambia, and Zaire (Congo). The United States continues to mine some copper in Utah, Montana, Arizona, Michigan, New Mexico, and Tennessee, but copper mining is being reduced due to environmental and economic pressures. Copper is used mainly as wire in electrical industries, as well in plumbing; construction; manufacturing alloys of brass, bronze, and beryllium; and as **electroplating** under other metals such as nickel, zinc, and chromium.
- *Gold* (Au) is found in small quantities in most countries but in major deposits in South Africa, Brazil, Russia, Australia, China, and Canada, as well as in the United States. It is mainly used for monetary purposes, that is, gold bullion, but also for dentistry and medicine, jewelry, and electroplating in the electronic industries. The oceans contain great amounts of gold, but it is not concentrated enough to be extracted profitably.
- *Lead* (Pb) is seldom found in a natural state but instead is found in combination with galena. It is mined in Canada, Mexico, Australia, Africa, Europe, Russia, and, to a lesser extent, in the United States. It is a cumulative poison if ingested or inhaled as dust in small amounts. The Environmental Protection Agency's (EPA) standard is zero lead in foods and less than 0.05% in house paint. Lead is an important and useful metal for many industries, including as acid-lead electrical storage batteries in automobiles; in solder, bearings, electronics (PCs), and crystal glass and ceramics; and as shielding for x-ray and nuclear radiation.

- *Lithium* (Li) is a soft metal that is the lightest of all solid elements and also the least reactive of the alkali metals (group 1A of the periodic table of chemical elements), but it will ignite in air at its melting point. It is obtained from spodumene, lepidolite, and amblygonite ores that are found in Canada, Africa, Brazil, Argentina, Australia, Europe, and the United States. It is an important metal used in many industries, particularly as an alloy to harden other metals and to hydrogenate liquid oils into solid fats as in hydrogenated foods. It is also used in the manufacture of soaps, greases, and rocket propellants; in vitamin A synthesis; as a nuclear reactor coolant; and in the production of **tritium**, used in thermonuclear H-bombs. Lithium salts (not the elemental metal) are prescribed to prevent and treat the disease manic depression; although it has some minor side effects, it is effective in preventing extreme mood swings.[15]
- *Manganese* (Mn) is the twelfth most abundant element on Earth. It is not found in a pure metallic state but instead as mineral ores of psilomelane, pyrolusite, rhodichrosite, and manganite minerals. These ores are mined in India, Brazil, Africa, and the state of Montana. One of its earliest uses was as a 20% alloy to harden the steel in iron railroad rails, making them last longer. Manganese is also used in the production of other metals and in the manufacture of hardened tools for cutting metals. Numerous compounds of manganese are important to a great many industries. For example, manganese dioxide (MnO_2), as the mineral pyrolusite ore found in Africa, is used as the black **depolarizer** in dry cells (batteries). More than a trillion tons of manganese nodules cover up to 50% of the Pacific Ocean floor and range in size from golf balls to bowling balls, containing about 30% Mn, 6% Fe, and 1% of Cu and Ni, plus a few other metals. They originated from undersea volcanic activity and are not profitable to harvest with present technologies.[16]
- *Nickel* (Ni) is an important industrial metal found in Canada, Cuba, New Caledonia, Indonesia, and the Dominican Republic. Mainly used as an alloy metal in the production of stainless steel, it is also used for electroplating, as fuel cell **electrodes**, in the making of strong permanent magnets, and in the hydrogenation of vegetable oils into solid fats.
- *Silver* (Ag) is mined primarily in the state of Nevada, but there are less extensive deposits found in other parts of the United States, as well as in Canada, Peru, and Russia. Thirty percent of all silver used in the United States comes from Nevada. It is used mainly in photography, as an alloy and a **catalyst**, for solder, in jewelry, and in electronics due to its high electrical conductivity. Its importance as a coin metal has waned due to its relatively high cost, and it has been replaced by other metal alloys that provide the same commercial purpose.[17–19]

Rocks (Petrology)

Rocks are naturally formed aggregates of one or more minerals that are consolidated. The major rock-forming mineral groups are feldspars, quartz, micas, olivines, pyroxenes, amphiboles, clays, and calcites.[20]

Ores that contain metals as well as minerals may also be considered rocks, since rocks are formed by compacted sediments (sandstone), volcanic activity (basalts), and heat (igneous). Rocks are found in the outer spheres of the solid geosphere, that is, the oceanic and continental crusts, the lithosphere, and the upper mantle.[21] Rocks are constantly, but slowly, being changed from one type or class of rock into another, for example, metamorphic rocks. This transformation is referred to as the "rock cycle." But before we take a look at the cycle, let's look at the various ways in which geologists classify rocks for identification purposes. Rock classification systems are based on one or more characteristics of the rocks being studied. The three basic types of rocks are classified as *igneous* (rocks formed from melts), *sedimentary* (rocks formed by sediments), or *metamorphic* (rocks that metamorphosed [changed] from their original material) (see Figure 4.4).

Each of these three major types of rocks also includes subclasses that are based on one or more of the following characteristics: the rock's chemical (elements and mineral) composition; its origin or so-called parent material; the depth and temperature at which the rock was formed; the temperature and rate at which molten rock material cooled; the rock's texture (grain size); its particle, color, specific gravity, hardness, formations and deformations, foliation, fracturing (cleavage), and solubility in acid or water; the geographic location and land structures where it was found; and, finally, its general appearance.

Figure 4.4
Examples of basic types of rocks formed by weathering, heat, and pressure.

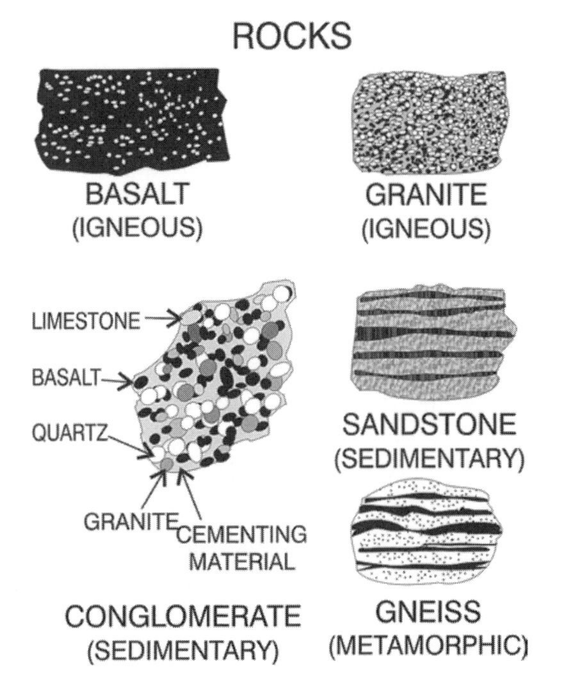

Igneous Rocks

Igneous rocks are formed when hot molten magma begins to cool as it approaches or reaches the Earth's surface. The rock's chemical composition, as well as its physical structure, is dependent on when and how fast the magma cools. The cooling rate determines both the size of the resulting mineral crystals and the rock's texture, the main characteristics that are used to identify igneous rocks. Although color is not a major factor in the classification of rocks, it can be helpful. Four basic properties are used to classify and identify igneous rocks.

1. *Phaneritic* rocks are characterized by crystals of 1 to 5 mm in length. Examples are granite, diorite, gabbro, and peridotite.
2. *Aphanitic* rocks have microscopic or very fine grained crystals. Examples are rhyolite (with some quartz) and andesite (no or little quartz). Aphanitic basalts (ferromagnetic minerals) are formed when magma melts at about 600°C to 1200°C and cools as it reaches the surface to form lava.
3. *Porphyritic* rocks have crystals within crystals visible to the unaided eye. Some basalts can be porphyritic.
4. *Glassy* rocks are melts that formed smooth, solid glass with no evidence of crystals. Obsidian is an example.

Igneous rocks are formed by the solidification of molten magma just below or after it breaks through the Earth's crust as a result of volcanic activity. The word "igneous" comes from the Latin words *igneus* or *ignis* meaning "fiery" or "fire." The magma that forms igneous rocks is a hot mixture of chemical elements that may or may not contain minerals. Igneous rocks comprise 95% of the Earth's crust and are classified by their texture, or grain size, which is dependent on the depth (pressure) and temperature at which they were formed and the rate at which they cooled. Basic factors for the classification of igneous rocks are whether they are *intrusive* or *extrusive.* Another factor considered for classification is their *silica* content.

Intrusive Rocks

Intrusive or plutonic rocks are formed by magma intruding into existing rock structures beneath the Earth's surface. These rocks were named after the Roman god Pluto, who was the equivalent of the Greek god of hell (Hades), and the name refers to the formation of rocks coming from deep in the Earth, generally from ~100 to ~350 kilometers (km) (62 to 220 miles [mi]), where they originate as magma. Since the crust is a poor conductor of heat, it insulates the magma that comes to rest deep in the Earth where it cools slowly over hundreds of thousands of years. Thus, magmatic materials have time to grow from medium, to coarse, to large-grained crystals depending on the time it takes the crystals to cool. The size of the crystal grains of igneous intrusive rocks is larger if the magma cooled slowly. Plutonic igneous rocks are characterized by crystal grains you can see with the unaided eye, thus permitting identification by simply observing the rock's texture. Granite and similar granitic rocks are examples of intrusive plutonic igneous rocks. They are the most abundant rocks in the continental crust and primarily contain crystals of quartz and feldspar.

The processes of radioactivity deep in the Earth produce heat that flows and is transferred from one place to another by either conduction, radiation, or heat convection cells in the mantle responsible for the circulation of magma that intrudes into solid rock formations closer to the surface. Two general classes of intrusive activity are *concordant* and *discordant* intrusions. Concordant igneous rock formations develop where magma intrudes and crystallizes parallel to the layers in the rocks that it encounters. Discordant intrusions cut across (not parallel to) existing rock formations as they intrude into the rock and fill rock crevasses.

- *Batholith* intrusions are large and dome shaped. They are generally discordant to concordant. Examples can be seen in the granite cones of the coastal range of mountains running from Alaska through western Canada.
- *Lopoliths* are intrusions that form saucer-shaped concordant formations. They range from about 100 km (60 mi) in diameter to about 8 km (5 mi) thick. Examples are found in the Bushveld area of South Africa and the Northwest Territories of Canada.
- *Laccoliths* are concordant igneous intrusions that have a flat base with a domed roof. They are much smaller than batholiths and lopoliths. The Henry Mountains in Utah are an example of laccoliths.
- *Dikes* exhibit intrusive discordant cross-cutting features. They cut across the layers of rocks beneath the dike.
- *Sills* are intrusive and concordant. They generally occur as sheets that are thin as compared to their length and width. Sills can be hundreds of miles long. The Palisades, located along the banks of the Hudson River that run through the states

of New York and New Jersey, are an example of sill magma intrusions.

- *Plutons* are intrusive discordant rocks formed by magma that crystallized at or before reaching the surface. They are irregular in shape and are composed of coarse-grained interlocking crystals.

Another type of intrusive igneous rock has a *porphyritic* texture in which large crystals are imbedded in a mass of smaller grains. These intrusive rocks are the result of large crystals that were formed underground at high temperatures and were erupted as lava as they became mixed with smaller grained crystals that formed around the larger ones.[22]

Extrusive Rocks

Extrusive rocks are also the result of volcanic activity, but they result when magma is cooled rapidly and is extruded onto the surface during volcanic fissure eruptions. The magma that comes to the surface and flows is called "lava." Extruded magma may occur on either the continents or the seafloor. When lava cools, it can become solid in just a few days, or it might take months or even years. Therefore, cooling lava does not have time to form large crystals. Extruded igneous rocks are either fully or partly glassy. Sometimes they contain crystals that can be seen with the unaided eye, and sometimes the crystal size is so small it can only be seen with the aid of a microscope. The sizes of crystals and grains, as well as glassy composition, are dependent on the chemical composition and temperatures at which the rocks formed and cooled.

Basalt is the most common type of extrusive igneous rock. It is generally formed by the relatively rapid solidification of lava, and its grain size is much too small to see with the naked eye. The ocean crust consists mostly of basalt. It also makes up some of the large, elevated volcanic plateaus on the continents. Obsidian as an igneous extrusive rock can be either glassy basaltic or rhyolitic (comes from Greek word *rhyax* meaning "stream of lava"), which commonly exhibits lava flow textures. The more rapid the cooling of the flowing lava the more likely the texture will be glassy.[23–25]

Other Classification Factors

Geologists classify rocks by examining the texture and grain/crystal sizes and by identifying the rock's mineral and chemical composition. If you drill into the Earth's crust, the temperature rises about 30ºC for every kilometer in depth. At ~100 to ~350 km (62 to about 218 mi), it is hot enough to melt rocks (i.e., magma). Temperatures at these depths range from ~900ºC to ~1,400ºC, which will melt rocks depending on the pressure exerted on it by the gravitational force of overlying rocks and on other factors, such as the melting rock's chemical composition, the amount of water in the magma, and its density. Magma is so hot that it expands and becomes less dense than the surrounding cooler rocks. Thus, magma rises into and through the denser rocks of the lithosphere and crust, similar to the way a hot-air balloon rises in the cooler air surrounding it. If the magma is extruded, it is called lava that hardens as extrusive igneous rock.[26] When they are in the field, geologists use the textures and grains of igneous rocks to aid in identification. After geologists describe the textures of rocks, they then identify the types and proportions of crystals and minerals found in the rocks.[27] Table 4.4 shows a comparison of grain size with the type of textures for igneous rocks.

The type of igneous rocks that are formed depends not only on how fast or slow the magma cooled but also on the amount of silica mineral content, which may

Table 4.4
Textures Compared with Grain Sizes for Igneous Rocks

Textures	Grain Size
Porphyry	Large grains mixed with fine grains
Coarse grains	Greater than 5 millimeters
Medium grains	1 to 5 millimeters
Fine grains	Under 1 millimeter
Aphanitic (can't see with naked eye)	Very fine grains, dull looking
Glassy (no visible grains, even with microscope)	Glossy/glassy looking

be either (a) *saturated*—igneous rock that contains the maximum amount of silica and maintains equilibrium; (b) *supersaturated*—where the magmatic solution contains more silica solute than is normally present at equilibrium; or (c) *undersaturated*—igneous rock that contains much less silica than normal. Two examples are feldspar and olivine. The silica content determines how fast lava flows. For instance, silica-rich lava is more viscous than silica-poor lava and flows more slowly, as did the lava from the Mount Saint Helens eruption. Also, the opposite is true: silica-poor lava flows more quickly, as does the lava from the volcanic activity in the Hawaiian Islands. Just as the percentage of silicates in magmatic material partly determines the flow rate of magma and lava, so does the percentage of water affect its consistency. Water lowers the temperature at which magma solidifies. For instance, dry magma may solidify at ~700°C, whereas magma with just 10% water will remain liquified at ~600°C.[28]

The primary elements found in most magmas are oxygen (O), silicon (Si), aluminum (Al), calcium (Ca), potassium (K), sodium (Na), iron (Fe), and Magnesium (Mg). Igneous rocks can be divided into types based on their chemical composition.

- "**Felsic**" (derived from the words *fel*dspar and *si*lica—rich in O, Si, Al, Ca, Na, and K) is the geological term given igneous rocks composed of the minerals quartz, feldspars (plagioclase and alkali feldspar), feldspathoids (nepheline and leucite), and mica. These primary minerals are light in color because they contain little iron or magnesium. Thus felsic rocks are also referred to as "leucocratic" rocks. Granite, one of the most common felsic rocks, is formed by both igneous and metamorphic processes. Granite resists weathering in dry climates but not in tropical and wetter climates. Granite is commonly the core rock of large mountain ranges, domes, and shear cliffs. Geologists refer to granite as a "basement" or "bed" rock because it is the foundation of all the continental landmasses.
- "**Mafic**" (derived from the words *ma*gnesium and *f*errous *i*ron—rich in Fe and Mg as well as varying amounts of O, Si, Al, and Ca) is the term used for igneous rocks consisting of minerals such as olivine, pyroxenes, calcic-plagioclase, amphiboles, and biotites, all of which are darker in color. Thus, mafic rocks also are referred to as "melanocratic." The dark color of mafic rocks can be used to distinguish them from felsic rocks (felsic rocks are generally lighter in color). Basalt is a mafic, intrusive rock composed of plagioclase feldspar and pyroxene. Basalt is a major constituent of the Earth's oceanic crust, which in turn overlies a deeper layer of the ocean crust

composed of gabbro/diorite, a coarse crystalline form of plutonic rock.[29]

- "Intermediate" is the term applied to rocks whose composition is somewhat in between felsic and mafic. If they are intrusive (magma solidified underground that intrudes into existing rocks), they are called *syenite, monzonite,* or *monzodiorite*; if they are extrusive (igneous rock formed by magma eruption on the surface of the crust), they are called *trachyte, latite,* or *andesite.*

A color index based more on shades of light and dark than on actual colors has been developed to assist in separating and identifying igneous rocks. Felsic rocks have a color (shade) index of less than 50 (quartz = clear to white; potassium feldspar = white to pink; plagioclase feldspar = white to gray). Mafic rocks have a color index of more than 50 (pyroxene = green; olivine = dark green; biotite = gray to black). Igneous rocks with a color index of more than 90 are termed **ultramafic** and vary in specific composition. Unfortunately, the colors of igneous rocks are not a one-on-one correlation with their chemical makeup.[30–31]

Sedimentary Rocks

Sedimentary rocks are produced by the accumulation and lithification (forming of hard rocks) of various rock fragments. Sediments become lithified by the processes of compacting and cementing the unconsolidated sediments into solids. Sedimentary rocks are generally formed as layers called "strata" and consist of skeletons of sea organisms and fine particles eroded from other rocks. In other words, sedimentary rocks result from the consolidation and lithification of deconstructed rocklike particles to form layered sediments.[32]

Sedimentary rocks also can be formed by the precipitation of solids from a solution to form what is known as "chemical sediments." Sedimentary rocks are mostly found near the surface of the Earth's crust on coastal areas. However, as the ocean floor spreads, sediments are subducted underneath continental plates where they descend into the upper mantle. The forces of nature, such as gravity, wind, water, temperature differentials (freezing and thawing), and organic decay are responsible for the mechanical and chemical conversion of rocks into gravel, sand, clay, and ions dissolved in water. Unconsolidated material resulting from this process is referred to as "sediment." Most of this eroded material is transported to the oceans and is deposited into layers that are, over time, compressed into layers of sedimentary rocks.[33]

Sedimentary rocks may be thought of as deposits or layers found only on the continents and ocean basins—not deep inside the Earth's lithosphere or mantle. About 80 to 90% of the Earth's land and ocean surfaces are covered with sedimentary rocks, but since they are deposits of particles from other rocks located on or near the surface, they comprise only 5% of the crust's total volume and less than 1% of the Earth's total volume. The sedimentary rock layers are about 5 or 6 times thicker on the continental landmass than they are on the ocean basins. This is due to the way the continents and oceans were formed. The layers of sedimentary rock on the seafloor are thinner than on the continents due to the spreading of the seafloor outward from the mid-oceanic ridges, followed by the subduction of ocean crust under continental plates over millions of years. One exception where the sediments are somewhat thicker in the oceans is the point at which major rivers enter the sea and dump large amounts of sediment to form huge, thick deltas, for example, the Mississippi River Delta and Nile River Delta. Most of the rock that is exposed on the continents is sedimentary.[34–35] Examples are shale (which accounts for ~70% of all sedimentary rocks), sandstone, some forms of limestone, chert, siltstone, mudstone, and small amounts of coal.

As with igneous rocks, geologists have developed classifications to identify and study sedimentary rocks. The basic categories of sedimentary rocks used for classification purposes are (1) clastic, (2) organic, or bioclastic, and (3) chemical.

Clastic Sedimentary Rocks

Clastic terrigenous sedimentary rocks are the result of rocks that have been eroded and turned into tiny particles (called "clasts") that were transported, deposited, and lithified (solidified). Clasts may range in size from clay to silt, sand, pebbles, and cobbles. They were transported from the site of origin by running water, mudslides, glaciers, wind, and gravity. Clast size and density determine how and where particles are deposited and eventually what type of sedimentary rocks they will become. Silicate minerals or rock fragments are the major components of sedimentary rocks. The following are examples of rocks classified as *siliciclastic* sedimentary rocks.

- *Conglomerates* are sedimentary rocks composed of lithified cobbles, gravel, rubble, and pebble particles ranging in size from 2 to 65 millimeters (mm) (1/16 to 2.5 inches) in a fine-grained matrix (see Figure 4.4). Clasts in many conglomerates are cemented by calcite or silica. The pore size and the degree of compaction determine the type of conglomerate formed. Spaces, or voids, in conglomeratic rocks can act as reservoirs that contain underground water supplies, but most aquifers are located in sandstone rather than in less porous conglomerates.
- *Sandstone* is composed of sand consisting mostly of quartz particles that are the remnants of other rocks. These quartz particles are just a fraction of a millimeter in diameter and are cemented within a fine matrix to form the sandstone. Beach dune sand is eroded sedimentary rock (mostly quartz) that has been rounded, with the sharp edges worn off, whereas building sand found in commercial sandbanks consists of sand with sharp edges that are needed when mixed with Portland cement, gravel, and water to form reliable construction concrete.
- *Limestone* is a fine to coarse grained sedimentary rock with more than 50% carbonates, such as calcite [calcium carbonate ($CaCO_3$)] and, to a lesser extent, dolomite [$CaMg, (CO_3)_2$]. Limestone often includes fossilized shells of marine life. It is composed of the consolidated equivalent of calcareous sand and mud and/or shell fragments. Along some tropical coastal areas where clastic sediments do not accumulate (because there are no rivers to deposit the sediments), limestone/coral reefs are built by small organisms and thick limestone beds are formed. These large limestone beds are called "carbonate platforms." They may extend for many kilometers. Two examples of carbonate platforms are the Florida Keys and the Bahama Islands located on the southeastern continental shelf of North America. Not only is limestone the most widely distributed carbonate rock, it is one of the most important economically. Through the use of heat, limestone undergoes calcination, yielding lime and other products such as Portland cement, which is mixed with sand and water (and possibly gravel or small stones) to make concrete.
- *Siltstone, mudstone,* and *shale* consist of silt and clay particles that are less than 1/250th of a millimeter in size. When undergoing metamorphic pressure, these

clay particles can form harder forms of sedimentary rocks, for example, slate. Mudstone and shale are rocks consisting primarily of micas that form layers that can be easily split. Silts are tiny pieces of quartz and feldspars. Shale may also contain the altered organic remains of plants and animals and is the primary source of oil and gas. (More on this in a subsequent section of this chapter.) Shale and siltstone comprise about 70% of all clastic sedimentary rocks. Many sedimentary rocks are actually alternating layers of various thicknesses of sandstone and shale that have formed by the successive layering of mud/clay, sand, and other sediments—one on top of the other. Table 4.5 contains a summary of the characteristics of clastic sedimentary rocks.

Bioclastic Sedimentary Rocks

Organic, or bioclastic, sedimentary rocks are mainly composed of the remains of plants and animals.

- *Chert* (flint) is one class of rocks formed by deposits of trillions of skeletal remains of tiny marine organisms (diatoms and radiolarians) that are composed of silica rather than calcium carbonate. Some cherts, however, are formed by the precipitation of silica dissolved in ground water to form flint nodules.[36] Since flint was formed in layers, it was possible for prehistoric people of the Paleolithic era (~10,500 to ~2,500,00 years ago) to flake off layers (**knap**) from flint rocks in order to form spear points, arrowheads, and cutting tools that retained durable sharp edges.
- *Coal* is the best known organic sedimentary rock. It comes from fossilized plants that lived millions of years ago. The forces of heat and pressure, as well as chemical changes that take place when there is a lack of oxygen, resulted in the significant conversion of the original organic matter into sedimentary rock. Usually when plants and animals die, oxygen will, in a very short period of time, chemically oxidize and break down their organic matter into basic chemical components (e.g., CO_2, H_2O, and CH_4). During the coal-forming period (Paleozoic era, ~300 to ~600 mya) on Earth, there was an abundance of plant life in swampy areas. When these plants died, they fell into the water where there was little oxygen. Layer upon layer of decaying plant matter was progressively

Table 4.5
Characteristics of Clastic Sedimentary Rocks

Cast Sediment: Diameter	Description	Name
Cobbles: 65–260 mm	Medium/larger rounded stones	Cobblestone
Gravel/pebbles: 2–65 mm	Rock fragments, generally rounded	Conglomerate
Gravel/rubble: 2–65 mm	Rock fragments are angular	Breccia
Grit/Sand: 1–2 mm	Grains visible, quartz (SiO_2)	Sandstone
Silt: 0.02–1.0 mm	Fine grained, too small to see	Siltstone
No particles (<2 microns)	Fine mud, mostly clay minerals	Mudstone
No particles (<2 microns)	Thin layers, mostly clay minerals	Shale/splits

Note: Clasts are individual grains or fragments of sediments produced by erosion.
Sources: *Sedimentary Rocks,* accessed 2001, http://www.home.expix.net; Graham R. Thompson and Jonathan Turk, *Earth Science and the Environment* (New York: Harcourt Brace, 1992), 49.

buried. The exclusion of oxygen and great pressure resulted in the cementation of these layers, followed by even more layers of dead plant material that eventually were compacted into masses of peat. Peat in turn was covered with sediments that further compacted the organic matter. This pressure, plus increasing temperatures, drove off gasses and concentrated the carbon in the peat, eventually converting it into the form of combustible sedimentary rock we know as coal.[37–38]

- *Oil shale* is another organic-rich sedimentary rock. In this case, petroleum is interspersed in tiny pores between particles in the rock. Pools of petroleum exist in many areas on the Earth in underground sand deposits as well as in other types of fractured rock. (More on coal, petroleum, and natural gas as nonrenewable resources in a later section of this chapter.)

Chemical Sedimentary Rocks

Chemical sedimentary rocks are usually grouped into two general divisions, carbonate and noncarbonate. Carbonate sedimentary rocks are composed primarily of chemical sediments from inorganic sources [e.g., the carbonate ion $(CO_3)^{2-}$]. Two examples are calcite and dolomite.

- *Calcite,* calcium carbonate ($CaCO_3$), precipitates from seawater and is the major mineral in limestone. Many marine organisms, such as **foraminifers**, clams, oysters, and corals (as well as a few saltwater plants), take up the dissolved carbonate ion and combine it with calcium to form their shells. As they die, the shells fall to the ocean bottom where waves and currents break them into small pieces. These tiny calcite shell fragments are deposited in deeper water where they are then lithified into sediments called "bioclastic limestone." Very fine shell fragments, when cemented together, form a different sedimentary rock, known as "chalk," which is light in color and somewhat softer than limestone or chert. Bioclastic rocks are formed by precipitates of inorganic carbonate. Thus, they are also considered seawater-originated chemical sediments. Sometimes the shells and skeletons remain as visible fossils. Over the eons, massive deposits of marine limestone eventually became raised mountains as some of the continents formed (e.g., the Appalachian Mountains in the eastern United States). Mountain building (orogeny) can also occur when and where the Earth's structural plates collide, and the land at the boundaries where these plates meet is subjected to folding and other deformations. The result is the heaving up of massive blocks of marine limestone and other types of rocks that form mountains.
- *Dolomite,* calcium magnesium carbonate $CaMg(CO_3)_2$, is also a form of carbonate. Dolomites are classified as limestones with more than 50% by weight of the mineral dolomite in them. They may have as little as 10% calcite. Dolomite is formed when the original calcite is replaced in seawater with magnesium carbonate. Some geologists refer to this type of limestone as "dolostone," but the term "dolomite" is the name given the mineral and rock.[39–40]

Noncarbonate chemical sediments are not as common as the carbonate examples, yet they are economically and socially important. Noncarbonate sedimentary rocks are composed of compounds other than calcium carbonate ($CaCO_3$). Some examples are given in the following list:

- *Phosphates* (any inorganic compound containing the phosphate group PO_4^{3-} ion, e.g., phosphorite rocks and phosphatic nodules on the seabed)
- *Silicates* (cherts, flint, and opal)
- *Ferriferous* rocks (ironstone, banded iron formations)
- *Evaporites* (salts, halites, e.g., common salt NaCl)

The majority of sedimentary rocks result from the erosion and transport of other rocks, minerals, or chemicals that have been deposited and altered. The forces of water and wind are processes responsible for the formation of a great variety of structures exhibited by sedimentary rocks. A few examples of these structures are (1) stratification or beddings; (2) ripple marks or parallel ridges, similar to sand dunes; and (3) cracks in mud and shale.

Metamorphic Rocks

Metamorphic rocks are defined by their name. The term derives from the Greek words *meta* and *morph,* as in "metamorphosis," which refers to change in shape, such as the stages of an insect's development from the larva to a pupae and then into an adult. Rocks also change as a result of natural forces.[41] The process of metamorphism of rocks may be defined as "any process imposing mineralogical, chemical, or structural changes upon solid rock in response to extremely different conditions and forces than those under which the rock originates."[42] As with igneous and sedimentary rocks, there is more than one way to classify metamorphic rocks. In general, they are classified as either *recrystallized* rocks that are formed under high temperatures (and sometimes high pressure), or as *cataclastic* rocks formed by mechanical forces (pressure).

1. Metamorphic rocks can be thought of as the *recrystallized* product of parent rock particles going through intensive structural and/or chemical changes due to the forces of temperature, pressure, and, to some extent, chemical alteration. An important distinction between igneous rocks (formed from melted rock matter), sedimentary rocks (formed by erosion of rocks), and metamorphic rocks is that, for metamorphic rocks, the forces of heat, pressure, and chemical action deep within the Earth alter the mineral content and/or structure of preexisting rocks without melting them. Although most metamorphic processes take place below sedimentary rocks, all types of rocks—igneous, sedimentary, and even metamorphic—can go through the process of metamorphism. For example, clay that becomes buried under great masses of overlying sediments and is heated to at least 600°C (1,110°F), while also being under high pressure, forms a new stable mineral known as feldspar.[43] Minerals that form rocks are generally considered stable, often unchanged over many millions of years. Under certain conditions, however, some minerals are structurally altered when subjected to gradients of increasing pressure and temperature or when water transports chemicals that seep through the rocks, which changes the rock composition by either the deposition of new minerals or leaching of its original minerals. More important are the ranges (gradients) of temperature and/or pressure that are exerted on parent rock material and that cause stages of metamorphosis to occur, thus forming great varieties of new types of rocks, including marble (from limestone), quartzite (from sandstone), slate (from shale/mud), and schist and gneiss (from igneous rocks).
2. *Cataclastic* rocks are not always classed as real metamorphic rocks because they were not subjected to the high temperatures and pressures that cause actual recrystallization. In other words, they did not actually go through complete metamorphic processes. Instead, rock particles and other debris were

simply pressed together and cemented into a mixture as a conglomerate rock. The materials from which cataclastic rocks are formed were mechanically sheared and broken down by tectonic faulting and folding at plate boundaries, thus undergoing pressure causing rock deformation. This process is known as *dynamometamorphism* (kinetic or dynamic metamorphism).[44–45]

Heat is the most important of several forms of energy responsible for the metamorphism that changes igneous and sedimentary (and some metamorphic) rocks into different metamorphic rocks. Heat accelerates almost all chemical reactions, and although metamorphism is not entirely a chemical activity, high temperatures deep in the Earth are mainly responsible for this process (but pressure is also responsible for the metamorphosis of rocks, and as per the gas laws, increasing the pressure also increases the temperature within a constant volume.) As previously mentioned, temperatures increase about 30 to 40°C for every kilometer in depth, and at about 10 km (6 mi) temperatures reach ~200°C (400°F), which is enough to metamorphose rocks. Heat energy is generated within the Earth's spheres by several processes and systems. One is the decay of radioactive **isotopes** of elements in the mantle and lithosphere. This is a natural process in which the atoms of one class of unstable radioactive elements are transformed into atoms of other, more stable elements. A small amount of leftover mass from this spontaneous decay of radioactive elements is converted to heat energy deep in the Earth. The three main heat-producing radioactive elements in the Earth are uranium, thorium, and potassium-40. Heat is also generated by the friction created between bodies of rock that slip past each other at faults (as during earthquakes) and at boundaries where one or more tectonic plates meet. Rocks are also heated when molten magma forces its way toward the surface through dikes (cuts or cracks in massive bedrock) and by the friction between rocks in the lithosphere and crust as they slide over or past each other. For instance, when an ocean plate dips (subduction) below a continental plate and the slab is pulled down into the mantle by gravity, a great amount of gravitational heat is generated. This heat melts magma near the surface above the descending slab, resulting in volcanic activity near the boundary zone (coastal area) of the continent. Internal heat, along with pressure, alters the texture, density, proportion, and composition of minerals in rocks that are undergoing metamorphosis. The temperatures required for these changes to occur range from a low of ~150°C (~300°F) to more than ~1,000°C (~2,000°F). The higher temperature ranges are responsible for the formation of igneous rocks. Two examples are the pluton granites, formed at ~650°C (~1,200°F), and volcano basalts, formed at ~1,000°C (~2,000°F).

As important as heat is for the process of metamorphism to take place, it is not the only factor. Scientists measure pressure in solids, liquids, and gases by referring to the normal air pressure at sea level, which is ~760 mm of mercury or ~14.7 lb/square inch, or 1 atmosphere at 100 meters (m) above sea level. This standard is then converted into bars, where one bar is equal to normal pressure. Or for measuring greater pressures, the standard is converted into metric units of kilograms per square centimeter. There are three types of pressure recognized deep inside the Earth.

1. *Lithostatic* pressure (meaning "rock staying in place") is created by the gravitational weight of the layers of rock found in the lithosphere.
2. *Confining* pressure occurs when rocks push on each other from all directions. This pressure is exerted on rocks confined in an area

and results in alterations in the rock's density and size but not shape. Lithostatic pressure can also be of a confining nature.

3. *Directed* pressure takes place where tectonic plates meet at boundaries forcing rocks to fracture and fold to form mountains and faults. The rocks are deformed in the direction of the pressure exerted by the moving plate(s).

Deep rock is pressed as it is acted on by the gravity pulling the rocks above it downward. This confines and compresses the deeper rock, thus increasing its density, which in turn reduces the rock's pore size, thus reducing its ability to absorb more fluid material. This increased pressure can also alter the chemical composition of the rock, especially if some water is present, which there is in abundance throughout the lower mantle. A research report in *Science* magazine stated that "Earth's mantle may store about five times more H_2O than the oceans."[46]

The different levels (gradients) of heat and pressure are responsible for the different textures and structures of rocks. These different textures are used to help classify metamorphic rocks. Two such classes are *foliated* rocks and *nonfoliated* rocks.

1. *Foliated* rocks are the consequence of metamorphism where the parent rock material was subjected to low temperatures but great pressures. The term "foliated" refers to the layered or parallel grains in a metamorphic rock. Pressure, without high temperatures, also causes alterations due to crushing and shearing near the surface of the Earth. Some examples of foliated rocks are slate, phyllite, schist, and gneiss. Although the grain sizes in these examples vary, they all exhibit parallel plane structures. Their textures remind some geologists of wood grain, and they vary according to the amount of pressure applied under relatively low temperatures.

2. *Nonfoliated* rocks are produced by high temperatures when the heat from cooling magma makes contact with the parent rocks. Nonfoliated rocks appear as finer textured rock crystals with fine grains that are unstratified (not parallel). The texture of nonfoliated rocks is more like the surface of a paved road rather than the grain in wood. Heat results in metamorphic changes of limestone and dolomite to the rock known as marble. High temperatures also metamorphose basalt into igneous rocks, and shale into slate.

A number of conditions are responsible for the different foliations and textures of metamorphic rocks. Geologists have developed four classifications of metamorphism based on the movements of magma and the Earth's crust.

1. *Contact metamorphism,* also known as thermal metamorphism, results when very hot magma intrudes and comes in contact with existing rocks. The heat of the magma, not pressure, is responsible for contact metamorphism of surrounding rocks known as "country rock." Because rocks are poor conductors of heat, the greater the distance from the source of magma, the less magmatic heat; thus, contact metamorphic alterations become less over greater distances of the surrounding rocks as the magma cools. In other words, the rocks closer to the intruding hot magma will be changed more than those at a greater distance. Also, the water content in the magma helps determine how close the contact rocks were to the magmatic heat source. Heat drives off the water, so the less water, the closer the rocks were to the hot magma, and the greater will be the extent of the rocks' metamorphic alterations. Contact metamorphism is limited to the relatively small areas where intruding plutons spilled magma to form sills. Also, at times, hot magma spreads out over large areas, resulting not only in extensive contact metamorphism of the surrounding rocks but also changes in their mineral content.

2. *Regional metamorphism* has formed the common type of rock found over larger regions of the Earth, more common than contact metamorphic rock. Unlike rocks that are the product of contact metamorphism, regional metamorphic rocks are the result of extreme pressures rather than high temperatures. Regional metamorphism occurs when magma and tectonic plate movement push up through the crust, forcing the roots of mountains to rise into mountain ranges. The heat supplied by hot, near-surface magma, as well as the heat from crust movement, creates additional new regional metamorphic rocks. Regional metamorphic rocks may be foliated types and are often found along with igneous rocks in mountains. Examples of regional metamorphism are found in central Canada, the Appalachians of New England, and the Cascades in the Pacific Northwest. Three common types of regional metamorphic rocks are *migmatite* (a mixture of metamorphic and igneous rocks), *schist, slate,* common *quartzite* (from sandstone and limestone), and some types of *marble.*
3. *Burial metamorphism* is sometimes classed as a subset of regional metamorphism. As the name implies, burial metamorphism occurs as sediments from rivers and coastal deltas sink and are buried with new sediments that are added on top of them over millions of years. Pressure and heat at a depth of about 10 km (6.2 mi) change the material in the lower layers of sediment by recrystallizing their minerals. Burial metamorphic rocks are mostly nonfoliated, with the minerals arranged without deformation and in a random pattern. Burial metamorphic rocks are found as deep as 12 km (8 mi) under the Mississippi River Delta off the southern coast of Louisiana. Some examples of burial metamorphic rocks are *mica, slates, phyllites,* and *schists.*
4. *Dynamothermal metamorphism* also is sometimes classed as a subset of regional metamorphism. Dynamothermal metamorphism results from pressure exerted by compression created when converging plates override and trap rock beneath them. This occurs during mountain building as the lateral pressure forces some rocks upward, and some downward, resulting in folding and fracturing as the rocks are changed. Some of the rocks forced downward reach depths up to ~10 km (~6.2 mi) where they are subjected to great pressure and geothermal heat. Thus, the term "dynamothermal" [from "dynamic" (pressure) and "thermal" (heat energy)]. This process creates very large regions of metamorphic rock. Examples of dynamothermal metamorphic rock can be found in the European Alps, the Appalachians in the eastern United States, and the Asian Himalayas.

There are also several other lesser classes or types of metamorphism.

- *Hydrothermal metamorphism* is a chemical process of rock formation that involves water. The water may come from the magma, or it may have been driven from the rock undergoing metamorphosis, or it may be from a ground source. Most hydrothermal metamorphism occurs on the ocean floor.
- *Fault-zone metamorphism* is produced by the great frictional heat generated when two blocks of rocks (or plates) slide past each other. The heat generated by the grinding between the two slabs changes the crystal structure of minerals. The greatest recrystallization of the metamorphosed rock takes place close to the fault, where the heat is greatest. Fault metamorphism is sometimes considered as a type of *dynamic* metamorphism, where rocks are altered by the strain created by localized faults, or as *shock* metamorphism, resulting from impact created by high velocity meteorites striking the Earth.
- *Pyrometamorphism* results when extremely high temperatures are created in very local regions of rock. An example

is when a bolt of lightning strikes a rock or sediment and recrystallizes it. Pyrometamorphism also can occur through the heat generated by the burning of an underground coal seam. Nearby rocks can undergo metamorphism by the accumulated heat generated from this source.[47–50]

It is important to recognize that the different types, classes, and processes of metamorphic rock are based on (a) the rock's appearance and (b) the conditions under which the metamorphic rocks were formed, rather than just their composition. While several types of parent rocks may be involved, their composition is important to their metamorphic changes. The following list provides a few examples of common metamorphic rocks and their characteristics.

- *Marble* is contact-metamorphized limestone. It is calcite and/or dolomite recrystallized due to exposure to moderate heat and pressure from deep burial. It is a nonfoliated rock with fine to coarse grains and little or no porosity. Marbles may differ in color according to the other elements or compounds present. Marble can be cut and polished to a high sheen, making it an excellent building and decorative stone.
- *Quartzite* can be either a *granoblastic* metamorphic rock, recrystallized sandstone, or *chert* composed mainly of the mineral quartz. It is nonfoliated and was formed by contact with hot magma in deep burial.
- *Hornfels* metamorphic rocks originate from a variety of minerals (*andalusite, mica,* and *quartz*) and parent stones (*shale, mudstone,* and *basalt*). They are formed by contact with hot magma while experiencing moderate pressure. Hornfels are nonfoliated, with equidimensional fine grain textures with no specific orientation. They may appear either as banded or spotted.
- *Migmatite,* from the Greek word *migma* meaning "mixture," is an intermediate metamorphic rock between granite and gneiss. It is formed from a mixture of magma and solid granitic rock material that contains *feldspars, quartz,* and *mica.* It is formed under high temperatures and pressures and is foliated with the alternating dark and light bands of different minerals.
- *Gneiss* is formed from several minerals, such as *feldspars, quartz,* and *mica,* under high heat and pressure along with tectonic deforming pressure. Some of the chemicals found in gneiss rocks are Si, O, Al, K, Na, Ca, and possibly Fe, Ti, Mg, and Mn. Gneiss is foliated by regional metamorphism and may be banded, **lenticular**, or granular in texture. It consists of dense, medium to coarse grains of dark and light bands of aligned minerals.
- *Slate* is a metamorphic rock formed at relatively low temperatures and pressure. Its parent rock material is *shale* and *mudstone,* while its minerals are *mica* and *chlorite.* Some clay minerals may be included. It is foliated and can be split easily into layers.
- *Schist* is metamorphic rock formed from different aluminosilicates, including *quartz, mica, kaynite,* and *garnets,* whose thin layers can be easily split or cleaved. Some of the elements found in schist are O, Si, Al, and Mg. Schist is formed under moderate heat by direct shearing pressure and tectonic deformations and is foliated with coarse, layered, often flaky, wavy textures and platelike structures. Mica, epidote, garnet, graphite, and hornblende are examples.
- *Phyllite* is an intermediate grade of metamorphic rock that is between a schist and

slate. Small crystals of white mica and chlorite are its basic minerals. It is metamorphized under direct pressure at low-to-medium temperatures. Phyllite is foliated with fine grains and is layered with a shiny, almost mirrorlike, wrinkled-wavy surface.

The Rock Cycle

The concept of a systematic cycle implies that it has no beginning and no end but continues from one stage or component of the cycle to the next, and so on, to be constantly continued. See Figure 4.5 for a general depiction of the rock cycle.

It is reasonable to suggest that the rock cycle started soon after land formed above the oceans several billion years ago. The force of gravity pulled all the particles of the nascent Earth together, compacting matter into spheres and layers of differing densities as this spinning globe was set into rotation around an axis. As gravity continued to compact matter, the original heavier material settled in the lower spheres while less dense matter formed the outer layers. With the addition of the heat generated from the decay of radioactive elements and residual heat from the original formation, significant amounts of this original matter underwent chemical as well as physical transformations. The internal pressure increased with depth, as did temperatures, creating heat convection cycles that also affected the distribution and characteristics of interior rock matter. Temperatures, as well as pressures, vary with depth. Therefore, great changes occurred over hundreds of millions of years. As the rocks melted, forming magma, their volume increased about 10%. Consequently, the melted rock material became less dense and arose as semi-plastic-like magma that intruded at the mid-oceanic ridges. It then spilled over both sides of the ridges onto the ocean floor, resulting in the spreading of the

Figure 4.5
An artist's depiction of the continuous recycling of minerals and rocks that takes place during the rock cycle.

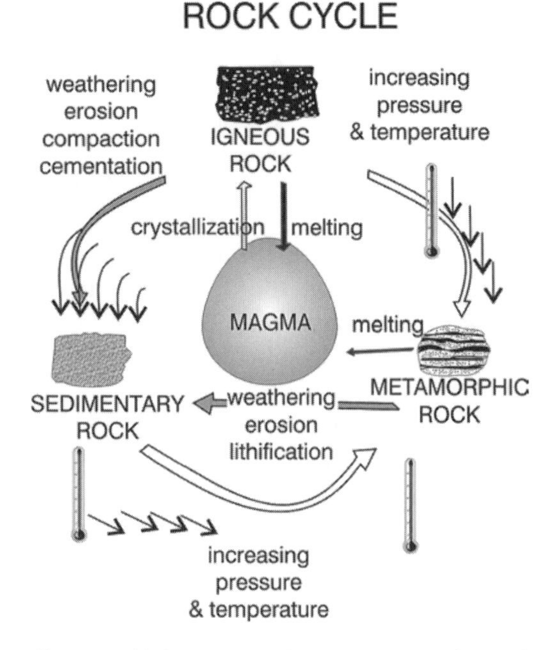

floor until the ocean plate met a continental plate, resulting in subduction of the ocean plate and formation of coastal troughs, volcanoes, and earthquakes. When two or more continental plates collided, they deformed their boundary areas, forming mountains. These mountains, in time, weathered and eroded to form sediments. The Earth's early crust consisted almost entirely of nascent extruded igneous rocks that, over time and due to geological processes and weathering, were converted to sedimentary rocks, younger igneous rocks, and deeper metamorphic rocks. The continental crusts were formed by a variety of processes, including cooled, solidified magma as igneous rock that, when exposed to weathering, eroded and formed sediments that compacted to form sedimentary rock. The global system of tectonic plate activity is the main process driving the system. The burial of sedimentary and young igneous rocks exposed them to great heat and pressure, which melted them and altered their chemistry and physical characteristics. The result of this process was the formation of metamorphic rocks that were then uplifted as igneous rocks that were again weathered and eroded, continuing the cyclic pattern of rock formation, deconstruction, and reformation. A somewhat similar process occurs with other types of rocks that are also part of the overall rock cycle.

Figure 4.5 is somewhat simplified for clarity. There are a number of crossover paths from one stage of the cycle to another, but in general the process is as follows:

1. Magma is formed in the Earth's mantle.
2. Some magma breaks through the crust, forming volcanoes and fissures, and some remains in the mantle and lithosphere, crystallizing slowly.
3. Rising magma cools on its path to the Earth's surface; it is called "lava" after it reaches the surface.
4. Extrusive igneous rocks are formed both on the surface and below the surface as magma cools.
5. Through weathering and erosion, igneous rocks are broken into particles from which sediments are formed.
6. Rock fragments and disaggregated minerals are transported and deposited at lower surface levels.
7. Lavas and deposited sediments are buried and lithified by gravitational compaction and cementation, forming sedimentary rocks.
8. Increasing temperature and pressure with increasing burial depth, along with alterations in chemistry, cause these rocks to undergo metamorphism, changing them into metamorphic and igneous rocks.
9. Metamorphic rocks (as well as some intrusive igneous rocks) are again melted and become magma.

The cycle of metamorphic rocks being uplifted, eroded, transported, formed into sediments, and again buried is a constant cyclic system that was repeated over and over in the past. The rock cycle will continue as long as the Earth remains a dynamic planet.

Nonrenewable Natural Resources

When viewed from the perspective of the universe, or even the solar system, the Earth is exceedingly small—just a minor speck in space. When viewed by a microbe or even an insect, the Earth is exceedingly large. Either perspective, small or large, implies that the Earth is of a specific size, with a given diameter, volume, and surface area composed of a given amount of matter. Thus, the Earth has finite amounts of that of which it is composed.

Definitions for Nonrenewable Resources

It is somewhat misleading to refer to *natural resources* in terms of "nonrenewable" and "renewable." Many of the nonrenewable resources are rocks and minerals, but not all of them are inorganic geological materials (e.g., petroleum). Nevertheless, differences between types of resources do exist. First, renewable resources are usually considered as those produced by solar energy in one form or another over relatively short periods of time. Some examples are the energy used during the process of **photosynthesis** (chloroplasts in green plants using the sun's light energy to convert CO_2 and water into carbohydrates and O_2) for growth of biomass. Heat, wind, and the water cycle, most of which are driven by solar energy, are also renewable resources. This concept of renewable is based on the assumption that solar energy will be around forever (at least another ~4.5 billion years). One example of a renewable geological resource is tidal energy driven by the moon and sun's gravitational forces interacting with Earth's gravity. Tides cycle on a regular basis, thus they could be considered renewable, but not in the sense of solar energy.

Nonrenewable resources are considered those substances originating on and in the Earth many millions or billions of years ago that cannot be regenerated in a reasonably short period of time. This is true, as far as this concept goes. However, just because these resources, whether chemical or physical, were formed eons ago does not mean that they will be lost if altered by either nature or living organisms, including humans. The law of conservation states that some physical amount (quantity) of an isolated system (e.g., Earth) is constant. These nonrenewable resources may be altered in their chemical or physical geological nature, or deposited in other locations in different forms, but they are not lost from the Earth. For example, when aluminum is extracted from its bauxite ore and is converted into metal for manufacturing useful items, the aluminum objects may, in time, be discarded, buried, recycled, or oxidized, which may change the object chemically and physically, but all of the original Al atoms are still here on Earth. The only natural resources ever truly lost from the Earth are the insignificant amounts of metals and minerals sent into deep outer space, but one might say that they still exist in the universe. Even low-orbiting satellites sooner or later return their nonrenewable resources to Earth, albeit in a chemical and physical form different from their origin. Some lighter elemental gases at the outer edge of the Earth's atmosphere may get knocked with enough energy by molecular motion to enable them to reach escape velocity (about 7 mi/second) and escape from the Earth forever. At the same time, tons of tiny micrometeorites and tiny ice particles are constantly being added to the Earth's mass from outer space. Alterations of the nonrenewable resources may occur naturally or by human intervention, but basically, humans do not increase or decrease the amount of matter on or in the Earth.

Since humans occupy only a tiny fraction of the outer crust of the Earth and the atoms of the Earth's elements are indestructible (with the exception of radioactive elements), man can alter and use, but not destroy matter. "[O]ne cannot 'exhaust' the Earth's supply of copper; the worst one can do is disperse it. And while energy can be degraded, humanity uses only a minuscule fraction of the vast amounts of nuclear energy continuously released in the core of the sun."[51] For example, when iron is smelted from its ore, it becomes concentrated as a metal that humans can use in the manufacture of any number of desired items. But once the item containing that iron is no longer useful, it may be converted by **oxidation** to a form different from the original ore or metal (namely, iron oxide or rust). Therefore, it can be argued, for example, that the nonrenewable natural resource located in the Earth's crust—iron ore—will be lost because it cannot be put back into the Earth in its original state, and once all the iron in its original form is converted for use by humans, no more original iron ore will be available. Obviously, no one knows when or if total depletion of iron ore will occur. However, it can also be argued that the element iron is still on Earth, just in a different form (i.e., conservation of mass).

Conservation of Nonrenewable Resources

Economic and social issues, as well as geological aspects relating to the importance of coal, oil, and gas, along with the mining of metals and minerals, are considered in this section. Technically, hydrocarbon energy sources are not geological in nature and were originally formed from renewable resources (plants and animals), but these processes occurred over such long periods of time that they will not be replaced as fast as they are used. Thus, they are usually considered nonrenewable.

Two major concepts are associated with conserving our environment and resources. One is *preservation,* or protecting, of the resource by not using it, thus saving it; the other is the *wise use* of natural resources, under which those nonrenewable resources (as well as renewable resources) necessary for a reasonable standard of living are intelligently used. It is impossible to preserve all of the nonrenewable (or even renewable) resources in today's world. This has been impossible from the outset of life on Earth because living organisms require the consumption of both renewable and nonrenewable natural resources. Not only do all life forms use some nonrenewable resources in their lifetime, but all living organisms, to some degree, also alter their environments. Even so, a number of people consider it wrong for people to use nonrenewable resources for the benefit of humans. As just one example, Ray Anderson, the industry co-chair for the President's Council on Sustainable Development, is reported to have said, " . . . it's a sin against the earth for humans to take one drop of oil, one grain of coal, one ounce of metal from the earth's crust."[52] Such a policy goal is unattainable in today's world, and it would have been impossible from the beginning of human civilization.

For our present standard of living to prevail, our requirements for nonrenewable resources for each human born in the United States during the early part of the twenty-first century—on the average and for all purposes (industrial, commercial, and individually)—are 3.75 million pounds of minerals, metals, and fuels over her or his lifetime. See Figure 4.6 for a detailed list of these natural resource requirements for the average American during her or his lifetime.[53]

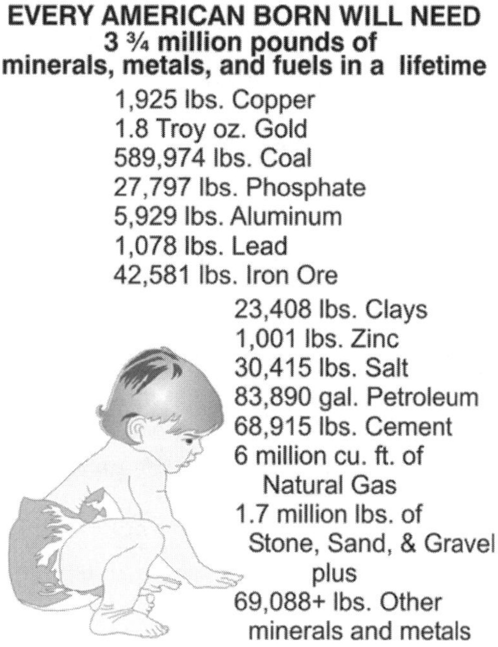

In addition, every year, 46,000 pounds of new minerals for all purposes must be provided, on the average, for every person in the United States to maintain our standard of living. The following list provides several examples of the required resources on a per capita yearly basis (these figures are based on the total amount of the resource used per year as compared to the total population in the United States):

Figure 4.6
An estimation of the resources used by one person in the developed world during a normal lifetime. No one person actually uses these amounts; the figures are calculated by averaging the resources used by everyone. (Figure used by permission of the Mineral Information Institute.)

Stone = 10,814 lbs.
Sand and gravel = 8,220 lbs.
Cement = 742 lbs.
Salt = 404 lbs.
Other nonmetals = 1,203 lbs. (est.)
Iron ore = 603 lbs.
Aluminum = 93 lbs.
Copper = 23 lbs.
Lead = 13 lbs.
Zinc = 12 lbs.
Manganese = 6 lbs.
Other metals = 10 lbs. (est.)
Petroleum = 7,520 lbs.
Coal = 7,821 lbs.
Natural gas = 8,164 lbs.
Uranium = 1/4 lbs.

In addition, no matter who builds the house, 14,105 board-feet of lumber, on average, is required to build a 2,075 square-foot two-story, three-bedroom home with central heat and air conditioning.[54]

Somewhat amazing is the amount of natural resources we use that are not readily available in the United States. Table 4.6 lists selected chemicals and minerals, the percentage of each imported (above 75%), and the countries from which they are imported to the United States. In addition to those listed in Table 4.6, 1–75% of more than thirty other important chemicals and minerals are imported by the United States.

Table 4.6
Chemicals, Minerals, and Metals Imported by the United States

Commodity	Percentage Imported	Countries Imported From
Arsenic (As)	100	China, Chile, Mexico
Niobium, columbium (Nb)	100	Brazil, Canada, Germany
Carbon, graphite (C)	100	Mexico, Canada, China, Madagascar, Brazil
Manganese (Mn)	100	South Africa, Gabon, France, Brazil
Mica, sheets	100	India, Brazil, Finland, China
Strontium, celestite ($SrSO_4$)	100	Mexico, Germany
Thallium (Tl)	100	Belgium, Canada, United Kingdom
Ytterbium (Yb)	100	China, United Kingdom, Hong Kong, Japan, France
Aluminum, bauxite (Al_2O_3)	99	Australia, Jamaica, Guinea, Brazil
Gemstones (ornamental)	98	Israel, India, Belgium, United Kingdom
Diamonds (industrial)	95	South Africa, Ireland, United Kingdom, Zaire
Asbestos (chrysolile)	95	Canada, South Africa
Tungsten (W)	94	China, Germany, Bolivia, Peru
Platinum-group metals	91	South Africa, Russia, United Kingdom
Fluorite, fluorspar (CaF_2)	88	China, South Africa, Mexico
Tantalum (Ta)	86	Australia, Germany, Canada, Thailand
Tin (Sn)	84	Brazil, Bolivia, Indonesia, China
Barium, barite ($BaSO_4$)	82	China, India, Mexico
Cobalt (Co)	79	Zambia, Norway, Canada, Zaire, Finland
Chromium (Cr)	75	South Africa, Turkey, Zimbabwe, Russia, Finland

Source: *1995 U.S. Net Import for Selected Non-fuel Minerals* (Denver, Colo.: Mineral Information Institute, 1997).

The Mineral Information Institute provides some interesting statistics related to the great dependency humans, and in particular U.S. citizens, have on nonrenewable natural resources that are required for maintaining a reasonable standard of living. For instance, in 1995 there were 133,029,661 cars in the United States alone. (About 8 million or more new cars are added each year.) Each car weighs about 3,000 pounds and during manufacturing uses more than three dozen different chemicals, minerals, and metals. These are mainly various forms of steel alloy, consisting primarily of iron. In the 1990s it was estimated that, in his or her lifetime, an *average* (based on total consumption) middle-class American consumes more than 28,000 gallons of gasoline, since he or she drives an average of 700,000 miles; reads and discards more than 27,500 newspapers at a rate of seven trees per year; adds 110,250 pounds of trash to the nation's garbage heaps; and wears and discards 115 pairs of shoes. The Massachusetts Institute of Technology asked more than 1,000 Americans what essential invention they could *not* do without: 63% claimed they could not do without an automobile, 54% said light bulbs, 42% said telephone, 22% said television, 19% said aspirin, 13% said microwave ovens, and 8% said they could not do without hair-dryers or personal computers.[55] Obviously, manufacturing these so-called essential, everyday conveniences requires conversion of nonrenewable resources to more useful forms.

The four main sources of energy today, for the developed nations, excluding solar and wind energy, are petroleum, natural gas, coal, and nuclear. (It might be mentioned that there are millions of people in developing nations who still depend on wood, cow dung, and other biomass for their main sources of energy.) As mentioned, **hydrocarbon** resources (petroleum and gas) are not exactly the same as most geological minerals, metals, and rocks. Hydrocarbons are usually considered as nonrenewable resources because they are energy sources of carbon and hydrogen found in geological formations. On the other hand, coal, which also originated from organic sources eons ago, is considered rock. Both petroleum and coal form at such slow rates that they are now considered nonrenewable.

Petroleum

What is petroleum? Where does it originate? These are questions that have been asked for many years, and the complete story is still not known. The name "petroleum" is derived from the Greek words *petra* (rock) and *oleum* (oil), and at one time it was called "rock-oil" because it is found in rock formations, as well as seeping to the surface. Tremendous amounts of petroleum seep to the surface of the beds of oceans and lakes. Two well-known examples of land seepage are the La Brea tar pits in California and the discovery of oil seeps in Pennsylvania in the 1800s. In general, petroleum was formed many millions of years ago by dying organic plants and animals that settled to the seabed. As layers of sediment blocked out oxygen over millions of years, these layers of decaying organic mud were buried deep within the Earth, where pressure and heat converted them to shale (the source rock). At the same time, the organic matter was converted to a variety of liquid hydrocarbons, ranging from light oils such as naphtha to more viscous oils called "maltha" (tars). Petroleum also varies in color from light yellow to dark brown, depending on the types of hydrocarbons and other molecules it contains. Some petroleum deposits are located in areas that, at times, were seabeds that were raised as continents formed. The original organic matter was converted to petroleum at temperatures of about 125°C, and if it became hotter (above the boiling point), some of the petroleum turned into natural gas. This is the reason natural gas is often found above or near petroleum deposits. Many forces, processes, and systems are involved, including the degradation of original organic material by biological organisms, in the complicated process of changing ancient marine organic matter into crude oil and natural gas.

Over a period of many centuries and in many countries, petroleum seeping out of porous rocks deposited pitch, bitumen, and asphalt on the land's surface or in tar pits. Historically, this surface residue had a number of uses, including waterproofing and medicine, but the supply was limited and expensive. It was not until 1859 when Colonel Edward Laurentine Drake (1819–1880), an expert in drilling salt brine wells, used his drilling technique to sink a well at an oil seepage site located near Titusville, Pennsylvania. He was part owner of a firm that used the seeping petroleum for medicinal purposes. He believed that if he could reach the main source of the seeping oil, he could collect all he needed for his business. His well was only about 70 feet (ft) deep, but it soon produced more than 400 gallons of oil a day. This started an oil boom that led to the modern petroleum industry. Some of the first uses were the production of kerosene (sometimes called "coal oil"), used as a source of light and heat that was superior to candles, whale oil, and wood.[56]

Nevertheless, petroleum is a nonrenewable natural resource that is obviously limited in supply, since it takes millions of years to produce significant quantities. The location, drilling, extraction, transportation, refining, and marketing of petroleum products are gigantic worldwide industries.[57] The world, and in particular the industrially developed countries, is extremely dependent on petroleum as an energy source. Even though new sources are periodically discovered, along with new technologies and conservation methods, the time will come when oil will become a scarcity, to the point that new sources of other energy will be required to maintain current standards of living. One question that cannot be accurately answered at this time is, just how much oil is there in the Earth? The exact locations and extent of all petroleum reserves are not known, so, with some wise-use strategies, it is a fair assumption that we will have petroleum products available for many years to come.

Natural Gas

The chemical name for natural gas is methane (CH_4), which is the first member of the paraffin hydrocarbon family. It not only burns cleaner than other fossil fuels, but it is also colorless, odorless, tasteless, and lighter than air. Methane's molecular structure and chemical reactivity make it a greater greenhouse pollutant than carbon dioxide gas. Methane was formed by the decaying of ancient organic matter in swamps and marshes and can be located in pockets in proximity to petroleum deposits. The lighter gas is often found in domes above pools of petroleum, or it can seep underground to separate underground reservoirs, or it can escape from the Earth's crust into the atmosphere where it contributes to global warming. Methane, a natural hydrocarbon gas, can also be produced commercially. The chemical processes of **adsorbtion** and **absorption** are involved when hot steam is passed over hot coal. Methane can also be formed by catalytic processes and is often a by-product of petroleum refining by **cracking** using steam. At one time, before the use of natural gas became widespread and plentiful, the old gas house produced methane, which was referred to as "coal gas" or "water gas," by passing steam over hot coals. The commercial production of this nonpetroleum methane was limited, so it was mainly used for lighting and as a home cooking gas.

Large quantities of methane are produced in the digestive systems of animals, particularly by ruminant flatuations. It is also produced by the decomposition of biomass. Heat is also produced by this decomposition. Years ago, during winter months, farmers parked their Model T cars on top of the manure piles located in back of their barns. As the chemical reaction in the manure released methane, heat was also released, which kept the car's engine warm. The November 1999 issue of *Scientific American* magazine included the article "Flammable Ice" that describes the discovery of methane-laced ice crystals on the seafloor at ocean depths of ~500 m (~1,064 ft). Methane ice is formed by seepage around openings in the seafloor where there is bacterial decomposition of organic material that allows ancient methane ice structures to build up. Some of the methane also escapes as gas into the surrounding seawater. The temperature at this depth is just above freezing. As gas **hydrates** decompose, they support colonies of **tube worms** that grow around the vents for escaping methane in the seafloor. When brought to the surface, the ice soon melts, releasing the methane gas, which can be ignited and burned off as it melts. The authors of the *Scientific American* article speculate that this source of methane stores more energy than all the world's fossil fuel reserves combined. There are significant problems in trying to recover this source of energy, because these methane hydrate deposits are not only deep in the seafloor but also are fragile. In addition, if significant amounts of these hydrates were brought to the surface, the methane gas released would exacerbate global warming.[58] Who knows, but this deep-sea source of methane may become a major source of energy in the future. Natural gas is currently being produced by very slow natural processes deep in the crust, but at a rate that is much less than at which it is being consumed. Again, the supply is limited. Estimates of how long the supply of natural gas will last at its current rate of use range from ~75 to ~200 years.

Coal

Coal is also considered a nonrenewable natural resource, but it has differences from the other two main forms of fossil fuels—petroleum and natural gas. First, coal is a solid hydrocarbon mineral, more rocklike than petroleum or gas. Second, depending on the particular deposit, it may contain significant inorganic elements and minerals as well as organic matter. Scientists describe coal as a solid, combustible material formed from prehistoric plant life. It consists of large molecules in groups of polynuclear aromatic rings that are connected to oxygen and sulfur, as well as some other chemical compounds.[59] The origin of coal is better known than that of oil or gas. When plants die and other organic matter accumulates on the surface of the Earth, it is exposed to the atmosphere, where oxygen, water, and chemical and bacterial reactions decompose the biomass to form CO_2, CH_4, and water. The processes that formed coal are different. Coal forms from large quantities of plant material that was buried in stagnant swamps and then fairly rapidly covered with silt and more plant material. Layers of the plant material were buried as sediments were deposited during rising sea levels, followed by lower sea levels where new biomass was grown and died, to again be buried by sediment. Each deeper layer of plant material was depressed by the layer above it, thus forming individual seams of coal of different types. Burial prevented oxidation and the escape of methane from the porous biomass, while layers of sediment squeezed out water, CO_2, CH_4, and other organic acids. In time this process converted the organic mass into different grades of coal, depending on the depth at which they were buried and the weight of overlying sediment deposits.

Different grades of coal are basically formed in the same manner but under different conditions of pressure and temperature. Thus, they contain different levels of carbon and inorganic elements.

1. *Peat* was formed during the period of luxurious plant growth on the Earth ~280 to ~350 mya referred to as the Carboniferous period. At that period in geological history, the peat-forming areas of the world were lush tropical or semitropical. Today, however, peat is found in northern England, Scotland, and Ireland, places that, certainly, are no longer tropical. The cumulative weight of overlying deposits compressed the decaying plant matter, squeezing out the water and gases. But since peat was formed close to the surface, the overlying sediments did not compress it enough to prevent the extraction of all the water and gases or to halt all bacterial activity from acting on the biomass. In this sense, peat may be thought of as incomplete coal, because it contains only about 50% carbon. In some areas of the world it is cut out of the ground in chunks, dried, and then used as cooking or heating fuel. As it burns, it produces great amounts of odiferous smoke and fumes. As more overlying sediments are added to peat deposits, they undergo additional changes that convert them to other forms of coal.
2. *Lignite* coal is formed when bacterial action continues to break down peat, forming an organic mass under greater pressure. These changes result in the conversion of peat into several types of coal that are concentrated in deeper layers of sediment. Lignite is the lowest grade of coal formed in this process. It is a very soft form of coal in which more CO_2 and water were removed from the peat, resulting in a coal that is about 70% carbon and has only ~20% water and ~10% oxygen. As lignite is buried still deeper, however, more changes take place.
3. *Bituminous* coal is formed as deeply buried lignite undergoes increasing pressures and temperatures and bacterial activity that convert the lignite into a soft, but lush, black, rocklike coal. Great quantities of bituminous soft coal are found in many countries, and it is usually deposited relatively close to the surface, thus facilitating strip mining by removing the overlying rock and soil. In the United States, soft coal is found in the states of West Virginia, Kentucky, western Pennsylvania, Wyoming, and several other western and southeastern states. China has the largest known deposits of bituminous coal. Bituminous coal varies in its carbon content depending on its origin, but it can contain about 65% to as much as 93% carbon. It is a major source of heat for many purposes, including producing electricity. Its smoke, if not scrubbed of sulfur and nitrogen, can result in **acid rain** and other pollutants. At one time, soft coal was the major source of heat energy in homes and factories in Great Britain. Even though bituminous coal produces less smoke than does peat or lignite, during certain weather conditions its smoke caused extensive smog conditions that were harmful to people and animals. Because of use of alternate fuels, these conditions have improved. If soft coal is burned with a limited supply of oxygen, some of the volatile gases are driven off, leaving behind what is called "coke," which has greater carbon content than does the original bituminous coal. This makes coke ideal for use in blast furnaces that produce iron.
4. *Anthracite* coal is a jet-black, shiny, hard rock also known as "hard coal." It is commonly located at greater depths than lignite and bituminous coal, where additional layers of sediment have buried it, increasing the pressure and temperature significantly. Anthracite is also formed in the metamor-

phic regions of mountains where limbs or tongues of folded sedimentary rock dip deeply into the surrounding rocks. Thus, it cannot be easily strip mined, but instead deep pit mines must be excavated that follow the underground seams of coal for miles. Deep mines are also located in England and some other countries, but the distribution of hard coal is much more limited than is soft coal. Limited pockets of anthracite are located in the Appalachian Mountains of northeastern Pennsylvania and a few other sites in the United States. Although no one knows exactly how much coal exists in the Earth, it is known that there is much less anthracite coal than there is bituminous. Not only is anthracite the scarcest variety of coal, it is also the most prized for both home heating and industrial uses because it produces more heat per pound with less harmful waste products than does soft coal. As oxygen and the water content decrease and the carbon content increases to as much as 98%, with just a small amount of inorganic impurities, the anthricite coal produces a hot blue flame with little smoke and ash.[60–61]

Coal can be an important source of many compounds when the process of destructive distillation (pyrolysis) is used. Coal (a hydrocarbon) can be altered by **hydrogenation** to produce synthetic crude oil, gasoline, and other products. Coal tar, a viscous, semisolid liquid with a sharp odor, can also be produced by pyrolysis. Coal tar has many uses but is also a human carcinogen. Therefore, some coal tar by-products are no longer manufactured. The dye and pharmaceutical industries rely on coal tars. Additional processing of coal yields hydrocarbon gases, ammonia, naphthalene, phenol, creosols, pitch, and xylene. The process of converting coal into synthetic fuel oils and gasoline is too expensive for use in the United States. However, it is used in some European countries to supplement their supply of gasoline.[62]

The famous gas lights of the nineteenth century resulted from an invention by William Murdock (1754–1839). Aware that when wood, peat, and coal were heated a combustible gas was given off, Murdock applied the process that converted coal to coke by heating it in a limited supply of air. He then collected and stored the emitted gas at what was known as the "gas house" or "gas works." He then piped it into homes for gas lamps, to communities for streetlights, and to industries for manufacturing purposes. It was a major advance in lighting over the candle and the whale oil lamps of previous generations.[63]

Both coal and petroleum are classed as fossil fuels that are widely spread over the crust of the Earth in large deposits. They are found on almost every continent but often not in amounts great enough to be economically exploited. About 95% of all coal reserves are found in the northern hemisphere. China, Russia, Canada, Western Europe, and the United States have tremendous reserves of coal, while most of the petroleum reserves are located in the Middle East, Asia, Russia, and South and North America. But most of the Earth's coal is low-grade, soft coal that when burned releases sulfur and nitrogen compounds into the atmosphere, in much the same way as does the burning of gasoline in internal combustion engines. These sulfurous compounds, unlike carbon dioxide and methane, have a cooling effect on global warming, but excessive sulfur and nitrogen smoke from burning coal can add to acid rain. Technologies have been developed to scrub many of these pollutants from coal smoke before it is expelled into the atmosphere. Electronic precipitators that extract **particulates** in the smoke stacks of industries before the smoke becomes airborne also are used. Because the recoverable coal reserves of the world are estimated at a total of more than 8×10^{11}

tons, coal will continue to be consumed to provide the main source of energy for many nations for many hundreds of years.[64]

Nuclear Energy

Just what is nuclear energy and how is it used to produce electricity? Radioactive isotopes are elements whose atoms have a different number of neutrons in their nuclei than do other forms of that element. Thus, they have different atomic weights and may be unstable. The unstable heavy nuclei of atoms of radioactive elements give off radiation (energy) as they break down by fission into atoms of different, more stable elements. These new, stable nuclei have approximately the same *combined* atomic mass as the original radioactive nuclei. This process is known as **fission** and results in a very small amount of the original mass of the unstable nuclei being converted into energy (namely, $E = mc^2$).

Uranium is one of the heavier elements with fissionable isotopes. It is also a relatively scarce element, located in the Earth's crust as crystalline uranium oxide, such as the mineral UO_2. Since uranium is soluble in water, it dissolves out of weathered igneous rocks and is transported to underground sediments where it combines with other elements. These uranium deposits accumulate in sand and gravel beds of ancient streams, from which it can be mined rather easily. It is either found as *pitchblende,* which is a black, shiny, crystalline form of uranium oxide, or mined as *carnotite,* which are bright-yellow, grainy particles in which the uranium has combined with other metals such as potassium and vanadium.

The most abundant form of uranium is U-238, whose stable atoms comprise about 99.3% of all isotopes of uranium. Radioactive (fissionable) uranium-235 constitutes only about 0.7% of all uranium isotopes, and it must be separated and purified to become useful as a nuclear fuel. The gaseous diffusion separation method was used to produce enough U-235 for the first atomic (nuclear) bombs, including the New Mexico test bomb and the first of the two bombs used to hasten the end of the war with Japan in 1945. The second atomic bomb of WWII used plutonium, which is a fissionable byproduct produced in nuclear reactors. At the current rate of consumption, the world's supply of U-235 for use as a fuel to generate electricity is limited to about 30 years. Plutonium-239 can be produced in nuclear breeder reactors by converting the stable U-238 into Pu-239, in which excess neutrons can create, or breed, more fissionable material than they consume. Plutonium-239, although highly radioactive, can provide an almost unlimited supply of fissionable material that can be used to generate electricity. During the fission of a **critical mass** of U-235 or Pu-239, high-speed free neutrons are produced. Two and, at times, three new neutrons are released for each atomic nucleus that disintegrates. These neutrons are then absorbed by other fissionable nuclei of U-235 or Pu-239, which also undergo fission in an ever-expanding exponential geometric ratio, resulting in a chain reaction that releases tremendous energy. Breeder reactors can provide a long-term, nonrenewable alternative source of energy far into the future (see Figure 4.7).

If the concentration of radioactive nuclei is great enough, and the number of neutrons released is large enough, and there is adequate additional radioactive material available (i.e., critical mass), the reaction becomes a self-sustaining chain reaction. If this chain reaction of fissionable nuclei is allowed to proceed at its own pace, it becomes a nuclear, or atomic, bomb. Only 33 pounds of uranium-235 or about 10 pounds of plutonium-239 are required to

Figure 4.7
Fission is sustained by the release of extra neutrons produced by the natural splitting of the radioactive uranium atom. A chain reaction is possible if a critical mass of a fissionable element is present, resulting in a nuclear explosion. If a neutron moderator is used to absorb some neutrons, the chain reaction can be controlled and used as an atomic pile, or a source of heat to produce electricity.

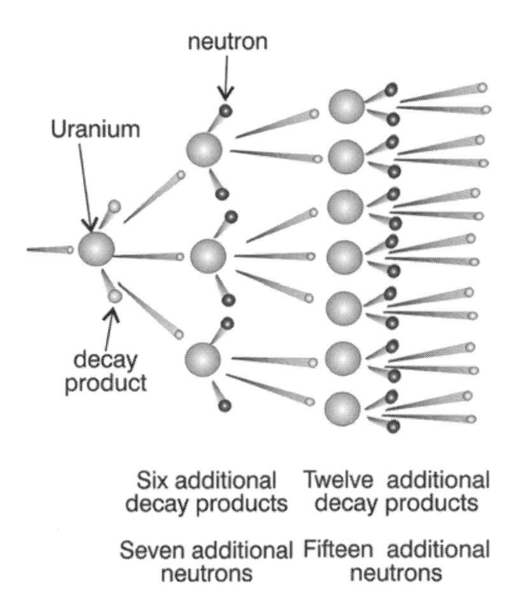

reach a critical mass for an uncontrolled fission chain reaction resulting in an explosion equivalent to several thousands of tons of TNT. The chain reaction can be controlled by using control rods that absorb some of the neutrons produced by the fissionable material. Much heat energy is still produced by this controlled nuclear reaction and that can be captured and used to generate steam to drive turbines that turn generators to produce electricity. The United States generates less than 20% of its electrical energy from nuclear sources, while France, which is short on fossil fuels, generates more than 70% of its electricity from the atom. In the future, nuclear power, along with several alternative energy sources, will most probably be used by the world to meet growing energy requirements.

In addition to uranium, other heavy elements release various forms of radiation. Most of them appear near the end of the periodic table of chemical elements and have relatively high atomic masses. Many elements with atomic masses beyond uranium have been artificially produced and have extremely short half-lives. Some of the following natural radioactive elements (given with their atomic masses) that appear on the periodic table are extremely scarce on the Earth: Technetium-99, Promethium-145, Polonium-210, Astatine-210+, Radon-222, Radium-226, Actinium-227, Thorium-232, Protactinium-231, Uranium-234, and Uranium-238.

Three basic types of radiation are produced by nuclear disintegrations: *alpha* (α) radiation, which are relatively heavy helium nuclei that can be stopped by a sheet of paper; *beta* (β) radiation, which are light, high-speed electrons that are more penetrating than alpha particles but can be stopped by aluminum foil; and *gamma* (γ) radiation, which originates from radioactive elements that are more penetrating than x-rays and is the most penetrating of all radiation originating on Earth but can be stopped by heavy lead shielding. True, radiation is accumulative and harmful or even deadly in excessive amounts, but it is also true that radiant energy is natural and exists in a great variety of electromagnetic waves that fill the world with of all kinds of radiation. See Figure 4.8 for an artist's depiction of the electromagnetic spectrum.

Each specific type of electromagnetic wave has a unique wavelength (λ) and frequency (ν). The varied characteristics of radiation assist in the study of many sciences, including geology; for example,

Figure 4.8
Visible light is radiation of colored light located near the center of the electromagnetic spectrum. Radiation that has mile-long wavelengths is located on the red side of light, whereas radiation that has progressively shorter wavelengths is located on the violet side of light's position.

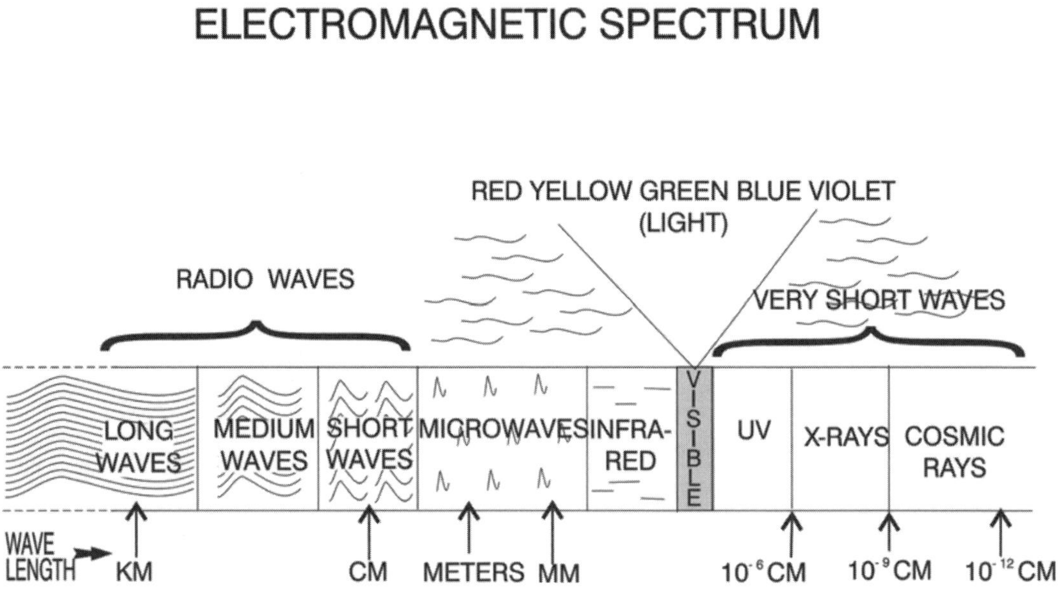

microwaves, infrared light rays, and x-rays are used to study crystal structures and so forth. In the following list, the characteristics of radiation are given from the shortest wavelengths and highest frequencies to the longest wavelengths and lowest frequencies:

1. *Cosmic rays* originate from outer space, have extremely high frequencies and short wavelengths, are very penetrating, and can be extremely dangerous but are partly filtered by the atmosphere. The Earth's natural magnetic field, which extends far out into space, diverts most of the cosmic radiation approaching the Earth's surface. Even so, people who live at high altitudes or who fly extensively in high-altitude airplanes or spacecrafts receive more cosmic radiation than do people who live at sea level. One theory is that cosmic radiation was responsible for some of the early genetic mutations that provided the evolution impetus for the **speciation** of organisms on Earth. The Earth's outer magnetic field diverts some cosmic rays to the polar regions, where the ionized particles form the aurora borealis or northern lights. Scientists also try to capture cosmic rays deep in mines to study their physical nature and geological and biological effects.
2. *Gamma rays* are generated from the decay of radioactive atomic nuclei. They can also be dangerous when the amount of exposure is excessive.
3. *X-rays,* both natural from outer space and those generated by radiological diagnostic instruments (x-ray machines), can be harmful in excessive amounts. Although similar to gamma rays, their wavelengths are longer and their frequencies are lower than are those for gamma rays. Both gamma rays and

x-rays are used to study and identify rock structures.

4. *Ultraviolet* (UV) rays, located at the short end of the sunlight spectrum, have a longer wavelength and lower frequency than x-rays. Excessive exposure to UV causes sunburn, skin cancers, and eye cataracts.
5. *Visible light* (white sunlight) radiation has a lower frequency and wavelength than does UV and is, thus, less harmful.
6. *Infrared* radiation, found at the long end of the light spectrum, is felt as heat that can cause burns.
7. *Microwave* radiation has much longer wavelengths and lower frequencies than does infrared radiation. Excessive exposure to microwaves may be harmful.
8. *Radio and TV* waves consist of much longer electromagnetic radiation wavelengths and have much lower frequencies than microwaves; thus, they are not harmful at ranges normally used for broadcast purposes.
9. *Electromagnetic fields* (EMFs) are not the actual electricity that surges through wires or the actual radio signals but instead are the fields generated around the wires themselves. This form of radiation has a very long wavelength (in miles, not millimeters) and low frequencies. The degree of harm caused by instruments that produce EMFs (e.g., cell phones, radios, TVs) is partly determined by how close one is to the source (inverse square law), as well as by the power of the source of the electromagnetic field.[65] Humans are exposed to radiation on a daily basis as part of natural phenomena, including radioactive gas (radon) and other radioactive elements in the Earth's crust. Contrary to what has been reported by the media, radio and electric radiation (what are usually referred to as EMFs) are not harmful in the amounts to which we are normally exposed. In addition to man-made EMFs, we live on a globe surrounded by a natural giant magnetic field generated in the deep Earth, as well as by several types of cosmic radiation (including numerous penetrating neutrinos) from the sun and outer space. Physicists, medical doctors, and epidemiologists agree that no reliable evidence or data support the belief that EMFs are dangerous.[66]

In extreme excess, just about any form of radiation with shorter wavelengths and higher frequencies can be dangerous. One example of a form of natural radiation that can be dangerous is radon (a radioactive gas) produced from the decay of the radioactive element radium, which is found in the Earth's crust. Depending on the geographic location, radon seeps into basements of homes in greater or lesser amounts. Radon (Rn-222) is located in group 18 (VIIIA), period 6 of the periodic table of chemical elements. That location places it as the only radioactive member of the noble gas group (inert elements). There are numerous radioactive isotopes resulting from the radium □ radon nuclear reactions, but most of the resulting isotopes, except radon, have short half-lives. There is some evidence that concentrations of radon gas in some homes may be responsible for cases of lung cancer.

One example of normal radiation that is both useful and harmful is UV light from the sun; overexposure to UV can cause skin cancers and eye cataracts. Another example is P-40, which is the radioactive isotope of the nonradioactive element phosphorus (P-39). These two forms of phosphorus exist together as isotopic forms in rocks and are an important ingredient in fertilizers and as a chemical in plant tissues. Some radioactive P-40 is ingested every time we eat plants that have absorbed this natural source of radiation from the soil. We can't avoid exposure to all forms of natural radioactivity found in the Earth or to electromagnetic radiation, whether natural or man made. But we can make intelligent use of radioactivity to assist us in learning more about the nature of the Earth and EMFs in ways that are both safe and beneficial.

A significant source for the centralized production of energy now available is nuclear power. The extensive use of nuclear energy to produce electricity could go a long way to reduce the use of fossil fuels and thus meet the Kyoto Protocol's plan to reduce global warming gases. Several countries are becoming more and more dependent on nuclear energy rather than on fossil fuels. According to the *Environmental News Service* there were 434 nuclear power plants operating in the world in 1999, down from 437 in 1997. The countries with the highest reliance on nuclear power in 1998 were Lithuania, 77.2%; France, 75.8%; Belgium, 55.2%; Sweden, 45.8%; Ukraine, 45.4%; Slovak Republic, 43.8%; Bulgaria, 41.5%; Slovenia, 38.3%; Japan, 35.9%; and Hungary, 35.6%. Eighteen countries (including the United States) relied on nuclear power plants to supply at least a quarter of their total electricity needs. The worldwide 1998 total nuclear electricity generated was 2291.4 terawatt-hours—more than the world's total from all sources in 1958. This is about 16% of the world's total electricity production.[67]

Western Europe has greatly increased the percentage of electricity it generates by nuclear energy as compared to fossil fuels. In contrast, since 1994 the United States has decommissioned several nuclear power plants, thus reducing its dependency on nuclear-generated electricity while increasing its dependency on electricity generated by fossil fuels. The basic nuclear fuels for generating electricity are obtained from the uraninite ore called pitchblende, which is essentially uranium oxide (UO_2). Uranium also exists in varieties as coffinite ($USiO_4$) and is found in Colorado, New Mexico, France, Zaire (Congo), Canada, South Africa, Australia, and Russia.[68]

At some point, it may become necessary for the United States to accept the fact that nuclear energy technologies are as safe, or even safer, than other technologies producing energy from fossil fuels. A case in point was the so-called accident at the Three Mile Island (TMI) nuclear power plant near Middletown, Pennsylvania. Although serious, the radiation was contained within the reactor building, and there was no meltdown as there was with the disaster in Chernobyl in the Ukraine.

> Estimates are that the average dose to about 2 million people in the area was only about 1 millirem. To put this into context, exposure from a full set of chest x-rays is about 6 millirem. Compared to the natural radioactive background dose of about 100–125 millirem per year for the area [in Pennsylvania], the collective dose to the community [Middletown] from the accident was very small. The maximum dose to a person at the site boundary would have been less that 100 millirem.[69]

People living at high altitudes, such as the mile-high city of Denver, receive more radiation (cosmic and UV radiation) in their lifetimes than people living near Three Mile Island at the time of the accident. Much research and development related to new technologies for building safe nuclear power plants has been halted. Even so, the United States and Western Europe have developed safeguards and regulations to minimize nuclear power plant accidents. Further fossil fuel shortages (and pollution) will sooner or later require people of all nations, and in particular the citizens of the United States, to address the problems of fossil fuel usage in the generation of electricity. Alternative energy solutions (wind, solar, tidal, biomass, etc.), by most estimates, can to some extent relieve the pollution caused by the burning of fossil fuel for energy purposes, but it is doubtful that these nonpolluting alternative systems will ever produce the amount of energy presently required by the developed, as well as developing, nations. One advantage of nuclear-generated electrical energy, as with most alternative energy sources, is that it does not generate greenhouse gases; however, nuclear energy does pose a waste-disposal problem.

At the time of this book's publication, not one death had occurred in the United States from radiation accidents in nuclear electrical generating power plants. The same cannot be said for energy production from power plants using fossil fuels. One reason is that, statistically, there are more, and older, fossil fuel electric generating plants than there are nuclear plants in the United States. Thus, the older, more numerous fossil fuel plants have a greater probability for accidents. Another reason is that nuclear power plants are better regulated and require more extensive safety precautions than do fossil fuel plants.

Mining

Ancient people made use of many different minerals and rocks that were found on or near the surface of the Earth. By scraping, pounding, and knapping (flaking) rocks such as flint, chert, jasper, obsidian, and quartzite, they created tools and weapons. They also learned how to use clay to make pots and vessels to hold a variety of substances. Once they learned that by heating the clay to a high temperature it would turn to a nonporous ceramic-type substance, which could then be used to hold water for cooking purposes, they improved their lives. They discovered how to mix mud with straw to produce bricklike building materials when dried and to use pretty minerals and metals for decorations. Therefore, it wasn't long before people were excavating deeper deposits of the minerals and rocks and, eventually, smelting metals from their ores. Evidence for this can be found in the famous historical mines of Cyprus, Egypt, and Mesopotamia. Later, the developments of the mining industries in Europe paralleled advancements of science and technology related to chemistry and metallurgy. Today, worldwide mining produces more than $500 billion worth of metallic ore each year, with another $700 billion of energy minerals (e.g., fossil fuels and nuclear energy).[70]

Most mineral/metal ore deposits exist because of the fundamental geological processes related to plate tectonics. A great variety of minerals are formed deep in the Earth. Both the temperatures and pressures of magma at different depths, as well as the cooling rate of magma as it intrudes and extrudes the oceanic and continental crusts, determine the types of crystals that are formed. Igneous rocks are metamorphosed deep in the Earth. When these rocks are raised and exposed on the continents, they are weathered, broken up, and eroded, forming sedimentary rocks, which in turn are subducted at plate boundaries and returned to the mantle to be recycled by metamorphic processes. Erosion on the surface, convection cells in the mantle, and the slow movements and subducting of tectonic plates all contribute to the system where minerals, ores, and rocks in the crust are recycled. Due to the nature of plate tectonics, most of the useful minerals, including metals and energy resources, are located in deposits that are not evenly distributed across the Earth. Although most minerals and metals are formed in the mantle, they are deposited on land in the following ways.

- *Hydrothermal processes* on land are responsible for the large total amount of ore deposits on the continents. These ores were formed when minerals and metals were dissolved by hot water and later deposited in veins within fractured rocks.

Almost all rocks contain some metals but in very low concentrations. When the dissolved minerals are deposited in veins, they can be very thin or concentrated to the extent that they can be profitably mined. Other ores, such as copper ores, are *disseminated,* meaning they are found as scattered, fine-grained, metal-bearing minerals interspersed throughout a body of rocks in quantities great enough to be considered profitable for mining.

Hydrothermal ore deposits are also found in areas on the seafloor where there were fractures in the ocean crust. These types of ore deposits are practically everywhere on the ocean floor, including the rift centers of spreading seafloor ridges and areas of active subsea volcanoes. The minerals and metals that were dissolved in hot seawater were deposited on the seafloor near the mid-oceanic ridge by magma from submerged volcanoes. Even today, it is possible to see black smokers that contain sulfide minerals that condense as the hot precipitates are cooled by ocean water. Rich deposits of copper, lead, zinc, gold, silver, and other minerals are formed by submerged volcanoes. Manganese nodules that range from golf ball to bowling ball size and are widely deposited on the Pacific Ocean bed are an exception to this process, because they are bacterial in origin and not created by normal hydrothermal activity.

- *Magmatic ore deposits* are formed when intrusive, magma-containing minerals are segregated into layers as they cool to form mafic rocks. These layered intrusions consist of crystals of metallic oxides in different concentrations than in the igneous rocks. One example is found in the area of Kimberley, South Africa, where the minerals olivine, phlogopite, limonite, serpentine, chlorite , magnetite, and perovskite are found. Many of these minerals contain significant percentages of carbon. This is the area where diamonds, an isometric crystalline mineral related to graphite, are found. Diamonds form in ultramafic volcanic rocks in high concentrations as deep, mineral-rich magma solidifies and rises into the upper crust in specific pipe-shaped deposits similar to a funnel.
- *Sedimentary ore* deposits of metal ores can be formed either by physical sedimentary sorting based on the different densities of the ores and the sand/gravel or by chemical precipitation of metals from rocks. An example of physical sorting is how gold is deposited. Since gold is heavier than mud, sand, and gravel, it drifts to the bottom of slow-moving streams and creeks, forming placer deposits. Historically, these deposits were panned; this method concentrates the ore by swirling water, sand, and gold dust in a pan and, because the gold is heavier than the sand, retaining the gold in the bottom of the pan as the water and sand are swirled over its edge. Today, huge high-pressure water hoses dig out the deposits, making it possible to process massive amounts of gold ore sediments and rocks. In addition to the sediments that are concentrated by gravity, the gold imbedded in rocks is recovered by chemical precipitation using cyanogen bromide (BrCN) as a reagent to extract the gold.
- *Weathering processes* are involved when rain and water dissolve soluble ions from soil and rocks as the surface of the Earth is eroded. These heavier elements then form residual deposits, for example, iron and aluminum deposits. Bauxite, the ore of aluminum, is formed in this manner.[71]

Before minerals can be used in the manufacturing of products, a number of steps must be followed to secure them.

1. *Exploration.* The mineral or energy source must first be located. Geologists have learned how to find and identify these concentrated deposits by observing surface structures, drilling test holes, and using modern electronic devices. The geological knowledge of how these minerals were formed is important in predicting the location of potential mineral deposits.
2. *Extraction.* This can take place on the surface or deep inside the crust. Both surface and underground mining have been employed for centuries, but only in recent times have the technologies for extracting (and refining) ores improved to the point where productivity greatly increased and prices per unit of extracted ore decreased. Both surface mining and deep, open-pit mining can cause great disturbance to the natural landscape. Therefore, surface strip-mining operations now restore the natural vegetation and landscape after the minerals are extracted. Mines that have been drilled and dug deep into the Earth for many years also can cause surface changes as the old, underground mines collapse. More than 2 million acres of land located in central Appalachia have settled into abandoned, underground coal mine shafts.
3. *Processing.* Very few of the minerals that are extracted are useful in their natural state. Most are imbedded in rocks that must be crushed and milled so the valuable mineral can be removed for further processing. As with extracting metals and minerals, both physical and chemical methods are used in processing them.
4. *Refining.* After processing, minerals must undergo various alterations to purify and concentrate them for further use. Refining can be accomplished by either mechanical, thermal, or chemical processes. Smelting is a major method of refining the crushed ores of minerals and metals. Low-grade iron ore can be ground into finer particles that can then be concentrated for further processing to extract the metal by smelting or chemical precipitation. A number of refining processes are also involved in converting crude oil into useful petrochemicals.
5. *Manufacturing.* Finally, the refined mineral resources, as raw materials, are transformed into an almost limitless number of consumer products.

Modern society, as we know it, could not exist without the mining industry. The welfare of people of all societies and nations is dependent on the mining and refining of minerals and metals. It is doubtful that the world could ever go back to the stone age, or the bronze age, or even the preindustrial iron age. To maintain our standard of living in the United States requires, on the average, the production of 46,000 pounds of *new* minerals, metals, and energy fuels for each person *each year.* These minerals provide our food, homes, schools, hospitals, factories, and the equipment and energy to operate these facilities.[72]

Options and Alternatives

The economic, social, and environmental concerns related to the use and misuse of the Earth's nonrenewable resources are of recent origin, as are our options and alternatives for the substitution and preservation or wise use of these precious resources.

Early humans, more concerned with survival, were pragmatic about the uses of natural resources. They were also unaware and unconcerned about the need to preserve or protect the environment. This posed no serious problem as long as the population was limited. Not many years ago, when there were only a few billion people in the world, it was observed that shortages of particular resources were occurring, or those resources were at least more expensive to secure and/or difficult to locate. Today there are more than 6 billion people, with an expected total of 9 or 10 billion people to be living on the Earth by the end of the twenty-first century. At one time, it was believed that it was not necessary to seek additional supplies or substitutes for the scarce commodities. Instead, the solution to the problem, as interpreted, was based on the belief that there were too many people in the world using the limited resources at hand. The concept of population control has been around for several hundred years and has become the **a priori** position of some current environmental activists, who believe, as do most scientists, sociologists, and the population at large, that there must be a limit to the increase in the number of humans, as there is with any living species. We just do not know what that limit is with any degree of accuracy. Most natural biological systems, of which the *Homo sapiens* species is a part, have a history of naturally imposed self-limitations on their numbers. At the same time, humans have great advantages over other species in that they have a high degree of intelligence and adaptability that may be used to further alter environmental conditions to alleviate the situation. Geologists and other scientists are becoming more knowledgeable about the Earth's structure, processes, cycles, and systems, and the approximate limits to growth for human populations as imposed by the finite geological limits of the Earth. There is little argument that at some point the distribution of worldwide natural resources will become even more out of balance between nations than it is today—and this is not just a social or political problem or solution: it is a geographic and geophysical reality.

In 1798, T. R. Malthus, in his book *An Essay on the Principle of Population,* claimed that this self-limitation to population growth was related to the food supply. His theory was simple: "Population, when unchecked, increases in geometrical ratio. Subsistence [food] increases only in an arithmetical ratio."[73] Modern day Malthusians have expanded this concept to include nonrenewable natural resources.

In 1972, the authors of the book *The Limits of Growth* predicted the imminent exhaustion of many of our major nonrenewable resources. They claimed that at exponential growth rates of the population, the world would deplete the supplies of gold by 1981, mercury by 1985, tin by 1987, zinc by 1990, petroleum by 1992, and copper, lead, and natural gas by 1993.[74] It's obvious that their predictive model has some problems.

Another story relates to the exhaustion of North African tin mines that were a source of an alloy metal for the manufacturing of bronze by the ancient Egyptians. Their problem was solved when explorers located deposits of tin ore on the southern coast of England. This discovery and exploitation of the new source of tin further developed the region through trade and commerce. At the same time in history (about 1000 B.C.E.), iron was far more expensive to produce than was bronze. However, with new technological improvements in metallurgy, iron became less expensive to manufacture than bronze, collapsing the bronze market.

These stories are really more about geology, economics, supply and demand, and availability than they are about the exponential population growth and limited amounts of these resources. One of the main problems is that no one knows, not even geologists or environmental scientists, the exact quantity, quality, extent, and location of the nonrenewable resources in the Earth. For instance, in the 1930s it was estimated that the world would run out of petroleum (and thus gasoline) in 15 to 20 years. After World War II, the estimate was revised to about the 1970s. Again and again the estimations and predictions of when we would run out of oil were revised. Why? After all, there must be a limit to the amount of petroleum within the Earth. The misconception was that we actually knew what the petroleum reserves were in all areas of the world; we did not, and we still do not know exactly how much oil (or any other nonrenewable resource) exists on or in Earth. The discovery of new petroleum deposits, the development of new technologies for extracting oil from deeper wells, the reclamation of oil from old wells, the use of deep horizontal drilling, the application of conservation practices, and the development of more efficient uses of petroleum are all factors that must be considered in any estimate of how long the oil supply will last.

Today, those who predict when we will run out of oil are more cautious in their estimations. Not too many decades ago, a whale oil crisis partially led to the development of the petroleum industry. Also, during World War II, Japanese control of the islands where the major rubber tree plantations existed led to the development of a synthetic rubber to replace latex rubber. Sometime in the future. facing the petroleum crisis will lead to improved technologies, substitutes, renewable alternative energy sources, and more efficient conservation practices. For instance, "[s]purred on by higher oil prices, industrial countries use 23 percent less oil to produce a dollar of output than they did in 1970"[75]; however, today we are using much more petroleum than we did in the 1970s. One thing is certain—petroleum and natural gas are limited, nonrenewable resources (at least in the short term). This does not mean we will be unable to use fossil fuels as energy sources in the future. Instead, they will become less important in the overall supply of energy used in the world. Not only will more efficient means of transportation and energy production be forthcoming; there also will be options and alternatives.[76] Who knows what role geology and geologists will play in a future crisis of limited nonrenewable natural resources? The more efficient technique of multidrilling opens new areas of known oil reserves, while at the same time it results in less environmental damage. With these new techniques it is now possible to recover about 5 billion barrels of oil under the Rocky Mountains, 10 or more billion barrels under a few acres of the Arctic National Wildlife refuge, and more offshore oil as well. This is in contrast to the Persian Gulf states, which have an estimated 600 billion barrels still in the ground (including more than 260 billion barrels in Saudi Arabia).

Petroleum production and reserves peaked in the United States in 1970. We consumed about 5.4 billion barrels of oil that year while producing just 3.6 billion, with reserves of 39 billion. By the year 2001 we consumed 7.1 billion barrels, and our domestic production dropped to 3.2 billion, while reserves sank to 22 billion. In the year 1977, 33.6% of our oil came from the Persian Gulf—3.7% from Iraq and 8.6% from Saudi Arabia. (Note that not all OPEC countries are located in the Persian Gulf region: Venezuela and Nigeria are also OPEC member nations.) By 2001, the total amount of U.S. oil coming from *all* OPEC countries dropped to 27.9%, with just 13% coming from the Persian Gulf OPEC countries. Since then, the United States has only slightly reduced its dependence on Persian Gulf oil, with non-OPEC countries now supplying more oil (Canada, 9%; Mexico, 7%; Norway 1.8%; United Kingdom, 1.5%; Columbia, 1.4%). Currently, the United States may be reducing its dependence on Persian Gulf oil, but overall the United States still imports, from all oil-producing states, almost 60% of the petroleum it uses. While this might seem oxymoronic, the United States also still exports a small amount of its petroleum.

The Petroleum Industry Research Foundation states that the consumption of oil in the United States has increased only 1.5%, whereas our GNP (gross national product) has risen about 3%. Looking at how we use our supply of oil provides an insight into how important it is for an advanced economy to maintain its standard of living. More than 68% of the petroleum used in the United States is for transportation purposes (56% of this amount for cars and SUVs), 24% is used for industrial purposes, 6% for home and industrial heating, and 2% for generating electricity. Natural gas has been substituted for home and industrial heating and coal and nuclear energy are increasingly being used to generate electricity, both of which decrease the rate of increase of our dependence on oil.

In his book *The Skeptical Environmentalist,* Bjorn Lomborg addresses, from an economic point of view, the issue of the use of alternative fuels to replace fossil fuels. He reports on three studies that examined all the costs related to electric production, including death from mining and illnesses; transportation and occupational hazards of production; and such consequences as acid rain, soot, waste gases, and so forth, on health and the environment. The conclusion of these studies was that even when currently considering these extra costs of using fossil fuels, they are still less than the costs for alternatives. Lomborg further states that there is little doubt that, in the future, renewable sources of energy used to produce electricity will be cheaper. He also states that this may be another reason not to worry too much about the long-term effect of global warming.[77]

Next to geological factors, economics is the main driving force for the availability of certain nonrenewable minerals, metals, and rocks. For example, as the price of an essential metal or mineral increases, several things happen: (1) methods for using the resource more efficiently are explored, (2) conservation measures are considered and possibly imposed, and (3) economical substitutes and viable alternative sources are sought. The implementation of one or all three of these alternatives is based not only on the availability of the resource but also on cost-benefit considerations. William Ashworth, in his book *The Economy of Nature: Rethinking the Connection Between Ecology and Economics,* claims that we do not have to choose between ecology and economics and that we can't have one without the other. He states that what is good for the marketplace is also good for the ecosystem, and vice versa.[78] His position can be extended to include natural resources as well, because they *are* the environment.

The following is an example of how people misjudge economics as it relates to the supply and demand of nonrenewable resources. Paul Ehrlich, a professor at Stanford University who published the popular book *The Population Bomb,* predicted a variety of environmental doomsday scenarios related to both renewable and nonrenewable resources as affected by population growth.[79] As the story goes, Julian Simon, an anti-Malthusian professor at the University of Maryland, whom Ehrlich attacked as a "cornucopian," engaged in a bet on the increase in prices in the commodities market for five specific metals over a period of ten years. The bet was that Simon was to sell specific amounts of copper, chrome, nickel, tin, and tungsten metals worth a total of $1,000. If the combined price of these five metals *rose* more than $1,000 ten years later (at 1980 prices), Simon would pay Ehrlich the difference. If the combined prices *fell* below $1,000, Ehrlich would pay Simon the difference. Ten years later (1990), the combined 1980 prices had fallen by more than 50%, making Simon's point that the supply of these nonrenewable resources was becoming more abundant rather than more scarce. The end of the story is that Ehrlich paid Simon $576.07, which was the drop in price for these five metals over the period of ten years.[80]

Recycling of nonrenewable resources often becomes a social concern that sometimes overrides economic considerations. The United States uses more than 5,500,000 metric tons of aluminum each year, and 50% of that is recycled aluminum. By recycling one pound of old aluminum it is possible to save eight pounds of bauxite ore, four pounds of other chemicals, and 6.5 kilowatt-hours of electricity. Aluminum is an example of successful recycling for several reasons. First, it is something consumers can do, see, and appreciate. Second, it provides a real economic cost-benefit because it takes much less electrical power to melt down aluminum cans and products than it does to extract aluminum from bauxite. In addition, because an aluminum can is worth 6 to 20 times more than any other packaging material, it is also the only packaging material that covers its own collection and processing costs at the recycling centers.[81] Home/community recycling is expensive and is not as effective as is industrial recycling at the manufacturing plant or as the proper use of landfills and **pyrolysis** incineration for disposal of used consumer products and wastes.[82] By far, most of the recycling of important metals is done at site by industry, not by homeowners (see Table 4.7).

Table 4.7
Important Metals Used and Recycled in the United States

Rank	Name of Metal	Percent Recycled
1.	Iron and steel scrap	100
2.	Platinum group metals	67
3.	Lead	65
4.	Gold	60
5.	Aluminum	50
6.	Silver	49
7.	Antimony	43
8.	Tin	35
9.	Tungsten	33
10.	Nickel	30

Note: About 10 other metals are recycled below 30% levels.
Source: *Recycling Metals* (Denver, Colo.: Mineral Information Institute, 1997), 25.

The recycling of most household products is not cost-effective, although some recycling may be environmentally effective. Several types of recycling, however, do improve the environment. For example, as mentioned, collecting and recycling used aluminum products provides a real economic cost-benefit. Also, using recycled plastic bottles to add to new plastic materials conserves petroleum, and one product that has been successfully manufactured by recycled plastics is park-benches. A questionable example is the recycling of newspapers, because collecting and processing them is not cost-effective, and removing the ink and other products from the paper itself pollutes water supplies. Economic incentives may be needed to encourage recycling as a means of conservation: for example, in some states a recycling fee is charged for the disposal of automobile batteries, or each time you buy a new automobile tire, a fee is added to the purchase price for disposal of the old tire. Although there are several possible uses of old tires (e.g., pulverized and mixed with asphalt they can be used for surfacing roads), not many of these ideas are cost-effective.

Economic incentives can also be used to promote substitutes for scarce nonrenewable resources. As essential metals and minerals become more and more scarce, substitutes can be found, often providing greater economic benefits as well as superior performance. One example is the replacement of copper wire with glass fiber optical cables or by use of wireless technologies. Not only is our dependency on copper reduced, but also a tremendous amount of energy is saved that formerly was used to smelt copper ore. In addition, the source of the replacement minerals (silicates) for fiber optics are less expensive, are in greater geological abundance, are more available, and are far superior for making telecommunication cables because of the increased amount of information that can be sent through them as compared to copper wires. A single, thin strand of fiber optics cable can handle 5 million telephone conversations at the same time.[83] Thus, even if fiber optics is more expensive to produce than copper wire, its use is more cost-effective because of its increased communication capacity.

Economic incentives offered by the U.S. government have been important in the development of viable sources of energy to replace petroleum and gas. In the western United States, wind-driven generators produce electricity at a cost almost competitive with electricity produced by fossil or nuclear fuels. A number of research and development projects are exploring the competitive production of solar electricity by using the heat of the sun directly to heat water for homes and swimming pools, as well as the direct conversion of solar radiation into electricity by solar cells. The development of fuel cells to power automobiles is in various stages of research and development. A Canadian company has produced several buses that use fuel cells instead of gasoline or diesel fuel that operate in the city of Chicago. Their model uses methane (natural gas) for the source of hydrogen and air for the source of oxygen for the reaction (O_2 + $2H_2$ □ $2 H_2O$ + energy). This reaction is the opposite of the electrolysis of water. One advantage of using fuel cells to produce electricity, besides replacing gasoline, is that their only pollution is water. If another source, other than natural gas (because methane contributes to global warming), could be found for the hydrogen for fuel cells, this would be an ideal energy system.

Some countries have developed geothermal sources of energy by tapping the hot water reservoirs in the Earth's crust. These plants produce hot water for homes and industry, as well as some electricity. Other areas construct dams to store the kinetic energy of water to run hydroelectric power plants, thus conserving some fossil fuels. The main limitation of these systems is that their location is not necessarily in close proximity to the areas where the electricity they produce is most needed. Another idea with local implications is using the high-to-low tides to generate limited amounts of electricity. These, and other systems, could be developed further to at least partially replace fossil fuels.

The article "The Power Plant in Your Basement, "in the July 1999 *Scientific American,* is an example of a viable alternative to extensive use of fossil fuels. The authors describe how a solid-oxide fuel cell can provide electricity, heat, and hot water to a home. The basement unit operates at ~800ºC (~1,500ºF). At this level, the system uses 90% of the cell's chemical energy not only to heat the house and hot water but also to produce 10 kilowatts of electricity. These units are expected to be on the market around the year 2004.[84] Four different types of fuel cells are now under research and development, with a few in the demonstration phase. The widespread use of self-contained home power plants will go a long way to solving the problems related to petroleum and gas supplies.

Another idea, not yet developed, is to use the nuclear wastes, the disposal of which causes so much concern. The vast majority of these wastes result from the dismantling of nuclear weapons and disposal of spent control rods from commercial nuclear electric power plants. Rather than burying these nuclear wastes, some effort and money might be expended to find ways to make use of this discarded source of energy.

In summary, although the world's population has increased, so has the standard of living for a greater *proportion* of people now living on the Earth than for any other period in history. This does not mean that we have eliminated people living in poverty or in want of a better life. In fact, one of the persistent challenges is to understand, and eventually correct, why vast numbers of the populations of countries rich in a variety of natural resources continue to live in poverty. If geology is not the main reason, then what are the reasons for more of these people not attaining a higher standard of living? It has been suggested that the main problems related to poverty in developing countries are mainly attributable to local politics, religion, and culture—not just geology.

However, all this does point in the direction of ever-expanding knowledge of how to intelligently adapt to our environment and a continuation of making wise use of our nonrenewable elements and resources in order to benefit a greater proportion of the world's population.

Notes

1. T. K. Derry and Trevor I. Williams, *A Short History of Technology* (New York: Dover Publications, 1960), 114–16.
2. Robert E. Krebs, *The History and Use of Our Earth's Chemical Elements: A Reference Guide* (Westport, Conn.: Greenwood Press, 1998), 90.
3. John Daintith, Sarah Mitchell, Elizabeth Tootill, and Derek Gjertsen, *Biographical Encyclopedia of Scientists,* vol. 1 (Bristol, England, and Philadelphia, Pa.: Institute of Physics Publishing, 1994), 7–8.
4. Isaac Asimov, *Asimov's Chronology of Science and Discovery* (New York: Harper and Row, 1989), 114.
5. James E. Bobick and Margery Peffer, *Science and Technology Desk Reference,* 2d ed. (New York: Gale, 1996), 553.
6. *Everything Is Made of Something* (Denver, Colo.: Mineral Information Institute, 1992), 44.
7. Mark J. Crawford, *Physical Geology* (Lincoln, Nebr.: Cliff Notes, 1998), 7.
8. Crawford, *Physical Geology,* 7.
9. Patricia Barnes-Svarney, ed., *The New York Public Library Science Desk Reference* (New York: Macmillan, 1995), 382.
10. Crawford, *Physical Geology,* 9.
11. Daintith et al., *Biographical Encyclopedia of Scientists,* 627.
12. Richard L. Lewis, *Hawley's Condensed Chemical Dictionary,* 12th ed. (New York: Van Nostrand Reinhold, 1993), 790.
13. *Encyclopedia Britannica,* CD-ROM 2001, s.v. "Earth Science: Crystallography."
14. Barnes-Svarney, ed., *The New York Public Library Science Desk Reference,* 383.
15. *Complete Home Medical Guide* (New York: The College of Physicians and Surgeons of Columbia University, 1985), 713.
16. Graham R. Thompson and Jonathan Turk, *Earth Science and the Environment* (New York: Harcourt Brace, 1999), 495–96.
17. Krebs, *The History and Use of Our Earth's Chemical Elements,* 90.
18. *Mineral Uses, A Study of the Earth* (Denver, Colo.: Mineral Information Institute, 1992), 41–43.
19. Krebs, *The History and Use of Our Earth's Chemical Elements,* 111–13.
20. Kenneth W. Hamblin and Eric H. Christiansen, *Earth's Dynamic Systems* (Upper Saddle River, N.J.: Prentice Hall, 1998), 64.
21. *Academic Press Dictionary of Science and Technology,* CD-ROM 1995, s.v. "Rocks."
22. Edwin Simons Robinson, *Basic Physical Geology* (New York: John Wiley and Sons, 1982), 210–20.
23. *Microsoft Encarta,* CD-ROM 1994, s.v. "Igneous Rocks."
24. Hamblin and Christiansen, *Earth's Dynamic Systems,* 83–97.
25. Crawford, *Physical Geology,* 18–19.
26. Thompson and Turk, *Earth Science,* 43.
27. Robinson, *Basic Physical Geology,* 221.
28. Thompson and Turk, *Earth Science,* 137.
29. Thompson and Turk, *Earth Science,* 47.
30. *Encyclopedia Britannica,* CD-ROM 2001, s.v. "Igneous rock: Mineralogical Components."
31. Thomas McGuire, *Reviewing Earth Science: The Physical Setting* (New York: Amsco School Publications, 2000), 29.
32. Bates and Jackson, *Dictionary of Geological Terms,* 454.
33. Sybil P. Parker, ed., *McGraw-Hill Concise Encyclopedia of Science and Technology,* 3d ed. (New York: McGraw-Hill, 1994), 1671–73.

34. *Encyclopedia Britannica,* CD-ROM 2001, s.v. "Sedimentary Rocks."

35. Thompson and Turk, *Earth Science,* 48.

36. Thompson and Turk, *Earth Science,* 51.

37. Thompson and Turk, *Earth Science,* 51–52.

38. John Tomikel, *Basic Earth Science: Earth Processes and Environments* (New York: Allegheny Press, 1981), 30.

39. Bates and Jackson, *Dictionary of Geological Terms,* 147.

40. Thompson and Turk, *Earth Science,* 52–53.

41. *The Oxford English Dictionary,* 2d ed., CD-ROM, s.v. "Rocks."

42. *Academic Press Dictionary of Science and Technology,* CD-ROM 1995, s.v. "Metamorphism."

43. Stanley Chernicoff, *Geology: An Introduction to Physical Geology* (New York: Houghton Mifflin, 1999), 183–84.

44. Parker, *McGraw-Hill Concise Encyclopedia of Science and Technology,* 1156.

45. Sybil P. Parker, ed., *McGraw-Hill Dictionary of Scientific and Technical Terms,* 5th ed. (New York: McGraw-Hill, 1994), 630.

46. Motohiko Murakami et al., "Water in Earth's Lower Mantle," *Science,* 8 March 2002, 1885.

47. *Rockdoctors Guide to Metamorphic Rocks,* accessed 2001, http://www.cobweb.net/~bug2/rock5.

48. Thompson and Turk, *Earth Science,* 59–65.

49. *Metamorphic Rocks* (Metamorphic Settings), accessed 2001, http://www.science.ubc.ca/~geol202/meta/metaset.

50. Chernicoff, *Geology,* 188–90.

51. Peter Huber, *Hard Green: Saving the Environment from the Environmentalists* (New York: Basic Books, 1999), 9.

52. Quoted by Samuel Edwards, "Ethics and Global Warming: Clergy's Support of Kyoto Treaty Based on Bad Science," *Valley Morning Star* (Harlingen, Tex.), 24 October 2000.

53. *Every American Born Will Need* (Denver, Colo.: Mineral Information Institute, 1997).

54. *Minerals Provided for Every Person* (Denver, Colo.: Mineral Information Institute, 1997).

55. *Must Have* (Denver, Colo.: Mineral Information Institute, 1997).

56. Asimov, *Chronology of Science and Discovery,* 335.

57. Thompson and Turk, *Earth Science,* 502–4.

58. Erwin Suess, Gerhard Bohrmann, and Jens Greinert, "Flammable Ice," *Scientific American,* November 1999, 77–83.

59. Lewis, *Hawley's Condensed Chemical Dictionary,* 290.

60. *Encyclopedia Britannica,* CD-ROM 2001, s.v. "Mineral Fuels."

61. Chernicoff, *Geology,* 574–75.

62. Lewis, *Hawley's Condensed Chemical Dictionary,* 290–91.

63. Asimov, *Chronology of Science and Discovery,* 251.

64. Parker, *McGraw-Hill Concise Encyclopedia of Science and Technology,* 413.

65. Lewis, *Hawley's Condensed Chemical Dictionary,* 989.

66. Joseph L. Bast, Peter J. Hill, and Richard C. Rue, *Eco-Sanity: A Common-Sense Guide to Environmentalism* (New York: Madison Books, 1994), 143.

67. *Environmental News Service,* accessed 2001, http://www.ens.lycos.com/ens/may99/1999L-05-06-06.

68. Lewis, *Hawley's Condensed Chemical Dictionary,* 1201.

69. *Three Mile Island 2 Accident Report: Health Effects,* accessed 2001, http://www.nrc.govOPA/gmo/tip/tmi/htm.

70. Hamblin and Christiansen, *Earth's Dynamic Systems,* 644.

71. Thompson and Turk, *Earth Science,* 493–97.

72. *Everything Comes from Our Natural Resources* (Denver, Colo.: Mineral Information Institute, 1992), 34, 47.

73. Quoted in *Isaac Asimov's Book of Science and Nature Quotations,* ed. Isaac Asimov (New York: Weidenfeld and Nicolson, 1988), 237.

74. Ronald Bailey, *EcoScam: The False Prophets of Ecological Apocalypse* (New York: St. Martin's Press, 1993), 67.

75. Bailey, *EcoScam,* 69.

76. Robert E. Krebs, *Scientific Development and Misconceptions through the Ages* (Westport, Conn.: Greenwood Press, 1999), 225.

77. Bjorn Lomborg, *The Skeptical Environmentalist: Measuring the Real State of the World* (New York: Cambridge University Press, 2001), 19. The three studies referred to are U.S. Department of Energy (Oak Ridge National Laboratories/Resources for the Future), Lee et al., 1995; The EU (DG XII, 1995); and The Empire State Electric and the NY State Energy Research and Development Authority, 1995.

78. William Ashworth, *The Economy of Nature: Rethinking the Connection Between Ecology and Economics* (New York: Houghton Mifflin, 1995).

79. Paul R. Ehrlich, *The Population Bomb* (New York: Ballantine Books, 1968).

80. Bailey, *EcoScam,* 53–54.

81. *Recycling Metals* (Denver, Colo.: Mineral Information Institute, 1992), 26.

82. Krebs, *Scientific Development and Misconceptions,* 238–39.

83. Ben Dobbin, "Small Town Gets Big Payoff," *Houston Chronicle,* Houston, Tex., 22 October 2000.

84. Alan C. Lloyd, "The Power Plant in Your Basement," *Scientific American,* July 1999, 80–86.

Earth's Inner Spheres

No doubt, ancient people explored deep caves, noticed the upheaval and folding of rocks and the formation of mountains, and experienced the Earth spewing out fire and ash, as well as the shaking of the land under their feet. These geological events must certainly have made them wonder what was below the surface dirt, soil, rocks, and water. Over the ages, many myths developed to explain what people thought was deep inside their land (see the section "Myths and Misconceptions about Earth's Interior" for more on these myths.)

Unfortunately, it was not possible to open up the Earth like splitting a piece of fruit to examine the inner core or pit and to directly examine what was there, so, several hundred years ago, geologists devised indirect means of determining the Earth's structure. They studied the nature of rocks and their formations, topological features such as **synclines** and mountains, and other observable surface features to arrive at tentative theories to explain what the various regions of the interior of the Earth might be like. During the twentieth century, particularly the latter part, vastly improved instruments and techniques were devised to obtain indirect measurements of the Earth's internal structure and processes. These efforts let to the current accepted global theory of plate tectonics and the natures of the different internal spheres of the Earth.

The following is a simple exercise to demonstrate how difficult it is to indirectly determine the shape and/or nature of something inside a closed container. Place a single simple, small, everyday object (e.g., a large marble, ballpoint pen, gum eraser, three large paper clips, or an uneven rock about 1 or 2 inches in diameter) in a shoebox. Tape the lid on the box so that it cannot be opened, and then ask friends to shake, rattle, and roll the box, but not open it or peek inside. After a few minutes, ask them to write a description and draw a figure inferred from their indirect observations. Then request they theorize as to the structure and identity of the object based on what they learned. This is somewhat like the problem early scientists had, and still have, in determining exactly how the interior of the Earth is structured, as well as how internal forces, processes, and systems determine the nature of the internal spheres.

The Mantle Sphere

The mantle is a major inner sphere of the Earth, comprising about 82% of the Earth's volume and 68% of its mass. It surrounds the Earth's central core, it is composed of hot rock material, and its temperature and pressure generally increase with its depth. When pressure is decreased and the rock material becomes less dense as it rises, it becomes molten magma capable of flowing to the surface or erupting as volcanoes. Seismic evidence indicates that the mantle extends from about 600 kilometers (km) (374 miles [mi]) to a depth of about 2,800 to 2,900 km (1,740 to 1,800 mi) and is composed of *peridotite* (a special type of igneous rock), which consists mainly of silicon, oxygen, magnesium, iron, and lesser amounts of other metallic and trace elements. The mantle increases in density with its depth. Near the upper (outer) part of the mantle the density is 3.2 to 3.5 grams (g) per cubic centimeter (cm^3), while the density in the lower part near the core is 5.0 to 5.8 g/cm^3.[1–2]

The term "mantle" is a combination of Middle English and French meaning "cloak." According to the Oxford English Dictionary (OED), the term "mantle" has been used for many centuries to describe areas that were either inside of or surrounding other objects, which aptly describes the Earth's mantle. The word was first employed as a geological term in 1895 to describe surface deposits of rock as *mantle borders,* folds, or margins, but it was not used until 1940 to describe the inner sphere of the Earth between the crust and core.[3]

The internal structure of the Earth is basically subdivided in two distinct layers (which also may be subdivided). Although the chemistry of the interior spheres is of some importance, the physical properties related to rigidity, viscosity, density, temperature, and pressure are more important in determining their composition and internal position. How the material in the layers reacts *mechanically* (i.e., how dense and how strong they are) determines the distinctions between layers, not their composition.[4] The deeper the mantle, the greater the density (see Table 5.1).

When defined by its *physical* properties, the mantle is referred to as the solid or semisolid *mesosphere,* extending from the asthenosphere to the core. These distinctions determine how one defines the mantle but not its importance as a major region that affects the dynamic nature and geological features of the Earth. Thus, the mantle may be defined as and considered a single sphere or as two spheres with an indistinct transition zone separating the layers.

Table 5.1
Comparison of Density to Depths of Earth's Spheres

Sphere	Depth of Sphere	Density
Crust	6 to 50 km	2.7 to 3.0 gm/cm^3
Lithosphere	50 to 200 km	3.2 to 4 gm/cm^3
Upper mantle	200 to 700 km	4 to 4.6 gm/cm^3
Lower mantle	700 to 3000 km	4.6 to 5.5 gm/cm^3
Outer core	3000 to 5100 km	10 to 11 gm/cm^3
Inner core	5100 to 6370 km	11 to 13 gm/cm^3
Whole Earth	0 to 6370 km	5.5 gm/cm^3 (overall)

Upper Mantle

Seismic waves travel at different rates through materials of different densities. Studies identified a discontinuity in the rate that seismic signals travel in relation to the density and other physical characteristics of material between different layers deep in the Earth. For instance, the major elements in the crust are oxygen-silicon-based compounds, whereas the compounds in the upper mantle, in addition to oxygen-silicon, include many compounds that are iron and magnesium based.[5] Although some geologists consider the lithosphere to be part of the upper mantle, a discontinuity was identified between the relatively shallow crust/lithosphere and top of the mantle at a depth of about 75 to 150 km (~50 to ~100 mi). Using seismic evidence for this region between the bottom of the crust/lithosphere and the top of the mantle, geologists named it the "Moho," after Andrija Mohorovicic (1857–1936), who first described it. Another discontinuity based on seismic-wave velocities was identified at about 250 to 350 km (~155 to ~220 mi) depth. Known as the *low-velocity zone,* this is a weak, heat-softened layer that lies just under the Moho and above the top of the mantle. The partially melted, relatively thin, low-velocity zone of hot material facilitates tectonic plate movement. It is also the region of the upper mantle called the asthenosphere. Several theories are related to the asthenosphere's role in plate movement. One is that the slow-flowing material carries the plates; another is that the plates slide over the partially melted material of the asthenosphere. This thin boundary between the semisolid asthenosphere and the rocklike lithosphere makes them physically distinct but not significantly different chemically. Their chemical similarity is the reason they are sometimes included as part of the upper mantle. This division, the *lithosphere-asthenosphere boundary,* is an important boundary that is partly responsible for the dynamic nature of Earth's plate tectonic system. This boundary, located at ~100 to ~200 km (~62 to ~125 mi) depth, is observed by seismologists who detect differences in seismic P- and S-wave velocity between layers, as well as by thermal distinctions and, more recently, by the use of **tomographic** techniques.[6] The boundary is an area of comparative weakness that has the capacity to yield to long-term strains and stresses that induce a semisolid state to the material but do not liquify it. This shallow region of the upper mantle is not only responsible for "floating" tectonic plates but also one place where magma is extruded to the Earth's surface. The physical properties of the boundary of the upper mantle are important factors related to earthquakes, seafloor spreading, and the flow of magma, as well as plate movement—which is about as fast as your fingernails grow.[7] As the asthenosphere's depth varies to about 250 to 350 km (155 to 220 mi), it gradually becomes a more rigid (stiffer) part of the upper mantle. The density of the upper mantle increases with depth. Conversely, the density decreases in the upper part of the mantle just below the lithosphere. In this region the temperature and pressure become precisely right for the rock material to melt, lose its strength, and transform into plastic-like flowing magma.[8] As previously mentioned, the magma originating near the asthenosphere is different in composition from the hot material in the lower mantle. Magma-rock is also formed in the upper mantle at ~40 to ~60 km (~25 to ~40 mi), where temperatures are increased as the pressure is reduced closer to the surface to the extent that rocks partially melt and flow as magma that becomes lava as it is extruded into the crust to form volcanoes. In 1993, a research team for the first time bored through the ocean's crust into the upper mantle from an ocean research vessel. They extracted a core of mantle-rock and determined that it is mostly peridotite, which is a coarse-grained metamorphic rock composed largely of olivine, pyroxene, garnet, and other minerals rich in calcium and magnesium that were formed under high pressure.[9]

If the crust, lithosphere, Moho, and asthenosphere are excluded, the upper mantle extends from a depth of about 350 to 600+ km (~200 to ~375 mi) below the Earth's surface. But there is no sharp discontinuity in density or other physical attributes at the junction between the upper and lower mantles. This is one reason why the mantle is often considered one huge internal sphere that makes up most of the Earth's mass rather than two separate mantles. Even so, there are some differences. Seismographs note these physical differences even though there is no region that sharply distinguishes the upper and lower mantles.

Lower Mantle

The lower mantle is located between the upper mantle layer and the core. The lower mantle is also sometimes referred to as the "mesosphere," or "middle sphere," which is, by far, the largest internal sphere in the Earth. It extends in depth from the bottom of the upper mantle at ~700 km (~440 mi) to ~2,900 km (~1,800 mi) where the Earth's core begins. In other words, the lower mantle is ~2,200 km (~1,365 mi) thick. Instead of a sharp boundary between the upper and lower mantles there is a more-or-less transitional zone between these layers where their chemical compositions, temperatures, pressures, and densities are somewhat similar. The weight of the overlying crust, lithosphere, and upper mantle is great enough to produce extremely high temperatures and pressures that subsequently are great enough to maintain the solid state of the lower mantle. At these greater depths, the lower mantle's minerals, crystal structures, chemical composition, and physical properties (e.g., density) become different from the upper mantle's properties.[10] Although the temperature in the lower mantle is hot enough to melt rock, the rock material mostly remains solid due to the great pressures. Because of the high pressures and temperatures in the lower mantle, the minerals and crystals formed at this depth are different from those formed in the upper mantle. Recent studies indicate that the lower mantle consists mainly of *perovskite* and *periclase,* which are compounds of magnesium, silicon, and oxygen.[11] In a recent article in *Science* magazine, researchers reported that by using ion mass spectrometry and other measurements, a great amount of water was discovered to be dispersed in various natural peridotic (rock-magma) materials in the lower mantle. These researchers estimated that the "Earth's lower mantle may store about five times more H_2O than the oceans."[12] This sounds like a great amount, but when the size (volume) of the lower mantle is considered, the water is sparsely distributed throughout the pores of rock material and crystals. Even so, mantle water has a great effect on the physical nature of deep, as well as shallow, rock and also influences the flow rate of magma as it approaches the surface.

The source of ~3,000°C or greater temperatures in the mantle is the decay of radioactive elements, as well as a small amount of residual heat left over from the formation of the solar system and the Earth about 10 billion years after the Big Bang.[13] This heat is responsible for much of the dynamic nature of the inner spheres that in turn creates geological changes on the Earth's surface. The Earth must physically find a way to rid itself of some of this internal heat. One way to do this is by circulating the heat to the surface, where it can be dissipated into the surrounding land, water, and atmosphere. **Rheology** is the study of how matter flows and the ways it is deformed. One important concept of rheology is exhibited by convection cells that are created by the flow of heat from greater depths to more shallow regions of the mantle. This is the main process related to the removal of internal heat that is also responsible for the dynamic nature of the Earth's inner spheres and some surface features.

Convection

Convection currents or cells are, more or less, cylindrical or spherical flow patterns in water, air, fluids, or semisolids that are driven by heat and/or density differences. Mantle material moves upward as it is heated, just as air does in the atmosphere or water in a hydrothermal system. Thus, the rising materials become less dense as they expand due to an increase in temperature and decrease in pressure. Note that heat is an increase in kinetic energy resulting in increased molecular motion that causes increased kinetic collisions between molecules that in turn increase the spaces between particles, thus expanding the particles of matter, which also results in a cooling effect (reducing molecular motion). Heat decreases a substance's density, whereas cooling increases its density. Note also that convection cells exist in closed systems where heated matter is raised from areas of higher temperature to areas of lower temperature. Because the basic amount of the mantle-rock (in a closed system) has not increased, the denser, cooled mantle-rock material descends along the outer side of the cell to fill in the areas from which hotter material is rising upward from the bottom and in the center of the cell. This results in convection cells in which the hot central area is rising and the outer, cooler regions are flowing down. Cooling is just the opposite of heating, that is, molecular motion slows down, thus reducing the spaces between particles and resulting in an increase in density as contraction takes place[14] (see Figure 5.1).

As mentioned, the science of rheology is the study of the flow and deformation of matter or suspended particles in a fluid. Thus, the internal heat of the Earth's inner spheres is responsible for driving convection cells that result in the extruding of lava into a spreading ocean floor that creates oceanic ridges, rift valleys, trenches, island arcs, volcanoes, and the movement of continents. Rheology and convection are also responsible for returning crust/lithosphere material back into the mantle by subduction at the edges of tectonic plates at the boundaries of colliding ocean and continental crusts. This churning and recycling of the crust/lithosphere material with the mantle material is in a constant state of flux.

Giant slabs of the ocean crust/lithosphere sink at the subduction zones as they meet the less mobile and thicker continental land masses. The subducted masses of crustal rock are pulled down by gravity all the way to the lower mantle to a depth of about 2,900 km (1,800 mi). The basic physical principle of conservation of mass is exhibited by the sinking of oceanic crust/lithosphere material, which is balanced by equal volumes of hot rock rising from deep within the mantle to the surface as it is extruded at the mid-oceanic ridge to replace the ocean crust/lithosphere material that was pushed down at the subduction boundaries of the plates. This process continues to drive the molten magma upward to be deposited again on the surface of the continental and oceanic crusts as lava. Geologists study the rock-magma material that resurfaced after many eons of this recycling

Figure 5.1
Heat (increased molecular motion) creates greater spaces between molecules, thus decreasing the density of the magma. The hot, less-dense, fluidlike substance rises toward the surface and becomes cooler, thus becoming denser. It then sinks toward the inner spheres of the Earth to again be reheated and continue the convection cycle.

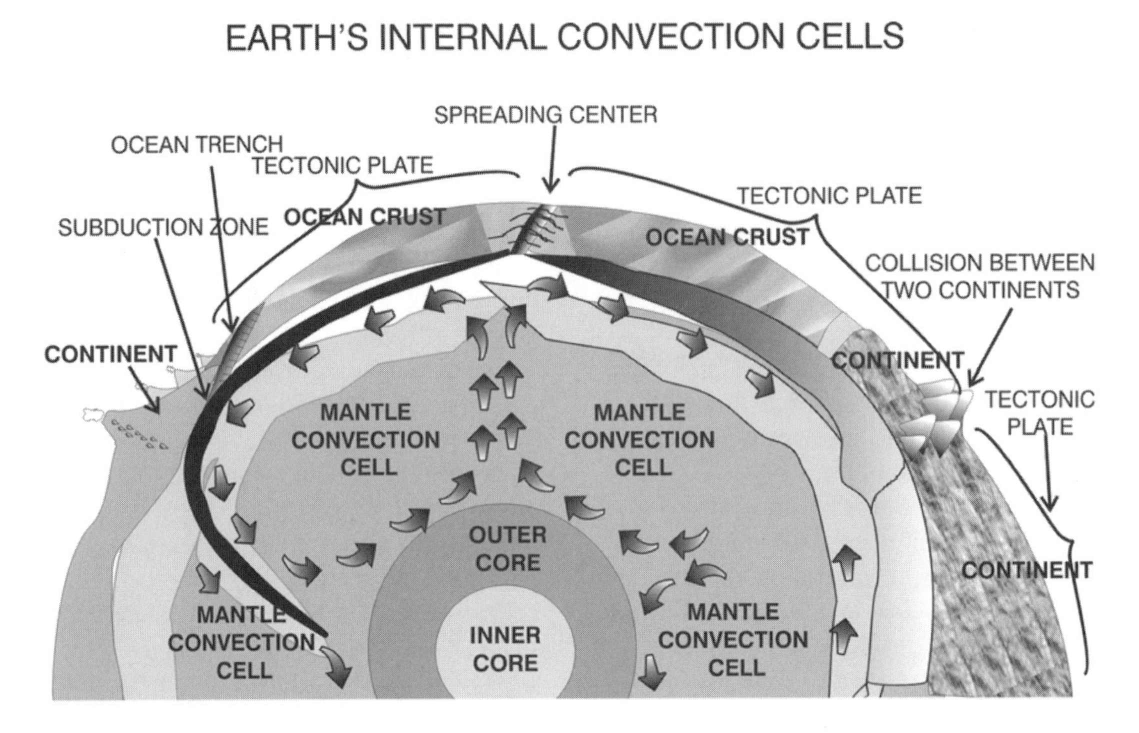

process to determine the chemical and physical properties of the mantle.

The Earth's Core

The core of the Earth consists of two distinctive spheres. The diameter of both spheres that make up the core is just over half the diameter of the Earth. (The Earth's diameter is approximately ~12,700 km [~7,900 mi], whereas the inner and outer core's diameters together are ~7,300 km [~4,550 mi]). In the last few decades, research has provided new information on the unique nature of these two inner spheres. A thin boundary layer separates the bottom of the lower mantle from the top of the outer core. This layer forms a distinct discontinuity of the chemical and physical properties between the lower mantle and the outer core, somewhat similar to the Moho discontinuity. This thin layer, located ~2,900 km (~1,800 mi) inside the Earth, forms an interruption, or a separation, where some mixing occurs between the lower mantle and outer core. It is known as the Wiechert-Gutenberg discontinuity (also called the Gutenberg discontinuity) and indicates a separation at the ~9.7 to ~10.8 g/cm^3 density of the core from the ~5.7 g/cm^3 density of the material at the base of the mantle. Thus, the mantle and core are separated by density as well as by temperature, pressure, and chemistry.[15] Although this discontinuity is only a few hundred miles thick, it represents a significant change in the state of matter from a solid to a more fluid nature. In addition to density difference, this distinction was detected by abrupt changes in seismograph readings. The velocity of seismic P waves decreases from 13.6 km (8.6 mi) per second to 8.1 km (5 mi) per second at this lower mantle/outer core boundary. While P waves slow down when they reach an area of reduced rigidity, the seismic S waves are not recorded at all because they are not transmitted from a liquid (core).[16] (See chapter 2 for more on seismic P and S waves.)

While the mantle/core boundary consists of highly compressed ultramafic rock that has a density of about 5.5 g/cm^3, it is soon changed to a substance that has a density of about 11 to 13 g/cm^3, which is a density more like a metal than rock. Iron and nickel are the only chemical elements common in the universe that have a density similar to these readings. Seismologists have detected a very thin zone (~10 km, or ~6.2 mi) of ultra-low seismic velocities at the base of the lower mantle where the speed of P waves is reduced. This zone is referred to as the ULVZ (*u*ltra-*l*ow *v*elocity *z*one) and indicates that the lowermost mantle material is partially melted. This situation corresponds to the hot spots on the Earth's surface where plumes of magma are extruded through the crust/lithosphere. In the discontinuity between the lower mantle and the outer core, some of the hot, liquid iron metal from the outer core forms plumes and hot spots into the lower mantle.[17] This indicates that the ULVZ is not a sharp, featureless boundary but instead one with a rugged topography similar to the surface of the Earth. An article in *Science* magazine described this region at the bottom of the mantle as a "mushy zone or a slurpy layer."[18] Continental size slabs of mantle rock, some as large as Alaska and 270 km (179 mi) thick, have been detected at the bottom of the mantle. These slabs exhibit what are described as upside-down mountains and valleys that form the ULVZ identified by seismology.[19]

The Earth's core is different from the other spheres of the Earth due to the extreme heat and pressure exerted on the elements and minerals of which it is composed. Scientists have learned a great deal about the core's chemical composition and physical properties by using several indirect (proxy) methods and computer models. The following list describes some of these methods and what has been learned by applying them.

1. The **moment of inertia** of the Earth indicates that there is a concentration of mass at its center.
2. Seismic P and S waves do not easily travel through the core, and those that do are refracted as if the central part of the core is composed of very dense material. The S waves are not transmitted through liquids, and because they bypass the outer surface of the core, this indicates that the outer core is liquid. Conversely, the P waves pass through the liquid outer core but are refracted from the inner core's outer layer, indicating that it is dense and solid.[20]
3. It is further deduced that the core is mainly composed of an iron-nickel alloy, because this is the major elemental composition of meteorites that were formed in the solar system at the same time as Earth and that continually land on Earth from outer space.[21]
4. Some other elements must also make up the core. Geophysical evidence indicates that the average atomic number of the material of the core is about 22, whereas iron is 26. This implies that elements other than iron are in the core, possibly sulfur-16, carbon-6, and cobalt-27 since they are found also in meteorites.
5. It is unlikely that silicon-14 or oxygen-8, which are found in the mantle, are also present in the core.[22]

Based on computer models and other indirect evidence geologists conclude the following about the nature of the Earth's core:

1. The core comprises about 16% of the Earth's volume but accounts for 32% of its mass.[23]
2. The temperature of the core ranges between 4,700°C (8,500°F) and 5,500°C (10,000°F).
3. The Earth's core begins just below the lower mantle, at about 2,900 km (1,800 mi) from the surface.[24]

There is seismic evidence that the core is divided into two distinct spheres. Another discontinuity layer between the outer core and inner core indicates a physical difference exists between these subspheres. In 1936, Inge Lehmann (1888–1993), using the P waves recorded on seismograms, suggested that a new discontinuity zone divides the outer liquid iron-nickel core from the solid inner core, even though both cores have similar chemical compositions. This separation, now known as the Lehmann discontinuity, is located at the liquid-solid boundary between the two cores.[25–26] More recently, improved seismological instruments and the use of computer modeling have confirmed this Lehmann discontinuity between a liquid outer core and solid inner core.

Outer Core

The outer core's diameter is estimated between 2,250 km (1,400 mi) and 2,500 km (1,550 mi). The best evidence for the existence of a liquid outer core is based on what seismologists call the S-wave and P-wave shadow zones. Scientists had discovered that no S waves were detected within an arc of about 154° across the Earth's surface from seismograph detections of an earthquake's epicenter to another seismograph detection instrument located on the other side of the Earth. This 154° seismic shadow, somewhat similar to a light shadow on the surface of the Earth, was interpreted as the inability of S waves to penetrate the liquid medium of the outer core that enveloped the Earth's very center. While P waves do penetrate liquids, they also cast a different type of shadow when they are *reflected* off the center core. They also change their speed as they curve as much as 103° from the source of an earthquake. In addition, when P waves are *refracted* inside the outer liquid core, they emerge at a 143° angle. This region between 103° and 143° forms the P-wave shadow of 40° that indicates the nature and dimensions of the outer core.

The main reason that the outer core is liquid instead of solid, as is the inner core, is because of the differences in pressure and temperature between the two cores. As the Earth cooled after being formed, the heavier molten iron (and some other elements) sank toward the center as they dissipated the Earth's internal heat outward toward the surface. Earth's outer core is the only inner sphere that is truly liquid. The average temperature of the outer core is ~5,000°C (9,000°F), whereas the inner core's average temperature is ~3,800°C (7,000°F). The liquid nature of the outer core is differentiated from the inner core by the slower speed of the S and P seismic waves. Evidence also suggests that at the Lehmann discontinuity boundary between the outer and inner cores, the lower region of the outer core is slowly solidifying just as its relative, the inner core. This most likely is the result of a continued slight decrease in temperature as heat is dissipated to the surface and, in part, the increase in pressure.

Inner Core

Although the inner core is smaller and solid, rather than liquid as is the larger outer core, it seems to have more intriguing characteristics than do all the other spheres. The inner core is not only the smallest but also the densest of all the Earth's spheres and has a radius estimated to be ~1,220 km (~758 mi). Many scientists consider the inner core to be ~1,300 to ~1,500 mi in diameter. Scientists confirmed that as seismic P waves pass directly through the entire Earth, they arrive on the other side earlier than was expected. As with sound waves, P waves travel faster in denser materials. This was confirmed when nuclear explosions, whose exact time was known, were detected and the speed of the P waves were measured as they traveled through the planet. This increase in the velocity of seismic waves as they passed through the planet proved that Earth's inner core is solid.[27]

It has only been in the past decade or so that geologists have determined that the solid-iron/nickel alloy inner core is *anisotropic,* meaning that solid metal crystals are not arranged randomly but have a directional texture similar to the grain in wood. This is known because seismic waves travel faster in one direction in the core than in the other. The evidence for this is being examined by supercomputers that model the crystalline structure of iron at this region of great pressure and temperature. Although three types of iron crystals are found in the Earth's crust, it appears that iron in the core is just one of these types of crystals, specifically the **hexagonal close-packed** (hcp). These crystals are formed under high pressure. The hcp crystals uniquely fit the seismic data for anisotropic or directional characteristics and, as a result, are all oriented in one direction, with the Earth's spin on its axis. (See Figure 4.2 for different forms of crystals.) Geologists have determined that if all the crystals point in the same direction, there is good reason to assume that the solid inner core is similar to one big iron crystal. They base their conclusion on the fact that the temperature-pressure at the inner core is ideal for the formation of this type of crystal and that a big, directionally oriented, central crystal can also explain anomalies of the Earth's magnetic field.[28] This polar orientation of the solid crystal inner core's iron also results in seismic waves moving faster through the core in the north-south direction than they do in the equatorial direction. This alignment is off 10° compared to the Earth's axis and is assumed to affect the orientation of the alignment of the magnetic north and south poles to be offset from the geographic north and south poles. Geologists have studied the alignment of the magnetic particles in some minerals and rocks that indicates that at some distant time in the past the orientation of the Earth's magnetic poles shifted several degrees from the true geographic north and south poles. In addition, in the distant past the magnetic poles actually flipped, and they have reversed their orientation several times over the past several hundred million years (see Figure 2.2 in chapter 2). The Earth's magnetism creates an external magnetic field that is detectable far out into the outer atmosphere and space. The Earth's magnetic field converges and is concentrated at the poles, just as the magnetic field around a bar magnet ends at its poles. This field is responsible for directing ionized particles and some forms of radiation from the sun away from the equatorial zone toward the poles. The aurora borealis, or northern lights, are the result of ions in the atmosphere being concentrated at the magnetic North Pole.

Geomagnetism

The Earth's magnetism was detected by the Chinese many centuries ago when they discovered that a lodestone (iron-rock), which contains some permanent magnetic properties, orients itself in a north-south position when placed on a block of wood floating on water in a bowl. It is alleged that Marco Polo brought this discovery to the western world. Slivers of iron that were magnetized by lodestones were used as compass needles by early navigators. In 1600, William Gilbert (1544–1603), an English physician, published his experimental investigations on the Earth's magnetism. In one of his major experiments he used a large, round lodestone over which he placed a compass. He noticed that the needle was deflected toward the center of the lodestone as well as to poles of the stone. He then demonstrated that a compass needle would not only orient itself toward the Earth's North Pole (Gilbert also originated the term "magnetic pole") but also dip toward the center of the Earth, and that at higher latitudes the dip was greater than near the equator. From this declination of the needle he concluded that the Earth was acting like a large bar magnet oriented with magnetic north and south poles.[29] Since that time, it is recognized that the magnetic nature of the Earth is more similar to an electromagnet than a bar magnet. (See Figure A.15 in the Appendix for a diagram of the Earth as a large magnet.)

Magnetism is caused by moving particles, such as electrons, that result in the ability of some substances to either attract or repel matter. Electrons that orbit the nuclei of atoms are spinning on their own axes; therefore, they create their own magnetic fields. Individually, electrons exhibit very weak magnetic fields. When they spin randomly, they cancel out each other, and thus this type of matter containing electrons exhibits no magnetism. On the other hand, substances in which the electrons mostly spin in one direction produce a magnetic field. When rocks containing iron (e.g., magnetite, basalts, gabbros, and hematite) are subjected to strong external magnetic fields, they align their internal magnetic particles in one direction. This is what occurs when a piece of iron is magnetized by rubbing it with a lodestone or another magnet. These magnetized rocks may become permanent magnets. They can also be demagnetized by the heat of lightning (greater than ~600°C, or ~1,100°F) or by weathering, and subsequently, they can again be magnetized. Wrapping a coil of wire around a soft iron rod or bar can form a different type of magnet, referred to as an electromagnet. As electricity surges through the coil of wire, its external field magnetizes the central iron bar. This process can produce strong magnets, but the magnetism in the soft iron core exists only as long as the electrons are flowing through the coil of wire.

Using magnetometers, instruments used to measure magnetic anomalies on seafloors and on land, geologists are able to examine magnetic rocks and detect the direction and other characteristics of the Earth's magnetic field, in both the past and the present. The Earth's magnetic field extends from the surface past the outer atmosphere for more than 60,000 km (37,000 mi). The Earth's magnetic lines of force that comprise its magnetic field are invisible but pervasive. The magnetic field encompasses, engulfs, and penetrates everything on and in the Earth, including rocks and living organisms. The internal magnet of the Earth causes this magnetic field to curve as it surrounds the Earth. The lines of force are directed toward the two magnetic poles of the Earth, where they strike the surface at a more perpendicular angle than at the equator.

Paleomagnetism is the historical geological record of Earth's magnetism. The Earth's magnetism has changed over geological history. For instance, when basalt lava cools, the crystals form magnetite rocks that, before becoming solids, align their iron particles with the magnetic field that existed on Earth at the time the crystals were formed. These rocks slowly break down, and particles drift to the ocean floor where they become aligned with the current field.

As they become imbedded in the seafloor and solidified, they maintain their alignment, even as the magnetic field changes. The Earth's magnetic field reverses (the magnetic North Pole alternates with the magnetic South Pole) every 25,000 to several million years. Since the magnetic particles in old rock are permanently aligned according to the time during which they were formed, they now provide a record of Earth's magnetic field and polar alignment for different times in history.[30]

Some questions still must be answered about the origin and nature of Earth's magnetism. For instance, scientists know, for at least three reasons, that the Earth itself cannot be a permanent magnet. First, high temperatures destroy the magnetic properties of permanent magnets. Therefore, the high internal temperatures of the core would destroy any type of permanent magnet, whereas, at the same time, the temperatures in the crust are low enough to permit some rocks to maintain permanent magnetic properties. Second, a great deal of evidence indicates that the magnetic polarity of the Earth has reversed over the past hundreds of millions of years. This has been determined by examining the shifting polarity of some formerly buried surface rocks and deep rock cores that exhibit magnetic pole reversals. Third, the magnetic pattern formed by zebra stripes on spreading seafloors as they extend from oceanic ridges (see chapter 2) indicates that the Earth itself cannot be a permanent magnet.

In the 1800s, it was determined that electricity flowing through a wire produced a magnetic field around the wire. This concept is based on the idea that electrons spontaneously moving through a conducting material can generate a magnetic field. This led to the concept of a theoretical system in the outer core known as a "self-exciting dynamo." It is well known that in modern dynamos a spinning electrical conductor (coil of wire) inside a strong magnetic field produces a flow of electrons with a magnetic field. On Earth it takes some force, such as water power or steam, to drive turbines that spin the conductors inside the magnetic field of electric generators (dynamos). The Earth's magnetic field generating system, although using the same scientific principles as man-made dynamos, is driven by the force of the rotating Earth that sets the outer core's liquid metal into motion, thus acting as a self-exciting dynamo. There is a big difference between man-made dynamos that require an external force and Earth's self-exciting dynamo. The natural rotation of the Earth provides the energy to sustain a self-exciting dynamo that generates a magnetic field, which in turn, produces a flow of more electrons, which in turn produces a stronger magnetic field, and so forth. This system will continue as long as there is adequate heat to maintain the liquid conductor (metal) in the outer core and as long as the Earth continues to rotate on its axis.

A similar explanation involves two types of motion of the liquid metal in the outer core. First, the liquid iron is circulated in vertical loops by heat convection. This is based on the fact that the bottom of the outer core (where it interfaces with the lower core) is much hotter than the upper region of the outer core (where it interfaces with the lower mantle). Second, as the liquid metals rise and descend in the convection cells, the spinning of the Earth on its axis deflects these liquid conductors, generating electromagnetism. In other words, heat causes fluidlike metal particles to rise, then cool and fall back down to the bottom of the outer core, while at the same time the metal particles are deflected sideways by the Earth's spin on its axis, thus generating electrical currents that give rise to magnetic fields.[31] This is referred to as the *geodynamo* model. This electromagnetic field is somewhat similar to a simple iron rod wrapped with wire that becomes magnetized when a low current of electricity is passed through the wire.

A third theory is based on the difference in the speed of rotation between the two cores. The powerful electromagnetic field produced by the liquid outer core drives the inner core. About a billion amps of current flow across the Lehmann discontinuity that separates the outer core from the inner core. This current, in relation to the strong magnetic field, causes the inner core to act as a rotor in a synchronous electric motor. Seismic evidence indicates that the inner sphere rotates faster than the outer sphere due to this electromagnetic influence. The **Coriolis force** (see Figure 7.2) and other forces also affect the rotation of the inner core. Although only about 10% of the Earth's magnetic strength is generated by the inner core, this is adequate to keep the inner core rotating eastward in sync with the liquid core above it and slightly faster than the Earth's surface spheres.[32–34] All of these theories for the origin of the Earth's magnetism involve intricate relationships among the chemical composition and physical features, such as temperature, pressure, and different rotational speeds for the two cores. In addition, there is rotational slackening of the outer spheres of the Earth resulting from tidal friction between the Earth's oceans and the moon, the Coriolis force, kinetic energy, and other forces. However, much still remains unknown about the inner spheres of the Earth and the nature of the Earth's internal magnetic forces.

In summary, the inner spheres, some solid, some liquid, and some mushy or semiplastic, are all influenced by physics, chemistry, and the laws of nature, and thus constantly undergo changes that enable the Earth to exist as a dynamic planet.

Myths and Misconceptions about Earth's Interior

From the distant past to the present people have fantasized and speculated about the internal structure of the Earth. Ancient lore and superstitions are updated periodically into more modern pseudoscience myths related to the internal spheres of the Earth.

Historically, many ancient civilizations, including the Greeks and Romans, associated the center of the Earth with the land of the dead. It is still considered the site of Hades or Hell by most modern western religions and civilizations. It was, and still is, known as the place of the afterlife where religiously or morally incorrect behavior of the living is punished. These ancient myths were incorporated into the Judeo-Christian tenets of Hell, Purgatory, and Heaven. A similar concept of a chance for redemption in heaven or damnation in the hell of the underworld is expressed in the Muslim Koran. Many other civilizations express the same myths about life and death and a special place for birth, virtue, and damnation located inside the Earth. For instance, Native American Hopi believe that Spider Woman controls the Earth below and uses the inner Earth as a womb.[35]

Another, more recent set of myths and speculations is related to the inner spheres of the Earth. These are all known as "hollow Earth" theories. A summary of just a few is given in the following list (see Figure 5.2 for an artist's conception of a typical hollow Earth).

1. In 1692 the English astronomer Edmund Halley (1656–1742), of Halley's comet fame, developed his theory of a hollow Earth by using mathematical calculations as related to slight variations of the Earth's magnetic field. He asserted that these variations were several separate fields, not just one that changes over time, and thus the

Earth had to be hollow with another sphere inside it in order to account for his calculations. Halley finalized his concept as four concentric spheres nestled inside one another, and he went so far as to insist that the inner Earth was illuminated by a luminous atmosphere and populated by people. He believed the northern lights (aurora borealis) were caused by luminescent gas escaping at the North Pole.

2. A Swiss mathematician, Leonhard Euler (1707–1783), revised Halley's model of the hollow Earth. He constructed a model containing multiple internal spheres surrounded by one large single sphere with a 600-mi diameter sun that provided the needed heat and light for the civilizations living in the center of the Earth.
3. In the eighteenth century, John Leslie revised Euler's hollow Earth model to include two suns.

Figure 5.2
An artist's combination of several proposed historical and mythical models for a hollow earth shows the polar openings to the interior land and sun.

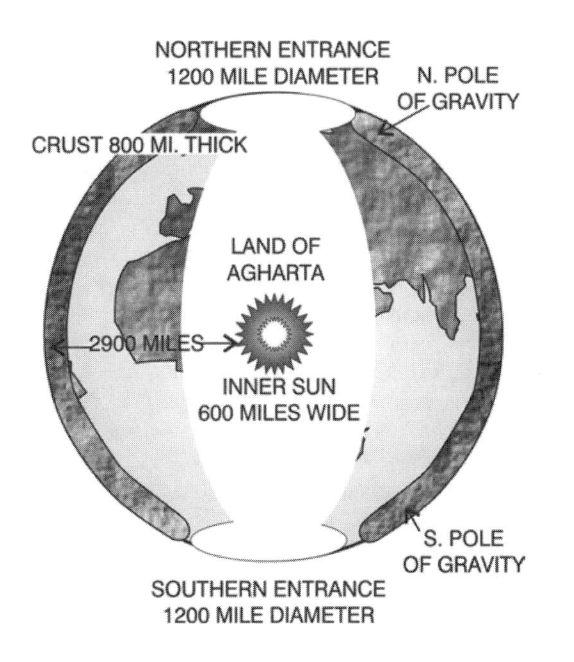

4. John Symmes (1742–1814), an American, alleged that the hollow Earth had a 4,000-mi-wide entrance (later revised to a 1,400-mi entrance by Raymond W. Bernard) at the North Pole and a 6,000-mi opening (also revised to 1,400 mi) at the South Pole. With the assistance of a newspaper editor, Symmes convinced the U.S. government to send an expedition to Antarctica to locate the hypothetical opening. They did not find an opening but did discover that Antarctica, unlike the North Pole, is a continent.
5. In 1846, Marshall B. Gardner claimed that the discovery of an extinct, frozen woolly mammoth was evidence that the Earth was hollow. He speculated that mammoths and other animals wandered in and out of the hole at the North Pole.
6. Late in the nineteenth century, Cyrus Read Teed (1839–1908), who later changed his name to Koresh and founded a cult in Florida (no connection with the David Koresh cult of Waco, Texas), proposed that people lived inside the hollow Earth where a central sun, which was half light and half dark, gave the appearance of sunrise and sunset.
7. In 1914, Hans Horbiger (1860–?) developed his theory of a hollow Earth from his dreams and his work with more than 500 ancient myths and legends about people living in the center of the Earth. He developed the cult called "WEL," which currently has an estimated 1 million followers in Germany and England.[36]
8. During World War II, Adolf Hitler (1889–1945) also believed in the hollow Earth theory. He sent an expedition to a Baltic island to locate an entrance into the Earth to find a way to photograph the British fleet across the hollow, concaved surface of the sphere. Claims have been made that at the conclusion of World War II some Nazis escaped into the hollow Earth.
9. One of the most astounding tales of a hollow Earth was reported in the diary of Admiral Richard Byrd (1888–1957), who flew 1,700 mi beyond the North Pole in

1947 where he claimed he saw a land of mountains, green vegetation, rivers, underbrush, and mammoths. In 1956, he wrote that he had penetrated 2,300 mi into an opening beneath the South Pole leading inside the Earth where he met advanced beings. His written records and personal diary of the Antarctic trip claim that he not only penetrated the Earth but met and communicated with the Master of Arianni, the domain of the inner world of the Earth, who warned him about the fate of humans on the surface of the Earth. His diary makes interesting, but unscientific, reading. It seems that Admiral Byrd was a strong critic of the atomic bomb and nuclear energy and may have been using the tales in his diary to convince others of what he believed would happen on the surface world in the future.[37–38]

The hollow Earth tales are similar to UFO stories and the Flat Earth Society's nonsensical claim that the Earth is not a spheroid. These are all obvious pseudoscience, not based on verifiable facts, and thus should not be advanced as having any scientific foundation. These myths have no place in increasing our understanding and factual knowledge of the Earth's geology.

Notes

1. *Structure of the Geosphere:* Mantle, accessed 2001, http://www.kheper.aus.com/aaia/geosphere/structure.html.
2. Kenneth W. Hamblin and Eric H. Christiansen, *Earth's Dynamic Systems* (Upper Saddle River, N.J.: Prentice-Hall, 1998), 13.
3. *The Oxford English Dictionary,* 2d ed., CD-ROM, s.v. "Mantle."
4. Stanley Chernicoff, *Geology: An Introduction to Physical Geology* (New York: Houghton Mifflin, 1999), 12.
5. Chernicoff, *Geology,* 41.
6. *Characterization of the Lithosphere-Asthenosphere Boundary,* accessed 2001, http://www.bham.ac.uk/IUGG99/jsa48.html.
7. *Earth's Journeys—Peel the Planet [Mantle],* accessed 2001, http://www.discovery.com/exp/earthjourneys/peel_mantle.html.
8. Hamblin and Christiansen, *Earth's Dynamic Systems,* 12–13.
9. Chernicoff, *Geology,* 309.
10. *Encyclopedia Britannica,* CD-ROM 2001, s.v. "Chemical Element: The Earth's Mantle."
11. Chernicoff, *Geology,* 310.
12. Motohiko Murakami et al., "Water in Earth's Lower Mantle," *Science,* 8 March 2002, 1885.
13. Edward J. Denecke Jr., *Let's Review: Earth Science* (Hauppauge, N.Y.: Barron's Educational Series, 1995), 99.
14. Robert L. Bates and Julia A. Jackson, *Dictionary of Geological Terms* (New York: Anchor Books/Doubleday, 1984), 110.
15. *Encyclopedia Britannica,* CD-ROM 2001, s.v. "Chemical Elements: Terrestrial Distribution."
16. Chernicoff, *Geology,* 310–11.
17. Ultralow Velocity Zones (ULVZ), *Investigating Earth's Lower Mantle with Seismology* (Berkeley Seismological Lab), accessed 2001, http://www.geo.berkeley.edu/seismo/annual-report/ar97.
18. Richard Monastersky, "The Mush Zone: A Slurpy Layer Lurks Deep Inside the Planet," *Science News,* accessed 2001, http://www.findarticles.com.
19. Chernicoff, *Geology,* 312.
20. Graham R. Thompson and Jonathan Turk, *Earth Science and the Environment* (New York: Harcourt Brace, 1999), 126.
21. Thompson and Turk, *Earth Science,* 126–27.
22. *Encyclopedia Britannica,* CD-ROM 2001, s.v. "Chemical Element: The Earth's Core."
23. Hamblin and Christiansen, *Earth's Dynamic Systems,* 13.
24. Thompson and Turk, *Earth Science,* 126.
25. Michael Carlowicz, *Inge Lehmann Biography,* accessed 2001, http://www.agu.org/inside/awards/lehmann2.
26. Andrew Alden, "A Guided Tour to Today's Mantle," About Geology, accessed 2001, http://www.geology.about.com/science/geology.libr.
27. Chernicoff, *Geology,* 114.
28. Ronald Cohen and Lars Stixrude, *Crystals at the Center of the Earth: Anisotropy of the Earth's Inner Core,* accessed 2001, http://www.psc.edu/science/Cohen_Stix/cohen.stix.html.
29. John Daintith, Sarah Mitchell, Elizabeth Tootill, and Derek Gjertsen, *Biographical Encyclopedia of Scientists,* vol. 1 (Bristol, England, and Philadel-

phia, Pa.: Institute of Physics Publishing, 1994), 346–47.

30. Chernicoff, *Geology,* 322–26.

31. Thompson and Turk, *Earth Science,* 127.

32. "Earth's Magnetism," *Scientific American:* Ask the Experts: Geology, accessed 2001, http://www.sciam.com/askexpert/geology/geology4.html.

33. *Core Spins Faster Than Earth, Lamont Scientists Find,* Lamont-Doherty Earth Observatory: Columbia University, accessed 2001, http://www.columbia.edu/cu/1996/0830/d.html.

34. *Science News,* accessed 2001, http://www.findarticles.com/m1200/20_15658037882.jhtml.

35. David Adams Leeming, *The World of Myth* (New York: Oxford University Press, 1990).

36. Robert E. Krebs, *Scientific Development and Misconceptions through the Ages* (Westport, Conn.: Greenwood Press, 1999), 83.

37. *The Museum—The Hollow Earth,* accessed 2001, http://www.unmuseum.mus.pa.us/hollow.htm.

38. *Hollow Earth: Agharta, The Land of Advanced Races,* accessed 2001, http/www.2eu.spiritweb.org/Spirit/Hollow-earth.html.

6

Biosphere: Envelope of Life

The biosphere may be defined as the sphere of the Earth where life exists. The main layers of the Earth that support life are the upper part of the crust with its thin layer of soil, the hydrosphere, and the lower atmosphere. Some forms of simple life have been found in extreme depths of the lithosphere, on the ocean bottoms, and at great heights of the atmosphere. However, most examples of major life forms and ecosystems are found in the upper crust (soil), near surface water, and the lower atmosphere (see Figure 6.1).

History and Background

In the past, geologists considered the Earth to be composed of two major classes of structures. The first class consisted of the major concentric regions called "concenters," while the second class of structures were subdivisions of the concenters, referred to as "envelopes." These early geologists considered the three great concenters to be the *core,* the *sima* region (lower crust of silica rock), and the upper *crust* (soil). At one time, it was believed that the three concenters did not mix with each other because each (sphere) was thought to be an independent and isolated mechanical (physical) system. Some geologists also considered the envelopes, or subdivisions of the concenters, to be a series of geospheres.[1] Geologists no longer accept these concepts. Today, they usually refer to the geosphere as all of the concentric spheres/layers that comprise the entire Earth.[2]

The accepted definition of the biosphere is "The part of the Earth and its atmosphere that is inhabited by living organisms."[3] It includes parts of the crust/lithosphere, hydrosphere, and atmosphere. The biosphere may also be referred to as the "ecosphere," which is defined as those portions of the Earth favorable for the existence of living organisms.[4]

The concept of the *biosphere* was first introduced in 1875 by Eduard Suess (1831–1914), a professor at Vienna University in Austria. He also coined the term "biosphere." Suess considered his biosphere to be a "life-saturated envelope" of the Earth's crust. His concept is still accepted, but several other, more comprehensive concepts of the biosphere have since been developed. One deals with the *physics* of living matter, another with the *organic* or living nature of the biosphere itself. Let's take a look at dif-

Figure 6.1
The biosphere consists of all the layers of the Earth where life resides. The Gaia (from the Greek, "Earth Mother") concept was developed as a hypothesis that proposes that the biosphere is in itself an organism, mutually beneficial and necessary not only to living organisms but also to the nonliving components of the Earth's crust, soil, water, and atmosphere. It is estimated that somewhere between 3 and 30 million species of plants and animals live in the biosphere, while only about 1.5 million have been identified. A symbiotic relationship exists between these multitudes of life and their natural environments that make up the biosphere.

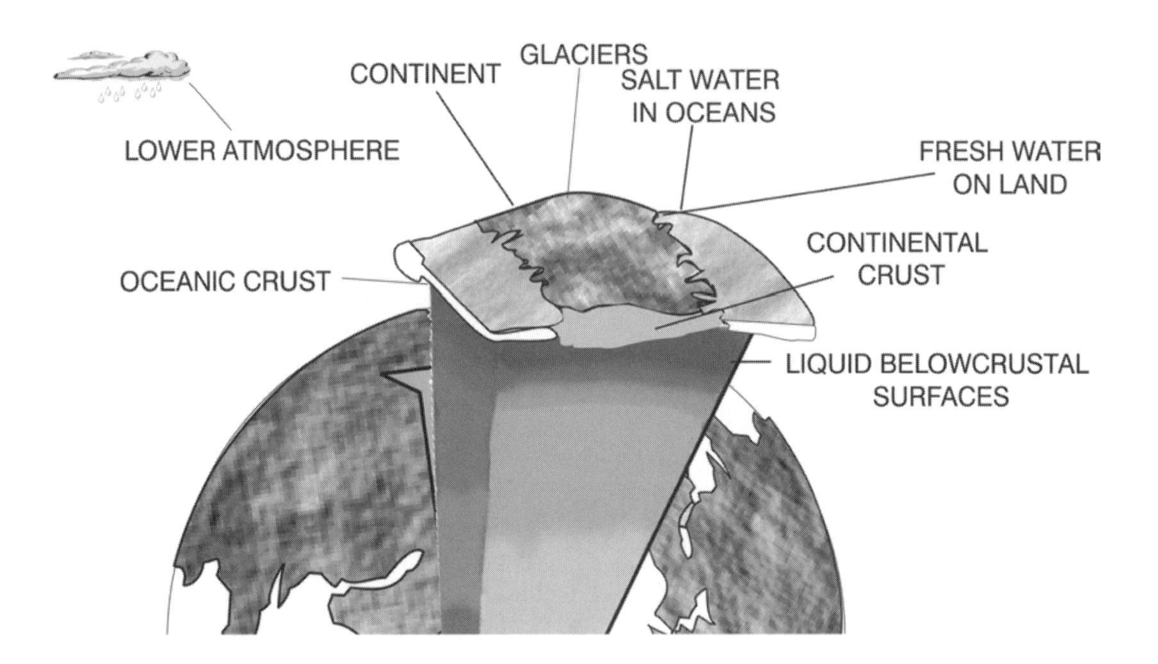

ferent concepts of the relationships of life to the geology of the Earth.

Edward O. Wilson, in his book *The Future of Life,* presents the biosphere from the points of view of different philosophies, from the mythical to the scientific.

1. The *poet* looks at the totality of life. Who are Gaia's children?
2. The *ecologist* explains that they are the many species of living organisms. And we must know the roles they play so that we might wisely manage the Earth's biosphere.
3. The *systematist* [classifier] might add, let's find out how many species exist in the world and who are their distinctive genetic relatives.[5]

These three levels of explanation of the biosphere are represented by several theories. First, the mythical concept of Gaia, if accepted as a metaphor, has some truth to it in the sense that some species exercise wide influences at the regional, as well as global, level. In contrast, however, it is based on a myth, a beguiling one to be sure, but a mythical theory not accepted by most scientists.

Gaia

James Lovelock (1919–) is generally given credit—as well as being criticized—for the concept of a living biosphere, a region where the Earth (geology) and life are intricately related as a single giant organism. But even he rejects the concept of the biosphere as a *superorganism* composed of many species that cooperate and balance the entire environmental system, similar to a colony of ants. At the same time, Lovelock and others do accept what might be called a weak version of Gaia, in which living species do exert some influence on the Earth, at least in the region where they exist, that is, the biosphere. Lovelock called this the Gaia theory because it can be used to make testable (and falsifiable) predictions related to biodiversity based on both ecosystems and the genetics of individuals and species.[6]

Lovelock's friend, the novelist William Golding, suggested that "Gaia" would be an appropriate term to describe the living nature of the Earth, since Gaia is the name the Greeks used for their Earth goddess.[7] The original Gaia hypothesis incorrectly postulated that life regulates and makes the Earth (geology) comfortable for itself. Lovelock corrected this hypothesis and restated it: the regulation that makes life possible is a property of the whole evolving system of life, air, ocean, and rocks.[8]

Biodiversity

No mater where life is found, or in what amounts, it is divided into three levels.

1. The main level of biodiversity is the *ecosystem,* consisting of general habitats such as rain and deciduous forests, oceans and coral reefs, ponds and lakes, valleys and mountains, and so forth.
2. The next level is the *species,* which consists of all the organisms that make up the ecosystem. Species range from the simple viruses, bacteria, and worms to higher plants and animals.
3. The next lower level of biodiversity for the division of life is *genetic.*[9]

The great variety of genes and their DNA are what make individuals, as well as species, unique "even though ALL living individuals and species share a portion of common genetic history because all organisms descended from a common ancestor (~3.5 billion years ago). [T]he rest of life is the body; we are the mind."[10] Genetically, humans are more like chimpanzees than chimpanzees are like other primates. We share ~98.7% of the same genes with chimpanzees. The main differences are brain size and the unique way human genes interact. In contrast, humans share only a small percent of the same DNA with fruit flies and bacteria.

The biosphere is further defined as the organismal self-maintaining environment that has a negative feedback regulatory system extending within 20 kilometers (km) (12.5 miles [mi]), or more, of Earth's surface (crust/soil, hydrosphere, troposphere) and comprises more than 30 million species.[11] Even so, life is found in the biosphere that consists of the Earth's outer membrane as a living, self-sustaining organism where the individual elements coexist in a **symbiotic** relationship with the physical envelopes.[12] At the same time, it is important to remember that more than 99.999% of the *total* Earth has no contact with living organisms nor does it provide a habitable environment for any form of life. Yet, a variety of life forms are found deep in the oceans, where some thrive in proximity to hot gas and sulfur vents on the seabed. Life is abundant near the surface of the ocean, particularly near coastal areas and to a lesser degree at great depths. Life is also abundant at various depths of land-based freshwater sources and in the Earth's crust, including the soil and upper rock layers of the lithosphere. Geologists have found as many as 10 million bacteria in each gram of rock at depths exceeding 2.8 km (1.7 mi) where temperatures exceed 75°C (167°F). They conclude that the limiting factor of how deep in the crust life can survive is not the increase in pressure but instead the increase in temperature. Bacteria are capable of growing in deep-sea volcanic vents where the temperature is 110°C (200°F) but can only survive for short periods at 140°C (285°F).[13] Life is found on the highest mountains/deepest oceans, the coldest/hottest, wettest/driest landmasses, as well as in the troposphere. The biological aspects of life are partially responsible for shaping the geology of the outer spheres. (All living things alter their environments, both physical and biological, just by living and dying.) Therefore, not only do physical forces shape Earth's geology, but biological forces also play an important role in the ecosystems of the Earth.

The Physical Biosphere

The Russian mineralogist and biologist Vladimir Ivanovich Vernadsky (1863–1945) first published his book *The Biosphere* in Russian in 1926. It was later translated into French and English. His view of life was as a cosmic phenomenon that followed universal scientific laws. In other words, life abided by the same physical laws as did all other matter in the universe. His three basic observations about the *physical* nature of the biosphere were as follows:

1. Life occurs on a spherical planet. (Vernadsky was one of the first to consider Earth as a self-contained sphere.)
2. Life makes geology. Life is not merely a geological force; it is *the* geological force. Virtually all geological features at Earth's surface are bio-influenced and are thus part of Vernadsky's biosphere.
3. The planetary influence of living matter becomes more extensive with time. The number and rate of chemical elements transformed and the spectrum of chemical reactions engendered by living matter are increasing, so that more parts of Earth are incorporated in the biosphere.[14]

Vernadsky spent his life investigating and writing about the physical and geological aspects of the mechanisms that form the "envelope of life" (biosphere). He related the Earth's position in the cosmos to the biosphere's geochemistry, its transformation of radiation energy (both cosmic and solar), its thermodynamics, and other mechanisms to conclude that "Life on Earth is, therefore, not an external or accidental phenomenon of the Earth's crust."[15] He was a firm believer that life is bound to the structure of the crust and is part of its mechanisms (i.e., physics), which assumes that all living matter is a single entity in the biosphere.

Although Vernadsky's theory of the biosphere was related to the physical geology of the Earth, it is not well accepted by present-day geologists, biologists, and other scientists because of its Gaia aspects. With some caveats and refinements, his theory of a physical/mechanical biosphere is accepted by many geologists and biologists.

Let's take a look at one of the most important outer layers of the Earth's crust.

Soils (Pedology)

Soils are composed of a great variety of minerals, chemicals, and organic matter that provide it with the ability to support plant growth. Furthermore, soil (sometimes incorrectly referred to as "dirt") is much thinner than other layers of the Earth. Even so, it is one of the major envelopes of the biosphere and is vital to living organisms. It is a mixture of weathered-eroded rocks, minerals, chemicals, air and other gases, water, and bits of dead plants and animals, and it is home to a multitude of living organisms.[16] In addition to having biological, chemical, and physical properties, soil is always changing. It is dynamic.[17]

In chapter 4 we discussed erosion as a form of weathering, including the transportation and deposition of eroded minerals and rocks. While soil is a mixture of weathered rock and organic remains of plants and animals usually found above bedrock, the type of soil is determined by the kinds of rocks from which it was formed by weathering processes that are affected by local climatic conditions. Now let's examine the physical and chemical nature of weathering that produces the basic raw material for soils.

Weathering

Soil is the product of the weathering of bedrock that breaks up at its surface into small fragments of rock, sand, and clay referred to as "regolith" by geologists. Soil composes the upper layers of regolith, but soil is much more than just fragmented rocks. It has organic matter distributed throughout and can support plant growth. Mass weathering of rocks and minerals occurs as a series of processes that are partly responsible for forming soils.

The two general types of weathering of rocks and minerals involve *physical* and *chemical* processes, both of which are an integral part of the rock cycle discussed in chapter 4.

Physical Weathering

A number of physical processes are responsible for mechanically breaking down rocks into smaller fragments without changing the chemical composition of the material that is fragmented. The major mechanical processes in the physical breakdown of rocks are given in the following list.

1. *Abrasion* is the most powerful mechanical weathering processes. Rocks are abraded by running water, waves, and wind. Larger particles are abraded and broken down into finer particles as they tumble over each other in streams, rivers, and beaches and as they scrape over each other on surface rocks of the landscape. One form of physical abrasion occurs as shoreline waves of lakes and oceans batter coastlines. The wave action, plus the sand and small rocks carried by the waves, erodes and undercuts shoreline cliffs, causing them to collapse into the surf. Two examples of this type of mechanical erosion, exacerbated by storms and strong winds, are the steady undercutting of the south shore cliffs of Lake Michigan in Illinois and Indiana and of the Pacific Ocean coastline of California.
2. *Gravity* is also responsible for a great deal of physical weathering. The pull of gravity causes the collapse of cliffs and overhanging rock formations. Gravity also dislodges rocks on mountains, causing their descent to lower levels. As the larger pieces tumble to lower levels, they are ground up into smaller pieces. In addition to gravity, there are several factors involved in mass mechanical weathering: (a) the steepness of the slope of the side of the mountain or cliff; (b) the type of rock; (c) the orientation of the rock layers in the mountains; (d) the angle of orientation of the rock layers at the bottom of the slope (talus); (e) the climate of the region (amount of snow, water, and vegetation); and (f) the frequency of earthquakes.
3. *Ice wedging* occurs when rain or melting snow collects in cracks and crevices in rocks and then freezes. (Crystal formation of all kinds exerts great per unit force that plays a major role in geology.) It is a well-known principle that when water freezes, it expands. Thus, freezing water in the cracks

of rocks acts like a wedge driven into a crack; as repeated freezing occurs, the cracks become larger, until the rocks are fractured into blocks and smaller pieces. The crystallization force of freezing water is very powerful. For example, freezing water creates a stress of 110 kg/cm^2, which is equal to hitting a wedge into a rock with a large sledgehammer. Ice wedging from seasonal freezing happens only when there is water present.

When high mountains are eroded by ice wedging, gravity pulls large chunks of rocks down the sides of mountains where they accumulate in distinctive triangular piles. Both gravity and moving water continue to move the broken rocks into ravines formed at the base of mountains and cliffs. These piles of blocks of rock and weathered debris tend to form cone-shaped deposits known as "talus cones." Talus cones are also formed when chunks of rock loosened by earthquakes accumulate at the base of steep slopes.

Another form of minor wedging sometimes occurs along the shorelines of bodies of saltwater where salt spray may accumulate in the cracks and crevasses of rocks. As the water evaporates, salt crystals grow. The slow but constant growth of salt crystals creates adequate force to split shoreline rocks. This type of weathering of rocks is called "salt cracking."

4. Two other factors related to mechanical erosion are *flows* and *slides*. During a flow, unconsolidated rock, sand, or soil can move down a slope slowly or rapidly, depending on the amount of water in the basic sand or soil. A slide, in contrast, is a coherent mass movement of material that occurs when either a fracture develops in a block of rock or unconsolidated material breaks loose and slides down an incline. Slides are usually more rapid than are flows as the fractured block descends to a lower level.

   Several other types of mass physical erosion of rocks, sand, or soil are *creep* (slow flow), *mudflow* (when soil is soaked and becomes a fluidlike mass of sediments), *slump* (the slide follows a concave surface), and *fall* (particles proceed down an almost vertical steep cliff).

5. *Sheeting* occurs when surface material is removed from bedrock, either by natural erosion or by human intervention, which allows the exposed rock below to crack into parallel sheets. Rocks buried deep in the Earth are under high pressure due to the load imposed by overlying material. When the surface material is removed, the pressure in the buried rocks is removed. This lowering of pressure causes rocks to expand, resulting in fractures that split the rocks horizontally. It has been reported that when stone is extracted in quarries, the pressure of underlying rock layers is sometimes released with such force that the rock explodes upward, creating new fractures in the rock below the bottom of the quarry.

6. Minor physical weathering processes include erosion of rocks caused by the expansion of plant roots, burrowing of animals, seasonal weather changes, and abrasion as wind and water carry particles that wear down rocks and round off the sharp edges of the smaller fragments and particles.[18–19]

As part of the rock cycle, weathered rock materials can form sediments that accumulate to great thickness. As they are buried, the weight of the overlying material compresses the particles while minerals form precipitates in the tiny spaces between the particles, cementing them together to form sedimentary rocks. If there is sufficient heat and additional pressure, these rocks can metamorphose into igneous rocks. These former sediments, now rocks, again reach the surface by mantle convection cells and the Earth's plate tectonic system (see the sections on plate tectonics in chapter 2). These recycled surface rocks are again weathered and fragmented, forming new sediments, repeating the dynamics of the rock cycle. Fragments from mass physical weathering and mechanical abrasion are of the same chemical and mineral composition as the original rock or mineral, although these fragments may be mixed with many other kinds of substances in the process of weathering.[20]

Chemical Weathering

Unlike the physical weathering of rocks, chemical weathering actually changes the chemical composition of minerals and rocks. These reactions are between the chemicals that make up the minerals/rocks and the natural chemicals in the crust, hydrosphere, and atmosphere.

The three basic types of chemical weathering are dissolution, hydration, and oxidation.

1. *Dissolution* takes place when the rock material is dissolved and becomes a solution. Chemists consider water to be a universal solvent, as it has proven to be a slow but sure way to dissolve most rocks over time. Water is a *polar* molecule, meaning that it has a slight negative charge on one side of its surface and an equal but opposite positive charge on the other side. These charges react with ions on mineral surfaces. Thus, the water molecule pulls off the ions from the mineral compounds to form solutions. For instance, when carbon dioxide is dissolved in water, a weak acid is formed that reacts with carbonate rocks such as limestone. The following chemical reactions occur:

$H_2O + CO_2$ □ H_2CO_3 □ $H^+ + HCO_3^-$
(water) (carbon dioxide) (carbonic acid) (hydrogen ion) (bicarbonate ion)

The reaction continues:

$CaCo_3 + H^+$ □ $Ca^{++} + HCO_3^-$
(calcium rock) (hydrogen ion) (calcium ion) (bicarbonate ion)

The calcium rock is finally dissolved by the carbonic acid's H^+ ion.

2. *Hydration* is a chemical reaction that occurs when water molecules attach to a mineral. This is not the same as dissolution, where one substance dissolves in water. Instead hydration occurs when water actually chemically combines with another substance. When water breaks into hydrogen and oxygen atoms, it forms as ions, that is, charged particles: H_3O^+ and OH^-. These ions chemically combine with silicate minerals in the crust to form new material. An example of chemical weathering is when the Na^+ (sodium ion) in one type of feldspar is replaced by the H_3O^+ ion of water to form clay minerals. Sooner or later the Na (sodium) is **leached** into the oceans, leaving the new weathered rock particles and soils deficient in sodium. At the same time, this process supplies the Na^+ ion that combines with the Cl^- ion to form the salt (NaCl) that is evident when seawater evaporates. A similar hydration process occurs with other types of minerals/rocks that contain K^+ (potassium ions), Ca^{++} (calcium ions), and ions of many other elements.
3. *Oxidation* takes place when the oxygen in the atmosphere or oxygen dissolved in water reacts with a mineral to form a new chemical compound. Oxidation is similar to the rusting of iron or burning of fuel where new compounds are formed. Minerals that have a high iron content, such as olivine, pyroxene, and amphibole, contain one or both iron ions (Fe^{++}, ferrous; or Fe^{+++}, ferric), which readily combine with atmospheric oxygen as follows:[21–22]

$2Fe_2SiO_4 + 4H_2O + O_2$ □ $2Fe_2O_3 + 2H_4SiO_4$
(olivine: mineral) (water) (oxygen) (ferric oxide: hematite) (silicic acid)

There are many other examples of chemical weathering of minerals and rocks due to oxidation, hydration, and dissolution. For instance, some plants secrete substances that both chemically and physically weather rocks. Example are lichens, which are symbiotic associations between a fungus and a photosynthetic green algae that colonize areas that are inhospitable to other plants, such as the surface of rocks or tree trunks.[23] Lichens break down the surface of rocks to extract nutrients that, over long periods of time, have a weathering effect on the rocks. Lichens are a mutualistic association between fungi and algae, two plant species that live as a single symbiotic organism. The algae make the food required by both organisms by photosynthesis, while the fungi provide protection, moisture, and minerals salts required by the algae. The lichen, as a single unit, makes acids that neither can produce alone and that weather the surfaces of rocks (and other surfaces they inhabit), thus aiding in soil formation.[24]

Products of Weathering

The end results of mass mechanical, physical, and chemical weathering are a great variety of soils. The stability of the mineral crystals that form soils is dependent on the temperatures and pressures under which the minerals were crystallized from melted silicates. Minerals that crystallize under higher temperatures (e.g., nonsilicate minerals, such as olivine, hornblende, and calcic plagioclase) are *less stable* and more susceptible to weathering. Conversely, silicate minerals that crystallized under lower temperatures (e.g., micas, feldspars, and quartz) are *more stable* and resist chemical weathering. This factor determines that the most stable mineral crystals are also the most abundant found in the Earth's crust, for example, quartz. Table 6.1 lists several minerals found in soils that range from the least to the most stable as far as weathering is concerned.

The strength of the metallic-oxygen bond for different elements determines how stable or unstable a mineral may be. This factor also determines which minerals are the most abundant in the Earth's crust (e.g., silicon, titanium, iron, and aluminum) and are thus less mobile throughout soils. At the same time, those minerals (e.g., manganese, sodium, magnesium, potassium and calcium) that are more easily broken down by chemical weathering are more mobile throughout the environment. Sulfur and carbon are also soon removed from surface rocks by chemical weathering. In other words, the least stable minerals are the most mobile. This matters because the more mobile, less stable elements are the most important for supporting life.[25] These more environmentally mobile elements required for life are constantly recycled between and among the three major spheres of the biosphere—the crust (soil), hydrosphere, and atmosphere. This an example of the dynamic relationship between geology and biology.

Table 6.1
Mineral Crystal Stability

Olivine	Calcic Plagioclase
Augite	Intermediate Plagioclase
Hornblende	Alakalic Plagioclase
Biotite	
Potassium Feldspar	
Muscovite	
Quartz	

Source: W. G. Ernst, ed., *Earth Systems: Processes and Issues* (New York: Cambridge University Press, 2000), 120.
Note: Minerals are listed from lesser to greater stability.

The products of physical weathering are particles formed when different types of rocks fracture and break up into smaller pieces and are subsequently abraded. The fractures may be along a plane of weakness, or they may be shattered or disintegrated to form angular rock fragments that are further weathered into globular shapes of finer particles. They may be separated into shells or sheetings similar to the layers of an onion. This is referred to as **exfoliation** and is responsible for the rounded shapes of large granite boulders or mountains.

The fragmentations that form sediments and soils vary greatly in size. Therefore, these fragments may be named and classified according to size (see Table 6.2).

Hydrolysis

Hydrolysis occurs when water reacts with minerals and when the water molecules become part of the new crystal's structure. The most important mineral that undergoes weathering by the process of hydrolysis is feldspar. It is the most abundant mineral in the Earth's crust that forms clay, an important ingredient of most soils. A number of different types of clays are formed as sheets of very fine grained silicates such as micas (less than ~2 micrometers, or 2 thousandths of a millimeter). Clays are structured in sheets that determine the types of chemicals involved in the original weathering and, thus, the nature of different types of soils. The following is the chemical sequence for feldspar's hydrolyzation to clay:

$$2KALSi_3O_8 + 2H^+ + H_2O \; \square \; AL_2Si_2O_5(OH)_4 + 2K^+ + 4SiO_2$$

(feldspar) (hydrogen ion) (water)
(clay mineral) (potassium ion) (silica)

Table 6.2
Products of Weathering

Name of Fragment	Diameter of Particle (cm)
Clay	0.00001–0.0004
Silt	0.0004–0.006
Sand	0.006–0.2
Pebbles	0.2–6.4
Cobbles	0.4–25.6
Boulders	greater than 25.6

Source: Edward J. Denecke Jr., *Let's Review: Earth Science* (New York: Barron's Educational Series, 1995), 166.

Regolith

As bedrock slowly weathers into smaller and smaller fragments and decomposes into sand or clay, the resulting thin layers of loose, broken-down chunks and particles of rocks form a material called regolith. The word "regolith" is derived from the Greek words *rhegos* (blanket) and *lithos* (stone), that is, "rock blanket." Regolith is defined as

> The fragmented and unconsolidated rock material, whether residual or transported, that nearly everywhere forms the surface of the land and overlies the bedrock. It includes debris of all kinds.[26]

The fragmented rock particles resulting from physical and chemical weathering cover most of the Earth's land surface. When rock particles such as gravel, sand, and silt are produced by weathering processes, they are referred to as "residual regolith." If the gravel, sand, and silt are deposited by streams, wind, or glaciers, they are referred to as "transported regolith." When the top few meters of regolith contain a mixture of fine particles of minerals and organic matter (humus) it is referred to as "soil." Transported regolith is responsible for much of the dynamic nature of soils that are constantly being mixed and moved by the forces of nature.[27–28]

Soil Formation

A number of basic factors are involved in soil formation. Some of these are (1) the nature of the parent rock; (2) climate, both temperature and precipitation; (3) rate of vegetative organic growth and decay; (4) time; and (5) topography of the land surface (steep cliffs have little topsoil). These factors are responsible for a great variety of soil types. Because soils have many overlapping characteristics, this creates a problem in developing theories for soil formation and/or viable classifications for soils. To overcome this, geologists have developed a system that is based on the measurement of the different properties of soils to describe their characteristics.

In 1960, the Seventh American Approximation System was introduced to provide more precise definitions of soil groups and general **pedology** terms. Using this system, soils are grouped according to observable inherent factors. Some of these identifying factors, as well as many technical pedological terms, are their horizons or layered structure; their source of parent rock material, for example, silica content; the climate of their geographic regions; their ratio of clays to humus; the degree of retention of chemical elements compared to those dissolved out in solution (percolation and leaching); and their color and texture.[29]

Trying to determine "how deep is soil?" is somewhat like asking "how deep is snow?" It all depends on what, where, and when the measurements are taken. The depth of soil is determined by the various layers of soils that are being measured, the climate (and weather) of the geographical location of the soil, and the period of geological history during which measurements are made. Soil can vary in depth from zero to whatever the depth of the underlying bedrock happens to be. One example is in the state of Wisconsin where bedrock is covered with gravel, sand, silt, and clay up to 120 meters (400 feet), but only a few feet of these deposits are considered topsoil. If topsoil, which is the layer of humus composed of rich organic matter, is considered, the depths vary greatly according to climate and geography, as well as other factors, including the types of plant and animal life in the topsoil. Additional factors are the effects of weathering, erosion, transport, and deposition of the deconstructed rock materials that form soils. Time is a major factor in determining the type, as well as the maturity, of soils. Older soils are considered mature if they have developed profiles exhibiting distinct horizons.

Soil Profiles

The nature of the parent rock material that created the regolith fragments is mainly responsible for the origin, as well as the nature, of soil. In other words, the parent material is the original source of minerals and rocks that make up soil, which is basically formed from the eroded and unconsolidated sedimentary rocks and clay materials. In addition, soil has been altered by a great variety of processes, but, time-wise, soil can be altered in just a few decades rather than the hundreds of thousands or millions of years required for rock formation and other geological events. The age of soil is measured by determining when the rock was first exposed to the surface, when it eroded, and when the alluvial and humus material was deposited. Young soils usually have a more undifferentiated profile than do older accumulations of clay deposits. They are also less acidic than older, deeper soils, which exhibit a higher acid content (lower pH value). Therefore, soils of different ages exhibit different stratifications and thus have different horizontal profiles.

Horizontal Nomenclature for Soils

Not all geologists use the exact same nomenclature or definition when describing the profiles of soils, but they all agree that great variety exists in the physical structure and chemical composition of the different horizontal layers of soil (see Figure 6.2). Three distinct layers, or horizons, are visible for soils located above bedrock in the temperate zones. They can be distinguished by their colors and textures.

The thin top layer is often referred to as the "O" (organic) layer. This is the thin surface layer where most of the organic matter that supports plant life is located. The next main layer is referred to as the "A" layer, or "leached horizon," since it is a mixture of organic humus, sand, silt, and clay. (Sometimes this horizon is divided into three sublayers: the "A_o" organic leaf-mold layer; the "A_1" or "A_h" dark, humus-rich layer; and the "A_2" lighter colored layer.) Together the "O" and "A" layers make up what we call "topsoil," which is the most fertile region of soil that supports plant growth. The next layer below the "A" layer is called the "B horizon," which is lighter or reddish colored subsoil. The characteristics of the "B" horizon depend on the types of chemicals dissolved in the water traveling downward from the topsoil. The "B" horizon soils are composed of fine clays and **colloidal** particles leached from the topsoil. The "C" horizon is a zone of broken, decomposed, and partially weathered bedrock, which overlies the unaltered bedrock, which is called the "R" horizon. As rain drips through the "O" and "A" horizons of topsoil, dissolved minerals, chemicals, and ions are carried to lower layers. This process is referred to as "leaching," which is the main reason soil horizons change from brown, to gray, to lighter colors including red.[30–33]

Figure 6.2
A typical soil profile exhibits several layers (horizons). Just above the "A" horizon of topsoil there is an "O" layer that is a thin deposit of litter consisting of humus and a few minerals. The "A" horizon is the fertile topsoil rich in organic matter and minerals. Typical topsoil is teeming with trillions of bacteria, fungi, algae, protozoa, worms, insects, and mites. The "B" horizon is a transition zone that consists mostly of clays and some organic matter, including roots, but less animal life. The "C" horizon is weathered rock that ranges from small to larger chunks. Some deeper roots penetrate into the smaller rocks at the top of the "C" layer. The "D" horizon is bedrock. It may be exposed if the top layers of soil are eroded away.

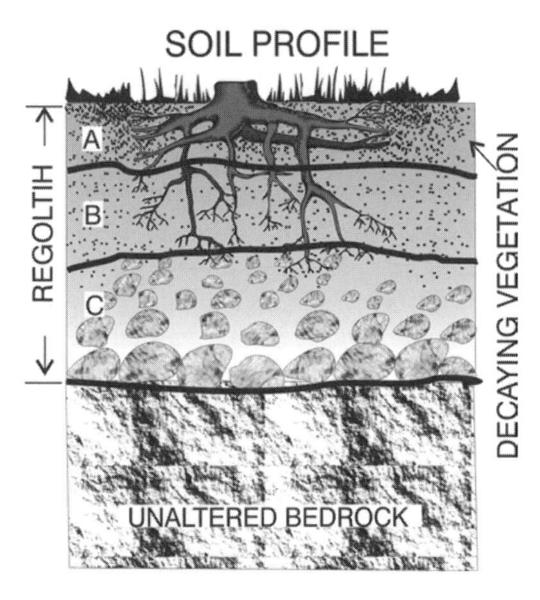

Soils are also classified according to somewhat subjective observations. Some examples are (1) *podsoils*—a horizon of ashy colored soils on top of darker richer soils; (2) *chemozems*—black soils, rich in organic mater with high calcium content; and (3) *krasnozems*—reddish colored soils with an undifferentiated horizon. In addition, soils can be classified according to their acid-base content as *acid, neutral, nitric,*

magnesic, or *calcic.* Or they can be classified according to their clay content, as *allophanic, illitic, kaolintic,* or *superkaolintic.*[34] Some other names used to identify soil types are given in the following list.

- *Pedalfer* soils are found in the "B" horizon and are high in aluminum and iron. *Hardpan* is a pedalfer layer of soil usually found in the "B" horizon and is formed by precipitation of salts through evaporation. Hardpan is cemented firmly with calcite or quartz and is often impermeable to water and difficult to break up. Hardpan is mainly found in western states of the United States, where it is called "caliche" and used for building gravel roads.
- *Podzol* is soil found in forests and areas where leaves and needles of trees decay to form a thin mat layer of organic matter overlying the "A_2" zone. Podzolization occurs in the cool forest regions of Canada and the eastern United States from New England to Florida.
- *Pedocals* are highly alkaline (high pH readings) soils with little or no humus and are formed in arid regions where subsurface evaporation of water crystallizes as calcium or lime deposits. *Alkali* soils are also pedocals, with a salt content high enough to kill plants. They are found in some southwestern states of the United States.
- *Laterites* are highly leached tropical soils, often with high concentrations of aluminum (as bauxite ore), gold, silver, iron, nickel, and copper. Laterite soils are found in the near-equatorial regions of Jamaica, the central areas of Africa, northern South America, and the southeastern United States. Crops raised in laterite soils require great quantities of water.
- *Loess,* a nonstratified and buff-colored, impregnable soil, may be either porous or clay rich. It is composed of calcareous silts and wind-blown dust deposited in northern Europe, eastern China, and the Mississippi Valley of the United States following the last Ice Age.
- *Loam* is a rich soil composed of organic matter mixed with some clay, silt, and sand. Loam soils are found in the great plains and farming areas of most continents.[35–37]

Soil Biology

According to most cosmologists the basic, lighter elements in the universe were formed shortly after the Big Bang, while the heavier elements are assumed to have formed later, particularly during the formation of the stars, including our sun. Earth formed at about the same time as the solar system, approximately 4.5 billion years ago, from leftover matter in the solar system as the chemical elements necessary for life arrived on Earth via meteorites that impacted it. These impacts provided more than 90% of our volatile matter, such as water vapor, carbon dioxide, ammonia, and other essential elements and molecular compounds. These chemicals interacted, forming a primordial atmosphere different from today's air and the liquid water of the oceans. How these organic molecules combined to form simple life forms about 3.5 billion years ago continues to be studied. Lynn Margulis and Dorion Sagan, in their book *Slanted Truths,* speculate that some type of self-making, self-maintaining **autopoiesis** resulted in organic compounds forming simple, single cellular life that, in time, evolved into multicellular forms of life.[38] Even many millions of years ago, rain, glaciers, and other forces caused the weathering of minerals and rocks into fragments that were transported and deposited as sediments and silts that formed early types of soils. It is not known how or if primeval soils were required as nutrients for early forms of life, but it is assumed that more advanced life forms took up residence in the oceans, as well as in soils, to obtain a supply of nutrients. The importance of the outer crust of the biosphere to the origin and maintenance of life is expressed in what is referred to by some scientists as Vernadsky's law. It states

> The migration of chemical elements in the biosphere is accomplished by either the direct participation of living matter (biogenic migration) or it proceeds life in a medium where the specific geochemical features (oxygen, carbon dioxide, hydrogen sulfide, etc.) are conditioned by living matter, by both that part inhibiting the given system and that part that has been acting on the Earth throughout geological history.[39]

In essence, this statement refers to a boundary between inert matter and living organisms, which were never produced by inert matter. (Living matter, by metabolism and other systems of life, alters the molecular structures of compounds but never the physical structures of elemental atoms.) Although Vernadsky's law is not accepted by all scientists, most scientists do agree that the cyclic life and death of bio-organisms recirculate atoms through the biosphere—life is always generated by life.[40]

Soil is composed of more than inert mineral/rock matter and recycled organic molecular compounds from dead plants and animals; it is also composed of many trillions of microorganisms and animals that live in its depths. We know that today's soil is teeming with life. A handful of fertile soil contains up to 6 billion bacteria.[41] Just one tablespoon of fertile soil contains more microorganisms than the number of people on Earth. An acre of mature topsoil contains many different species of life, and the total weight of all the living organisms in the top six inches of an acre of soil can range from 5,000 pounds to as much as 20,000 pounds.[42] In addition to **prokaryote** and **eukaryote** bacteria, there are other types of protozoa (e.g., paramecia, rotifers), many species of worms (nematodes, earthworms), insects (ants, termites, beetles and their larvae), small reptiles (snakes, skinks), and small mammals (shrews, moles, gophers, etc.). All of them eat organic matter or each other, defecating and dying while enriching and mixing the soil in the process. Earthworms are a good example. They carry dead leaves and other organic matter deeper into soils as they burrow down to lower levels. In the process they not only mix the fine soil particles but also convert the organic material into **humus**. Contrary to popular opinion, earthworms do not eat dirt; instead they survive on organic matter in the soil. Earthworms also aerate the soil, allowing water and oxygen to penetrate the topsoil, making it **friable**. Earthworms move from 1 to 100 tons of soil per acre per year.[43] Other small animals and insects burrow into soils to seek grubs, roots, and other nutrients or to construct living quarters. In doing so, they mix up the particles that compose soil. As far as animals, including humans, are concerned, soil and its nutrients are a main requirement for the growth of their basic food, namely, plants. Even carnivores that subsist on meat are, at some point in the food chain, dependent on plant eaters for their food. Humans are one of the few larger **omnivore** species that can survive on either plant or animal tissues.

Soil is dynamic in the sense that it functions as a part of ecosystems within the biosphere, where it enhances biodiversity by recycling biomass. In other words, soil itself is a recycling system within a larger system. In addition to sustaining plant and animal life, it regulates water by absorbing floodwater that is later released. The upper layers of soil clean and filter out impurities and kill potential disease-causing organisms, as well as potential pollutants. Soil also cycles and recycles nutrients and essential elements, and it is used to support human habitations and other structures. Soil assists in regulating climate by taking up carbon dioxide, which helps reduce global warming. Microorganisms convert methane, nitrous oxides, and other greenhouse gases into harmless by-products. Total greenhouse gases could be reduced by 16 to 42% through improved soil conservation.[44]

Issues

A number of controversies are related to humanity's long-time interactions with soil.

Soil as a Renewable Resource

When humans subsisted as hunters and gatherers, they probably had little, if any, concern with how soil was formed and used. About 10,000 years ago, agriculture became more than a garden hobby of hunters and gatherers. It was about that time, after the last Ice Age, when humans in the Middle East systematically used soil as a means of producing a sustainable supply of food. In the past thousand years or so, the natural flora and fauna landscapes in Europe and Asia, and many years later in North America, quickly gave way to cultivated crops and domesticated animals.[45]

> Early Americans thought America was a land of limitless bounty, which shaped how we used our natural resources. Early settlers, the pioneers, and their successors used the soil, trees, the game, and all the other treasures of the new land with careless profligacy, confident that the supply was inexhaustible.[46]

We now know that planting and harvesting both plants and trees can be a sustainable method of farming. There are four trees planted for every three that are harvested in the United States. There are more trees now growing in our country than there were 150 or even 100 years ago. The timber industry realized many years ago that if it did not treat their forests as renewable crops, they would sooner or later be out of business. Many of the private timberlands are better managed as a sustainable and renewable crop than are some public forests. The U.S. Department of Agriculture and its experimental research stations provided farmers with information on how to practice sustainable crop farming while at the same time maintaining or improving the health of their soil. As farmland is used for nonagricultural purposes (urban and industrial development), the effective and efficient use of both current and unused farmland becomes important. More demands, both domestic and international, will be made on products from the 400 to 500 million acres of cropland in the United States. When demand for products from the soil increases, wise-use conservation must be, and can be, practiced, because soil is a renewable resource whose quality can be improved. Therefore, soils, at least in the developed world, will not be depleted as some environmentalists claim.

Soil Conservation

The U.S. Soil Conservation Service estimates that topsoil regenerates at five tons per acre per year. Research conducted and reported by the institution Resources for the Future estimates that the regeneration of soil is more likely twelve to sixty tons per year per acre, depending on the geographic location. A conclusion of the study was that high-yield crops that have more of their roots deeper in the soil may actually restore productivity.[47] This is particularly true for the roots of **legumes** whose nodules extract nitrogen from the atmosphere and deposit it in the soil. Other sources provide more conservative figures for the formation of soils. One states that "It takes between 100 and 400 years for one centimeter of topsoil to form."[48] (The rate of soil renewal depends on the nature of the bedrock, climate, and geographic location.) The important point is that soil is dynamic. Soil changes depending on how it is managed, and management practices affect the renewability of soil.

One of the least understood concepts about soil conservation is erosion. As previously mentioned, erosion of minerals, rocks, and soil is a natural ongoing process of a dynamic Earth. The National Research Council of the National Academy of Science says the practice of using erosion estimates as measures of soil loss is wrong. Soil is moved or displaced; it isn't lost. Soil that leaves one farm is far more likely to end up on another farm than to flow into the ocean. The research done by Resources for the Future also concluded that erosion of our soils (in the United States) at the present rate presents no threat to future agricultural productivity. Even if we did not improve agricultural practices, crop yields would drop only about 2% over the next century.[49]

Sustainable, renewable farming practices that have actually improved mature soils have been practiced for several hundred years by the Pennsylvania Dutch farmers. They nurture their land and put back as well as take from it. Similar excellent farming practices are found in other areas of the United States, as well as in England and other parts of Europe.

Farming in the United States has changed dramatically over the past fifty years. Farms are no longer subdivided into small plots, which means small farms are disappearing at a rapid rate as larger farm operations develop. Although this trend is a social issue and is seen as a negative development by some, it has the potential to improve both soil management and agricultural productivity. At the same time, improved practices have been implemented, such as the following:

- *Crop rotation* (alternating different crops each year or leaving the fields fallow for a season)
- *Planting green manure crops* (crops that are plowed under, which act as fertilizer returning important nutrients to the soil)
- *Contour plowing* (plowing horizontally around the contour of a hill rather than up and down the slope of the land; the plowing follows the contour of the land)
- *Strip farming* (planting different crops in strips, side by side, to prevent erosion—often used with contour plowing)
- *No till farming* (avoiding the deep plowing of the soil and instead scratching its surface to prepare for seeding crops)

All of these methods greatly reduce erosion. The good news is that soils in the United States and most of Europe are getting better. Since the 1960s, farmers have adopted many of the previously mentioned conservation practices. Today, in the United States, more than 100 million acres are farmed using conservation tillage and more than 30 million acres of fragile croplands, now enrolled in the Conservation Reserve Program, are seeded to grass or planted with trees.[50] Although erosion of topsoil is still a serious concern in most countries, the United States is making great strides not just to control erosion but to improve the soil as a renewable resource. The bad news is that much of the underdeveloped world has not implemented good farming methods. For example, in the Middle East and some other areas people raise large herds of goats that strip the bark off trees and eat the roots of grasses. This practice has contributed to massive desertification of formerly productive lands. It is estimated that there is adequate productive land (worldwide) to feed the present population of 6+ billion people, and even the projected ~9 billion by the year 2050, of the world. However, that projection applies only if political strife, war, and other cultural problems, which hinder some countries from even nearing self-sufficiency as far as raising food for their own populations is concerned, do not occur.

Biomass as a Source of Renewable Energy

Plant growth is dependent on both solar energy and the nutrients obtained from the Earth. All animal life is directly or indirectly dependent on vegetation, and by inference also soil and, to a lesser extent, the oceans. Therefore, issues related to soil are also related to biomass and its uses and welfare.

More than one hundred years ago the concept of *uniformitarianism* was developed. This basic principle states that geological process, systems, and natural laws today operate to modify the Earth's crust in much the same ways as these processes operated throughout geological time. It also states that geological events of the past can be explained by similar events occurring today. As the saying goes, "the present is the key to the past." It might be added that this principle does not mean that all past and present geological events occur with the same intensity or at a uniform rate.[51] A century ago, proponents of this concept thought that the forms of living matter had changed (evolution), but the overall worldwide volume and weight of living matter had not changed throughout time. In other words, at one time it was thought that the total mass and volume of living matter was always the same throughout time. We now know that biomass, both in the oceans and on land, has increased over time. The amount of living plant and animal matter in the biosphere is equal to 10^{20} to 10^{21} grams, which would be much greater if the processes of exponential animal reproduction were not restrained. There are two kinds of living matter in the biosphere, *photoautotrophic* (green plants that use light to convert inorganic CO_2 and H_2O into organic molecules by photosynthesis) and *heterotrophic* (which are *chemotrophic* animals that obtain their energy and carbon by ingesting organic compounds, i.e., plants and other animals). The total *photoautotrophic* biomass on Earth equals ~740×10^{15} grams, whereas the total biomass of animals is only ~10×10^{15} grams.[52] Yet, more than 95% of all organisms on Earth are animals that live on chemical energy and fixed carbon from plants that grow in soil or seawater.[53] At one time it was thought that the total mass of green life in the oceans exceeded that of plant life on land. This has proven to be wrong. Actually, the total mass of organic carbon from plant life found on land is far greater than the organic carbon produced from green plants in the oceans.[54] The biomass on land, or more specifically life that depends on soil for nutrients and the sun for its energy (i.e., green plants), is the form of living matter that is referred to as one source of renewable energy. Another source of biomass is plankton and other plant life that exists in the oceans and is part of the food chain used by humans in the form of seafood.

Sources of renewable energy that could augment or replace nonrenewable fossil fuels were discussed in chapter 4. These renewable sources are solar, hydroelectric, wind, geothermal, and tidal/wave action. Photoautotrophism-produced biological material (plant biomass) could be another important renewable energy source. In a sense, crops and other sources of biomass grown in soil for the production of fuel and energy are really stored solar energy, because they convert sunlight into organic carbons (carbohydrates). When woody plants, grains, agricultural and garden wastes, organic garbage, and so forth are burned, the oxidized carbohydrates release stored-up solar energy.[55]

An excellent example is corn, which can be used two different ways to produce renewable energy. It can be fermented to produce liquid ethyl alcohol that can be added to or substituted for fossil fuels (e.g., gasohol, a blend of 10% ethanol and 90% gasoline). Dried corn kernels and other grains also can be directly burned in special stoves that produce energy adequate to heat homes. These stoves have a hopper that can be filled with kernels of dried feed corn that are automatically fed to the firebox. The stove needs to be refilled only a few times each day. A professor at Pennsylvania State University stated that a bushel of dried corn produces about as much heat as five gallons of liquid propane. He claimed that corn had a wholesale value of about $2 a bushel in late 2000, while propane cost about $1.30 per gallon (it is assumed both prices have increased with inflation). In the year 2000, some farmers were able to drastically reduce their heating bills, by as much as $2,000 per month, by burning corn or other plentiful grain crops they grew. At the current higher prices for nonrenewable fuel oil and natural gas, renewable biomass such as corn, barley, and other excess grains are economically competitive fuels.[56] Whether grains *should* be used for fuel, rather than being shipped to countries with grain shortages, is a different issue.

Jeffrey Bair, in the article "Biker Runs Harley with Soybean Oil," reported another way to use biomass for fuel to power automobiles and motorcycles. He reported that a mechanic converted a motorcycle gasoline engine to a diesel engine that gets more than 100 miles per gallon by burning soybean oil. He said it sounds like a jackhammer (as do most diesel engines) and smells like a fast-food restaurant. Bair stated that any number of food oils derived from plants could be used as fuel in converted gasoline engines.[57] Renewable bio-food oils from plants grown in soil are economically competitive with nonrenewable fossil fuels and possibly less expensive than gasoline in some countries.

Plant biomass can be used in another way to produce renewable energy. As plant material, including animal manure that still contains some plant matter, decays and rots, it produces the gas methane (CH_4), which is released into the atmosphere. During the process, heat is also generated. An interesting story involves early twentieth century farmers using their manure pile as a parking area for their Model T automobiles to keep the engines warm during cold winter months. Methane gas is the major component of nonrenewable natural gas, but methane also can be generated in special biomass digesters, captured, and used as a renewable fuel. Some countries now use methane, both from natural gas and coal sources as well gas generated from biomass, to fuel automobiles whose engines have been converted to burn this CH_4 rather than petroleum gasoline.

Large municipal incinerators can produce adequate steam to drive large electric generating plants. They can do this by using the concept of high temperature pyrolysis, which completely oxidizes all kinds of waste materials by altering the chemistry of the waste material, including industrial, home, municipal, and toxic wastes, as well as waste biomass (e.g., tree trimmings, brush, weeds, paper, etc.). The oxidation process is almost complete in this high-temperature combustion, resulting in much less air pollution and residual ash than is produced by common burning, which at the same time greatly reduces the need for landfills. A pyrolysis plant is expensive to build, and the process requires great quantities of garbage and trash in order to maintain a constant 24-hour operation to be effective, but it could help reduce use of nonrenewable fossil fuels for the production of heat and electricity.

Notes

1. Vladimir Ivanovich Vernadsky, *The Biosphere,* trans. David B. Langmuir (New York: Springer-Verlag, 1998), 91.
2. Robert L. Bates and Julia A. Jackson, *Dictionary of Geological Terms* (New York: Anchor Books/Doubleday, 1984), 209.
3. Norah Rudin, *Dictionary of Modern Biology* (Hauppauge, N.Y.: Barron's Educational Series, 1997), 40.
4. Bates and Jackson, *Dictionary of Geological Terms,* 55, 159.
5. E. O. Wilson, *The Future of Life* (New York: Knopf, 2002), 12.
6. Wilson, *The Future of Life,* 11–12.
7. James Lovelock, *The Ages of Gaia: A Biography of Our Living Earth* (New York: Bantam Books, 1988), 3.
8. James Lovelock, *Gaia: A New Look at Life on Earth* (New York: Oxford University Press, 1995), 144.
9. Wilson, *The Future of Life,* 10–11.
10. Wilson, *The Future of Life,* 132.
11. Lynn Margulis and Dorion Sagan, *Slanted Truths: Essays on Gaia, Symbiosis, and Evolution* (New York: Springer-Verlag, 1997), 350.
12. *Scientific American Science Desk Reference* (New York: John Wiley and Sons, 1999), 316.
13. James K. Fredrickson and Tullis C. Onstott, "Microbes Deep Inside the Earth, " *Scientific American,* "Earth from Inside Out," special publication, 2000, 10–15.
14. Vernadsky, *The Biosphere,* 15.
15. Vernadsky, *The Biosphere,* 58.
16. Anna Claybourne, Gillian Doherty, and Rebecca Treays, *The Usborne Encyclopedia of Planet Earth* (Tulsa, Okla.: EDC Publishing, 1999), 12.
17. Mississippi State Extension Service, accessed 2000, http://www.ext.mssstate.edu/anr/plantsoil/soil/soilpapers/basic.
18. Kenneth W. Hamblin and Eric H. Christiansen, *Earth's Dynamic Systems* (Upper Saddle River, N.J.: Prentice Hall, 1998), 229–31.
19. Graham R. Thompson and Jonathan Turk, *Earth Science and the Environment* (New York: Harcourt Brace, 1999), 182–84
20. Thompson and Turk, *Earth Science,* 186.
21. Hamblin and Christiansen, *Earth's Dynamic Systems,* 232–35.
22. Thompson and Turk, *Earth Science,* 184–86.
23. Rudin, *Dictionary of Modern Biology,* 211.
24. John Mongillo and Linda Zierdt-Warshaw, *Encyclopedia of Environmental Science* (Phoenix, Ariz.: Oryx Press, 2000), 210.
25. W. G. Ernst, ed., *Earth Systems: Processes and Issues* (New York: Cambridge University Press, 2000), 121.
26. Bates and Jackson, *Dictionary of Geological Terms,* 425.
27. Hamblin and Christiansen, *Earth's Dynamic Systems,* 242–43.
28. Stanley Chernicoff, *Geology: An Introduction to Physical Geology* (New York: Houghton Mifflin, 1999), 140.
29. *Encyclopedia Britannica,* CD-ROM 2001, s.v. "Soils: Seventh Approximation System."
30. Thompson and Turk, *Earth Science,* 189–90.
31. Hamblin and Christiansen, *Earth's Dynamic Systems,* 243–44.
32. *Encyclopedia Britannica,* CD-ROM 2001, s.v. "Soil: Horizon Nomenclature."
33. Edward J. Denecke Jr., *Let's Review: Earth Science* (New York: Barron's Educational Series, 1995), 166–67.
34. Chernicoff, *Geology,* 141–47.
35. Mark J. Crawford, *Physical Geology* (Lincoln, Nebr.: Cliffs Notes, 1998), 59.
36. John Tomikel, *Basic Earth Science: Earth Processes and Environments* (New York: Allegheny Press, 1981), 84.
37. Bates and Jackson, *Dictionary of Geological Terms,* 301–92.
38. Margulis and Sagan, *Slanted Truths,* 348.
39. A. I. Perelman, *Geochemistry* (Moscow: Vysshaya Shkola, 1979), 215, quoted in Vernadsky, *The Biosphere,* 56.
40. Vernadsky, *The Biosphere,* 56.
41. Claybourne et al., *The Usborne Encyclopedia of Planet Earth,* 112–13.
42. Soil Resources on the Web, *Soil: A Critical Environmental Resource,* accessed 2000, http://www.swcs.org/t_resources_critical_fact.
43. Soil Resources on the Web, accessed 2000, http://www.swcs.org.

44. Soil Resources on the Web, *The State of the Soil,* accessed 2001, http://www.swcs.org/t_resources_state_facts.htm.

45. Philip Shabecoff, *A Fierce Green Fire* (New York: Hill and Wang, 1993), 32.

46. Stewart Udall, *The Quiet Crisis* (Salt Lake City, Utah: Peregrine Smith, 1988), 54.

47. Ronald Bailey, *Eco-Scam: The False Prophets of Ecological Apocalypse* (New York: St. Martin's Press, 1993), 47.

48. Thomas McGuire, *Reviewing Earth Science: The Physical Setting* (New York: Amsco School Publications, 2000), 65.

49. Bailey, *Eco-Scam,* 48.

50. Soil Resources on the Web, *The State of the Soil,* accessed 2000, http://www.swcs.org.

51. Bates and Jackson, *Dictionary of Geological Terms,* 546.

52. Vernadsky, *The Biosphere,* 72.

53. Rudin, *Dictionary of Modern Biology,* 179, 286.

54. Vernadsky, *The Biosphere,* 73.

55. Hamblin and Christiansen, *Earth's Dynamic Systems,* 654.

56. Dan Lewerenz, "Grain Furnace Usage Heats Up: More Planters Save Money by Burning Crops Instead of Fuel," *Houston Chronicle,* Houston, Tex., 24 February 2001.

57. Jeffrey Bair, "Biker Runs Harley with Soybean Oil," *Valley Morning Star,* Harlingen, Tex., 27 February 2001.

A view of Earth as it rises over the horizon of the moon. Taken by NASA astronauts. (Courtesy of NASA. Photo ID: AS08-14-2383.)

Above: A 200 m across steam and gas ring blown out of the vent of Mount Etna in 1999. (Photo courtesy of Dr. Jurn Alean, stromboli.net.)

Left: Mount Saint Helens volcano eruptions, May 1980. (Courtesy of the U.S. Geological Survey.)

Examples of common rocks. (Courtesy of the U.S. Geological Survey.)

Limestone

Conglomerate

Granites and Rhyolites

Metamorphic

Glauconite Sandstone

Examples of common crystals. (Courtesy of the U.S. Geological Survey.)

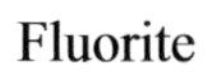
Fluorite

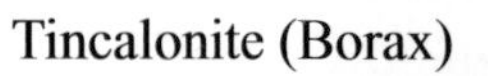
Tincalonite (Borax)

Barite

Calcite

Feldspar

Gold in Quartz

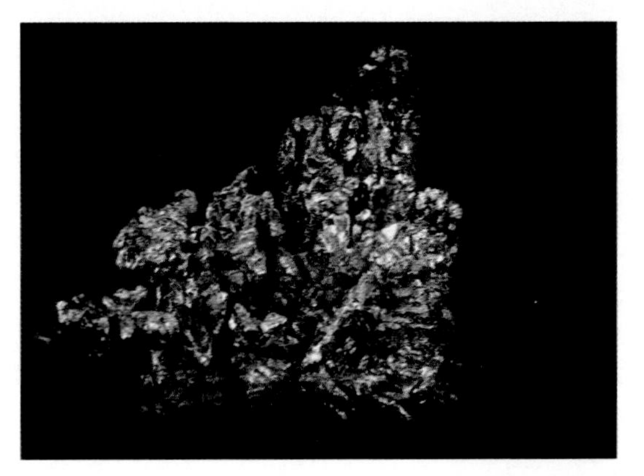

Silver

Gold

Views of different tornadoes. (Courtesy of the U.S. National Oceanic and Atmospheric Administration [NOAA].)

Various types of cumulus clouds. (Courtesy of the U.S. National Oceanic and Atmospheric Administration [NOAA].)

Cumulus

Towering Cumulus

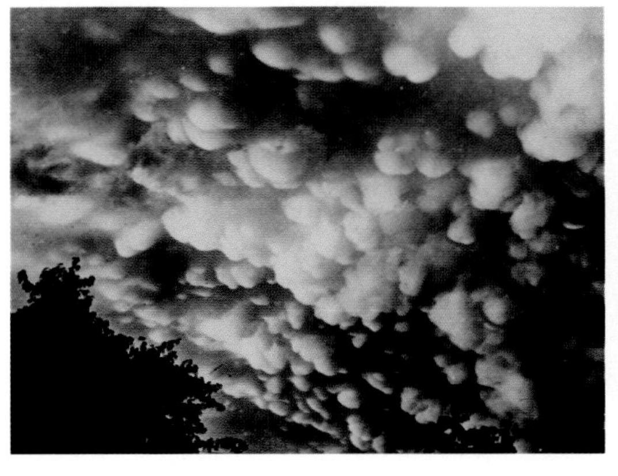

Mammatocumulus

Altocumulus

Cumulus Lenticularis

Altocumulus

Examples of storm clouds and lightning. (Courtesy of the U.S. National Oceanic and Atmospheric Administration [NOAA].)

Developing Storm Clouds

Developing Storm Clouds

Developing Storm Clouds

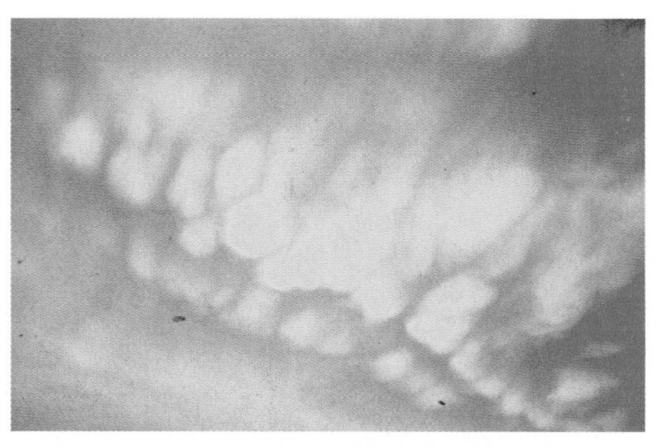

Developing Storm Clouds

Lightning during Storm

Lightning during Storm

7

Hydrosphere: Sphere of Water

The hydrosphere is the envelope of water in all its physical states or phases (surface water as liquid and ice, ground water in soil and rock, and gaseous water as atmospheric water vapor, including clouds and precipitation) found on the Earth's surface, underground, and in the oceans and atmosphere (see Figure 7.1).

The hydrosphere is a major system within the larger system of the biosphere. An intimate relationship exists in the entire biosphere between the soil/lithosphere, the atmosphere, and the hydrosphere. These three envelopes are required to sustain living organisms, with the exception of chemosynthetic microbes located deep in the crust that do not require an atmosphere. The hydrosphere makes Earth a unique planet in the solar system, although some of the other planets in the solar system, and a few moons, have water at their poles or beneath their surfaces. Some also have atmospheres with a variety of gases and densities, and some have primitive soils. For instance, according to researchers at the National Aeronautics and Space Administration (NASA), Jupiter's moon Ganymede exhibits signs of hidden saltwater under its icy crust.[1] But none of the other planets or moons in the solar system have the three envelopes of soil, water, and gases in the proportions or physical states necessary to sustain life as we know it. Earth is at the right distance from the sun to maintain a temperature for water to exist as a liquid, gas, and solid. If it were closer to the sun, water would soon boil and evaporate. If Earth were located at a much greater distance from the sun, it would be a frozen ice ball. Energy, mainly from the sun, drives the hydrological water cycle, as well as providing the energy required for life on Earth.

History and Background

Hydrology is the study of water, its history, its chemistry and physics, and its interactions with a variety of environments within the biosphere. Water is constantly circulated between and among the oceans, the landmasses, and the atmosphere. The questions What is its nature? and Where did it come from? have been subjects of study for centuries.

Origin of Water

There are several theories related to the origin of the water on Earth, as well as in the

Figure 7.1
The hydrosphere is the relatively thin layer of saltwater and freshwater that covers much of the Earth's surface. Water is necessary for life to exist, and the Earth's unique distance from the sun assures that water exists in all three states, namely, solid, liquid, and gas, and is cycled between these physical states.

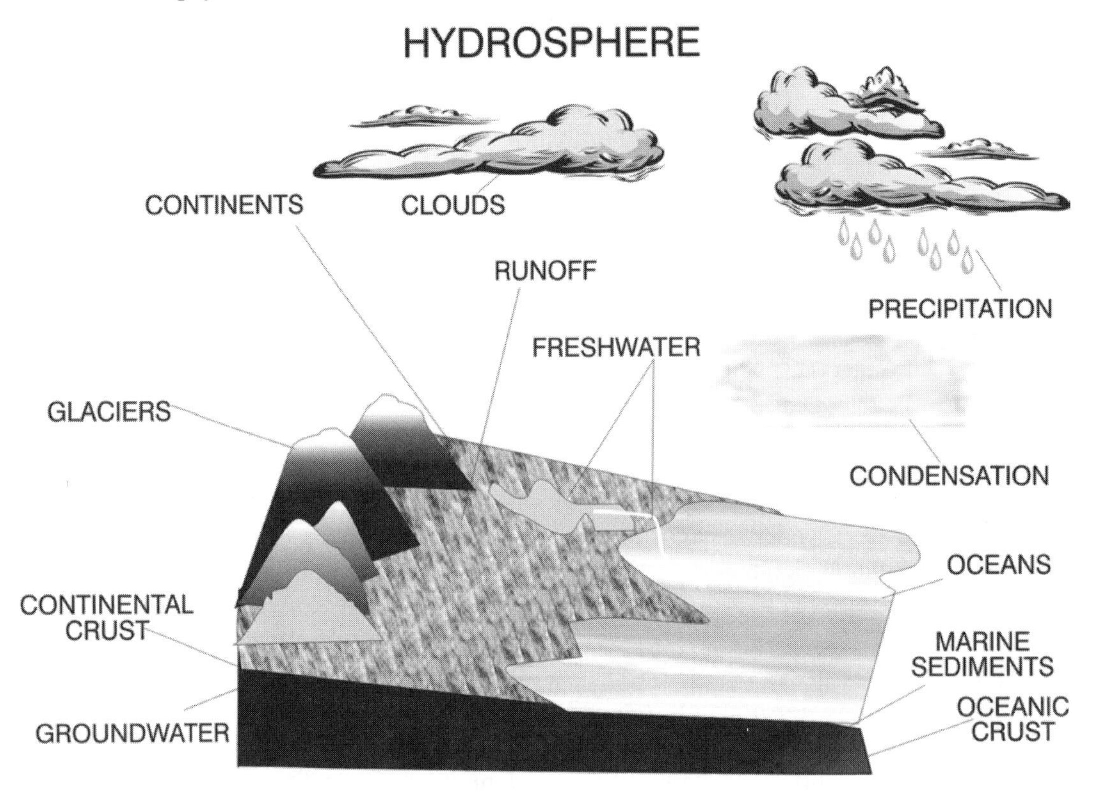

universe. They all agree that water on Earth arrived about, or shortly after, the time that the Earth was formed by **accretion** of planetesimals, for example, asteroids, meteors, and other stuff left over after the formation of the sun about 4.6 billion years ago. Several models are related to these theories.

- One source of Earth's water may have been contained in the rock material that accumulated and formed the early planet between 4.6 and 3.9 billion years ago. The dates for this model are based on the evidence that the oldest sedimentary rocks (~3.9 billion years old) must have required liquid water to erode rocks to form sediments that later formed sedimentary rock. Thus, water must have existed on Earth before ~3.9 billion years ago.
- Another model is based on the multitude of stony meteors that bombarded the newly formed Earth (~4 to ~4.6 billion years ago) and produced a thin layer of water-rich material. These meteors, called "chrondites," are composed of small spheroids of magnesium and silicates that were once molten, plus 1% water by weight. It is estimated that these early chrondites could supply much more water than was needed for the oceans.[2]

One bit of evidence for this meteor theory is based on a 1999 report that two rocks that fell in west Texas contained minute particles of salt in miniature capsules of water. Scientists found several minute samples of water sealed inside the salt crystals (NaCl) of what are known as the Monahans (Texas) meteorites, each weighing 1200 to 1300 grams [g] (about 2.5 pounds). The scientists say that these minute droplets of salty brine traveled through the solar system for millions of years before landing on Earth. They are not sure if this water was created in interstellar space and traveled into the solar system in comets, or if it was created by some process during planetary formation.[3]

- According to another model, ice comets in the solar system added water to the surface of the Earth during its early accretion phases. It was determined that some identified comets (e.g., Halley's and Hale-Bopp) could not have been the source of Earth's water because their water contains about twice as much deuterium as does the water in the oceans. (Note that deuterium, or heavy water, is composed of the heavier isotope of hydrogen, H-2. Because the hydrogen nucleus can have extra neutrons, it has three isotopes: H-1 [normal hydrogen], H-2 [deuterium], and H-3 [tritium]). However, in a May 17, 2001 news release from Goddard Space Flight Center, NASA reported observations of the breakup of the comet C/1999 S4 LINEAR just before it approached the sun. This comet, now called comet LINEAR, originated near the planet Jupiter. LINEAR's nucleus was ~750 to 1,000 meters (m) (~2,500 to ~3,000 feet [ft]) in diameter and contained ~3.3 billion kilograms (kg) (~3.6 million tons) of water. Astronomers were able to analyze the water vapor as the comet broke up when it approached the sun and discovered that the comet's water contained about the same amount of deuterium as does ocean water. Statistically, LINEAR-like comets that form near the sun are more likely to collide with the Earth than are those that formed at great distances; thus the percent of deuterium in the water of near-comets matches that of the oceans. Also, using telescopes sensitive to infrared light, astronomers discovered that LINEAR-like comets contain fewer volatile organic compounds (e.g., CO, CH_4, C_2H_6, and deuterium) than do comets that arrive from deep space.[4–5]
- A related, but controversial, model suggests that over the past billions of years the Earth was, and to some extent still is, constantly bombarded by small, icy comets commonly referred to as "dirty snowballs" (much smaller than LINEAR-type comets). These presumably pelted the ancient Earth with enough water from space to form the hydrosphere. Over the first millions of years, water may have been added to the atmosphere and Earth by numerous asteroids, meteors, and icy comets. Even today, small amounts of water from space in the form of tiny, ice-covered particles of matter enter Earth's outer atmosphere. It is speculated that these tiny snowballs may also have contributed to Earth's past, as well as present, water supply.
- Recently, astronomers discovered a hot, steamy interstellar cloud of gas located close to the Orion nebula. In 1997, a satellite operated by the European Space Agency detected a large concentration of water vapor in this gas that will soon freeze in outer space. The amount of water vapor in this Orion cloud is twenty times greater than has been measured in other interstellar gas clouds. Scientists

speculate that these giant nebula-like clouds of interstellar water vapor may have been the source of water for the Earth and possibly other planets.[6]

On the one hand, these various processes for explaining the origin of water on the Earth suggest a composite model that includes the accretion of large asteroids and meteors that formed the Earth between ~4.6 and ~3.8 billion years ago and later bombardments by icy comets. As mentioned, all of these sources could provide more than enough water to fill all the oceans and lakes on the Earth and more, since it is now known that the lower mantle contains about 5 times the amount of water that is held in all the oceans.

On the other hand, James C. G. Walker believes that icy comets were not important in providing water for the Earth's oceans. He mentions that, during accretion, the kinetic energy of colliding planetesimals was converted into heat, resulting in a very hot Earth, possibly too hot for ice to be a major source of water. Walker suggests that water was more likely trapped on the Earth in claylike minerals. He further states that, during the past 4 billion years, there has been (and still is) an exchange of volatile materials, such as water, between the Earth's surface and mantle (i.e., tectonic plate subduction and convection cells).[7]

Some Facts about the Hydrosphere

The hydrosphere is about 4 billion years old. Currently being debated is whether the total amount of water found on the Earth has remained virtually unchanged over the past billions of years or there have been periods when the amount of water increased (and possibly is still increasing) on Earth.

Although water covers more than 71% of the Earth's surface and contains ~1.3 billion cubic kilometers (km^3) (~326 cubic miles [mi^3]) of water molecules, the hydrosphere is one of the thinnest spheres of the Earth. However, it is the most massive layer of the biosphere. Even so, it contains only ~1/4500 of the total mass of the Earth (~1.3×10^{24} g). This is about one-half of the total mass of sedimentary rock and only 5% of the mass of the Earth's total crust.[8]

The two major components of the hydrosphere are the saltwater of the oceans and the freshwater of the lakes, rivers, ground water, ice caps, and water vapor.

Oceans

The following list provides some facts about the Earth's oceans.

- The saltwater oceans cover ~71.2% of the surface of the Earth but contain about 97% of the total mass of all of the hydrosphere. (About 2.15% of the Earth's water is frozen in ice caps and glaciers, which leaves about 0.65% freshwater in lakes, streams, ground water, and the atmosphere.)[9]
- The average depth of the oceans is ~4 km (~2.5 miles [mi]).
- The volume of all the oceans is ~18 times that of the continents.
- More than half of all Earth's seawater is found in the Pacific Ocean.
- About 3.5% per hundred (35 parts per thousand [ppt]) of the weight of the oceans consists of dissolved salts (NaCl and other salts). The salinity of seawater varies with the depth of the oceans in different areas around the globe. Saltwater is a consequence of millions of years of weathering and erosion of continental landmasses that resulted in dissolved salts being transported by freshwater to the oceans. As the ocean water evapo-

rates, the salt is left behind, increasing the concentration of salt in the oceans. At the same time, the freshwater from rivers and rain dilute the saltwater and maintain normal salinity. All the rivers of the world carry about 2.5 billion tons of dissolved salts to the oceans each year, along with a great deal of freshwater. Volcanoes on the seabed and movement of tectonic plates also contribute slightly to the degree of salinity of seawater. Contrary to expectations, this continuing inflow of salts from rivers to the oceans and the constant evaporation of seawater do not continually increase the average percent of salt in the oceans. Much of the salt is precipitated and is incorporated into the formation of a variety of minerals, leaving the salt level of the oceans in a steady state. Even so, the degree of salinity varies slightly with the geographic area where the measurement is made, the number of freshwater rivers flowing into a body of seawater, and the depth of a particular ocean. The salinity of seawater essentially has remained unchanged over eons of time.

- The oceans contain traces of about 70 of the 97 natural chemical elements. The most abundant in seawater are chlorine, sodium, magnesium, sulfates, calcium, potassium, bicarbonate, bromine, carbon, nitrogen, and strontium, plus traces of other elements.[10] The most important salts of these elements are NaCl, $MgCl_2$, Na_2SO_4, $CaCl_2$, and KCl.
- Seven main ions of the above named salts compose about 3.5% of seawater by weight. These ions are listed in Table 7.1 in order of percentage of weight in seawater.
- Several elements required for biological nutrition (iodine, iron, copper, and manganese, in addition to nitrate, nitrite, and phosphate ions) are found in a few parts per million (ppm) in the oceans.
- More than 90% of all life on Earth exists in the oceans. Most of these forms of sea life are very small types of plankton and so forth, which form the food base for other sea life. At the same time, the total biomass on the continents is much greater than the total biomass in the oceans.

Table 7.1
Salt Ions in Seawater

Name	Symbol	Percent by Weight	Grams of Ion/kg H_2O
Chlorine	Cl^-	1.93	1.4
Sodium	Na^+	1.07	10.8
Magnesium	Mg^{++}	0.1	1.3
Sulfate	SO_4^{--}	0.27	2.7
Calcium	Ca^{++}	0.042	0.4
Potassium	K^+	0.038	0.4
Bicarbonate	HCO_3^-	trace	0.1
Plus a small fraction of a percent of ions of other salts			

Sources: Kenneth W. Hamblin and Eric H. Christiansen, *Earth's Dynamic Systems* (Upper Saddle River, N.J.: Prentice Hall, 1998), 210; Graham R. Thompson and Jonathan Turk, *Earth Science and the Environment* (New York: Harcourt Brace, 1999), 314.

- There are 4.4 kg of gold in each square kilometer of seawater, which totals about 5.7 billion kg of gold in all the oceans. It is, however, much too expensive to extract.[11]
- The oceans contain a large amount of carbon dioxide—more than 60% of the amount of CO_2 found in the atmosphere. (The atmosphere contains about 0.03^{+}% CO_2.) Research is being conducted to determine just how effective the oceans are as a sink for excess carbon dioxide that is produced by natural processes, as well as by the burning of fossil fuels. Experiments are being conducted to pump excess carbon dioxide down to deep ocean floors, where it becomes stabilized under great pressure.[12]
- The source of ocean salts are from dissolved ions from river water and from interior volcanic action at rifts and ridges on the ocean floor.
- The density of ocean water is determined by salinity and temperature and varies with depth (pressure).
    1. At the surface, and to a depth of 50 to 200 m, seawater is mixed and ranges from 21 to 26°C in temperature.
    2. As depth increases, so does the density of seawater, as the temperature consistently decreases in the **thermocline**.
    3. Eighty percent of the mass of ocean water is found at great depths, where it is at its greatest density and lowest temperature (3.4°C) but with uniform salinity. This cold region below the thermocline extends to the surface at the polar regions.[13]
- Sea ice at the polar regions is a relatively stable form of water. Polar ice is an important determinant of climate temperatures, because it reflects more solar radiation back into space than does seawater. Sea ice freezes at about −2°C and covers about 15% of the Earth's surface.
- At the Arctic (North Pole), sea ice melts in the summer and refreezes in the winter, whereas at the Antarctic (South Pole) and on Greenland, ice is more permanent and covers about 7% of the ocean.[14]

Four main motions of the oceans affect climates by transferring energy on the Earth's surface from one region to another.

1. Surface currents are circulated by wind. A major surface current proceeds along the North Atlantic coastline and then eastward to join the North Atlantic Drift. This surface current maintains the mild climates of the British Isles and Scandinavian countries. This current is referred to as the Gulf Stream and is only one of several similar surface currents driven by the wind. Most ocean currents are driven by warmer water temperatures originating near the equator, which then circulate north. At the same time, cooler ocean water from the north drifts south toward the equator, thus affecting global climates.
2. Ocean currents are also driven by the Coriolis force due to the rotation of the Earth on its axis. The Coriolis force is named after the French physicist Gaspard-Gustave de Coriolis (1792–1843), who determined that the Earth spins as a unit on its axis. In other words, at the equator a point on the surface must move more than 25,000 mi in one 24 hour day, resulting in a surface speed of about 1,000 mi per hour, while the distance covered by a point (location) on the Earth becomes less and less as that point moves toward the poles. As the Earth spins eastward, the dynamics of the Coriolis force cause the oceans, as well as the atmosphere, at the equator to move in curved patterns in opposite directions for the northern and southern hemispheres. This movement causes circular paths, called **gyres,** that move clockwise in the Northern Hemi-

sphere and counterclockwise in the Southern Hemisphere[15] (see Figure 7.2).

3. Temperature and salinity (**thermohaline**) circulation move deep ocean water from one region to another.
4. Coastal upwelling also moves deep water to warmer regions.

All of these motions are related to heat energy that affects the pattern of global ocean circulation and, thus, regional and global climate.[16–17]

Estuaries are the places where freshwater from rivers meets the sea, and thus seawater is diluted at this junction. Freshwater is lighter than saltwater, so it tends to form a layer on top of the saltwater. This produces a boundary that is mixed during tides, creating brackish salt flats and wetlands. There is a special type of circulation at estuaries where water mixes the organic material from the rivers, providing nutrients to support marshes, seaweeds, **phytoplankton**, and **zooplankton**. These conditions at estuaries are responsible for much biological life productivity, particularly at the early stages of aquatic life.[18]

Freshwater

- Streams, rivers, lakes, and ground water comprise only ~0.65% of the total mass of the hydrosphere. Water vapor in the atmosphere comprises only about 0.001% of the water in the hydrosphere.

Figure 7.2
The spinning of the Earth on its axis creates a force on both the ocean currents and winds in the atmosphere. The Coriolis force has a profound effect on the Earth's regional and global climates.

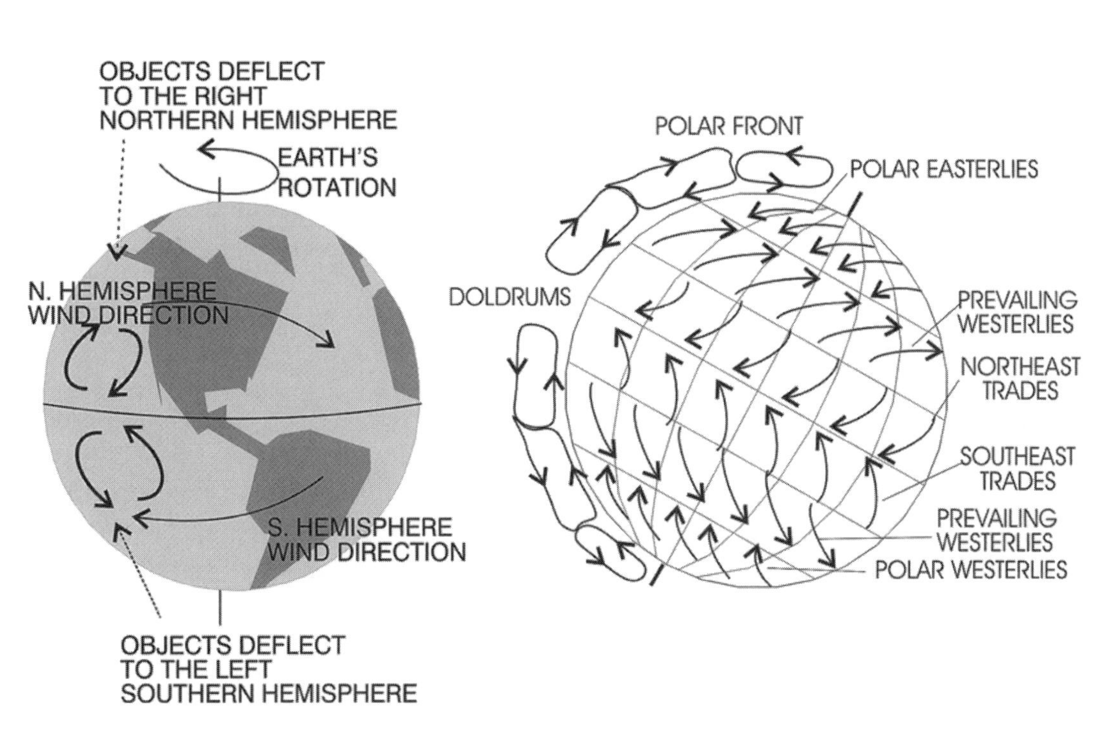

- Frozen water tied up in continental glaciers and polar ice sheets comprises about 2.15% of the water in the hydrosphere.
- About 10% of the world's landmass is covered with glaciers. At times in the past, most of the Earth was covered by ice, while at other times the glaciers receded to the point that the far northern regions were tropical.
- During the last Ice Age (~40,000 to ~10,000 years ago) glaciers covered much more land than they do today. The sea level was about 400 ft lower than current levels due to all the water frozen in continental glaciers.
- If all the glaciers melted today, the seas would rise about 260 ft.[19] (This is not likely to happen, even with concerns about global warming. One reason is that much of the water from melted ice is refrozen each winter at both polar regions.)

Flowing Water

Water flowing over the surface of the Earth is one of the main sources of erosion that continually changes the surface features of landmasses (see chapter 2). Precipitation in the form of rain, snow, hail, and sleet is the main source of freshwater required to support land-based life. The total precipitation on land is ~113,000 km^3, which after evaporation leaves ~41,000 km^3 of freshwater available for use each year. This is about the same amount that would flood the entire land surface with one foot of water per year. Unfortunately, most of this precipitation occurs in areas where it is not needed (the Amazon, Congo, and other remote rivers basins). Also, it is not spread out evenly over the year; in particular, in Asia about 80% of the precipitation on land occurs between May and October, resulting in about three-quarters of it being lost in floods and runoff. Land-wise, about 9,000 km^3 of precipitation is captured in dams and so forth, resulting in about 12,000 km^3 being stored for use. This equals about 5,700 liters (L) of water per day for each individual worldwide. The average daily use per person in Europe is about 566 L. This figure is higher in the United States, where on the average each person uses ~1,442 L/day, and the average per person/day use is much lower in many regions of Asia and Africa. Worldwide, per person use is equal to about 10% of all available freshwater.[20] Of course, the problem related to this is not that 90% of the available freshwater is unused by humans but that accessible freshwater (particularly potable water) is not always available where it is most needed. At the same time, use of freshwater has increased ~600% over the past century, mostly due to irrigation for agriculture. This per day use is expected to increase later in the twenty-first century to a worldwide daily use of about 2,000 L per person/day. Even so, this is less than 17% of the accessible and renewable freshwater.

Once precipitation reaches land, it flows from higher altitudes to lower levels, first in tiny rivulets that join to form runs, streams, creeks, and finally larger rivers that empty into the oceans of the world. Over the centuries, humans have learned how to capture and store much of this runoff before it flows into the oceans, but during drought periods these systems become inadequate to supply all the water needed for agriculture and other uses in some geographic areas.

A *watershed,* also referred to as a river basin or catchment area, is the total land area that contributes surface runoff water to a stream, river, or lake. The size of the watershed, the slope of the ground, the type of soil and vegetation, and how the land has been developed all determine how much water is available for agricultural, industrial, and municipal use.[21]

The term "stream" is used to describe water flowing within a channel. There are several other names for small streams, for example, brook, run, branch, creek, burn, bourne, and arroyo. Streams are called "tributaries" if they flow into, and join, larger bodies of water that become rivers. Rivers, sooner or later, usually flow into larger bodies of water such as lakes or oceans.[22]

Not all water from streams and rivers comes from surface runoff originating from precipitation. Some of it originates in ground water; at the same time, some water in streams and rivers seeps through the streambed, contributing to the supply of ground water. Even during periods of drought streams continue to flow as they are fed by springs from this stored ground water. Arroyos, (from the Spanish for "gutter") found in the southwestern United States, are deep, flat-floored gullies with steep banks that are dry stream beds most of the year, but during rainy seasons they drain off much of the floodwater from low-lying areas.

A number of factors determine how fast and where a stream will flow, as well as its effects on the landscape.

- The *gradient* is the steepness of the slope traversed by the stream. The gradient, or slope, of tributaries decreases exponentially downstream. In other words, the steeper the landscape, the faster the flow of water, and the greater the erosion and transport of sediment; as the slope decreases, less erosion occurs.
- The number of *stream tributaries* decreases downstream in a mathematical progression, whereas the length of the tributaries becomes progressively greater downstream. Thus, near the beginning of river systems there are numerous small streams, and so forth, that join the main channel. As this increased water supply progresses downstream, the number of small tributaries decrease, while at the same time these fewer tributaries become longer (and larger) as they join the main river before emptying into an ocean.
- The *competence* of the stream is the measure of the size of particles it can transport from one location to another. Obviously, fast-flowing streams can transport more and larger particles than can slow, sluggish streams. This is an important process that results in somewhat localized erosion.
- The *capacity* of a stream is how much sediment it can carry past a specific point in a given period of time. The stream's capacity is dependent on both its discharge rate and velocity. In other words, the steeper the streambed, the faster the flow of water and, thus, the greater the stream's capacity for carrying eroded sediments to either rivers, lakes, or oceans.
- The *discharge* is how much water is flowing in the stream. It is expressed as the volume of water that passes a given point in a specific time period (measured in cubic meters per second). As expected, the discharge rate is dependent on the velocity and rate of flow of the stream, which also affect its ability for local erosion and carrying capacity of sediments.
- The shape of the *stream's channel* and the nature of its bed and banks determine the amount of friction exerted on the flow of water and, thus, the discharge and velocity of the flow. The channel becomes progressively deeper and wider as the stream increases downstream due to increased flow where it discharges into a river or ocean.

- The *base level* of a stream is the limit to which the stream can erode its bed. Obviously, the larger the stream, the steeper its gradient, and the less its volume, the more its bed will be eroded, but if the base level has no gradient the flow will be limited, as will other factors responsible for erosion. In time, the stream may stop flowing.
- The *dissolved load* is the stream's ability to carry dissolved ions. This capacity is determined more by the stream's discharge and chemistry than its velocity.
- The *suspension load* is dependent on the type and size of particles, as well as their origin. Clay and silt particles are small; thus, they can remain suspended (as cloudy water) in even slow-moving streams.
- The *bed load* is determined by the volume and velocity of the stream's movement. During floods and periods of fast-flowing water, rocks and regolith rubble can be moved over the stream's bed. Sand and smaller particles are continually moved downstream as they are swept from one area of the stream's bed to be deposited at another location.
- *Deltas* are formed when sediments carried by streams and rivers meet the oceans and their loads are deposited in a flat, spread out area (something like the fingers of your hand when spread out in a dendritic pattern) that resembles the Greek letter Delta (Δ). The Yellow River of China deposits more than 1.5 billion tons of sediment in the ocean each year. The Indus and Ganges Rivers of India have both carried millions of kilometers of sediment from the Himalaya Mountains to the Indian Ocean to form deltas and large, spread-out, underwater deposits of silt of "alluvial fans." It is estimated that tectonic activity raises the Himalayas about 5 millimeters (mm) per year and erosion lowers the mountains about 0.5 mm per year. However, if in the future tectonic activity decreases, the erosion will continue. The Mississippi River deposits about 5 million tons of sediment each year to form the delta region at its mouth south of New Orleans, Louisiana. Another example of stream erosion is the riverbed at Niagara Falls, New York. The falls have eroded the streambed about 11 kilometers (km) (68 mi) up river over the past 9,000 years. The falls are slowly moving upriver as erosion continues, day-to-day.[23–24]
- As discussed in chapter 4, erosion by running water and the formation of valleys and canyons by rivers is dependent on three basic factors: (1) removal of regolith (rocks and rock debris); (2) cutting the channel deeper by abrasion of moving sediments, sand, and the regolith downstream; and (3) headwater erosion, where erosion is greater at the head of valleys rather than the sides of valleys as water concentrates and speeds up where the stream's channel starts. An excellent example of this type of stream (water) erosion is the Grand Canyon on the Colorado River that is 2 km (1.25 mi) deep.

Still Water

Compared to the relatively fast-flowing freshwater of streams and rivers, there are several classifications of relatively slow-moving water. *Lakes, ponds,* and *swamps* are generally defined as bodies of water constrained in a depression or valley whose beds are considered to be on impermeable ground. In essence, all lakes, ponds, and swamps are defined as landlocked, with no direct connection to an ocean other than an outlet steam or river. Some characteristics of lakes are given in the following list.

- Lakes cover 1.8% of the continents, and all the lakes in the world have a total surface area of about 2.5 million square kilometers.
- Most lakes have an inlet and outlet for water, but the flow is relatively small compared to the size and volume of the lake water; therefore, any flowing motion of lake water is seldom observable. In essence, the water coming into a lake (from all sources) enters faster than it can exit by an outflow stream, ground seepage, or evaporation.
- Small lakes are called "ponds," "lagoons," or "pools." Large lakes may be called "seas," for example, the Caspian Sea, which is the largest lake in the world.
- Lakes vary in depth from a few feet (e.g., Lake Pontchartrain in Louisiana, which is about 15 ft deep) to thousands of feet (e.g., the Great Lakes in the United States and Lake Ladoga in Russia, northeast of St. Petersburg).
- Not all lakes consist of freshwater. When water containing salty sediments enters a lake, but there is no exit for runoff, the lake tends to become more and more salty over the centuries as evaporation of the water leaves the salt behind. Two salty lakes are the Great Salt Lake in Utah in the United States and the Dead Sea located in the Near East, which has about nine times the salt concentration of seawater. Both of these lakes are devoid of normal aquatic plant and animal life.
- Lakes do not maintain their size over periods of time. The Great Salt Lake in Utah at one time was 14 times larger than it is now, and the ancient, very flat Lake Bonneville, also in Utah, was as large as Lake Michigan, covering an area of ~50,000 km^2 (~31,000 mi^2) and having a depth of ~300 m (~985 ft).[25] Today the dry, flat lakebed known as the Bonneville Salt Flats is so smooth it is used to test high-speed racecars. It dried up many years ago because surface evaporation was more rapid than the input of water from rainfall and ground water. These shrinking lakes are referred to as "pluvial lakes" (from the Greek for "rain" lakes), which were formed mostly during interglacial stages when less precipitation fell, and the pluvial lakes shrank to form small salt lakes or dry flats known as dusty "playas."
- In the 1980s, two lakes (Monoun and Nyos) in Cameroon, East Africa, released lethal gas, mostly carbon dioxide (CO_2), that killed many people. About 0.17 km^3 of CO_2 from the bottom of Lake Nyos welled up on August 21, 1986. This explosion of gas was overwhelming and was responsible for killing about 1,700 people, plus their livestock and dogs. The lake literally exploded as the gas and dissolved minerals from recent volcanic activity were violently released. In normal lakes this deep gas is slowly released from deep deposits and is mixed with surface water, where it slowly escapes into the atmosphere. But in these lakes the gas accumulates, remaining separate from the upper layers of freshwater until pressure builds up, which is periodically and explosively released. Scientists installed several large-diameter pipes in both lakes that allowed a self-sustaining fountain of gas-rich water to gush to the surface. By degassing the bottoms of the lakes, they hope to prevent future large gas eruptions and deaths.[26]
- In addition to being formed from ground water springs and water flowing into ground depressions, many lakes were formed by glacier activity. During the Pleistocene era, much of the precipitation in the Northern Hemisphere remained on

land as ice-forming glaciers many thousands of feet thick. These glaciers slowly moved south and melted over a period of 10,000 to 20,000 years, transporting great amounts of the Earth's surface material, including giant boulders from Canada to as far south as Ohio. Glacier activity also occurred in northern Europe, Asia, and the Southern Hemisphere. The world has just emerged from the last age of glaciers, about 9,000 to 11,000 years ago, and after the current period of warming it is predicted that the Earth will enter a new Ice Age; ice age glaciers and global warming periods seem to follow a long-term natural cycle, regardless of human activities.

- Melting glaciers deposited their loads of transported sediments that often damned up valleys to form lakes. A moraine is a mound or ridge of glacial drift (till) deposited by the movement of glacier ice. Moraine lakes are formed as the mass of till is moved, leaving behind a depression that becomes a lake. As glaciers moved south, great blocks of ice gouged out huge depressions and/or their massive (heavy) ice blocks left depressions as they melted, becoming moraine lakes and ponds that were continually filled by flowing water. Examples are the many smaller lakes in the lake regions of New York, Minnesota, and other northern states in North America. The movement of ice glaciers in Scandinavia, northern Europe, and Asia also created many lakes and ponds.
- The Earth's crust under the glaciers was first depressed by the weight of the ice during the Pleistocene glaciations. Due to the process "isostatic adjustment" (state of equilibrium), the glacier ice melted and its weight was removed, which resulted in the land slowly rising. About 13,500 years ago the ancient land area around Puget Sound in Washington state became submerged under water by the weight of ice. After the glacier ice melted several thousand years later, this submerged land area slowly began to refill. It has been raised to ~150 m (~483 ft) in altitude since that time. During the same period, the seafloor at the Hudson Bay region in east-central Canada rose ~300 m (~965 ft). These areas (and others areas formerly covered by glaciers) will continue to rise another ~80 m (~257 ft) as the land returns to the same level it was before the period of glaciers. These isostatic changes in the surface areas affect both the formation and decline of lakes.[27]
- As streams deposit sediment in lakes, the depressions fill up and, in time, become swamps, bogs, or fens. As vegetation takes over, they are transferred into meadows, or even forests, with streams often following the old channels. In time, most lakes become filled with mud, sediments, and vegetation or become swampy areas.[28–30]

Stored Water

Reservoirs (man-made) and *ground water* (natural) are both systems for storing freshwater. Although both types of storage areas use natural geological and topographical formations, humans can control the water level or even construct artificial water-storing basins. Nature stores freshwater underground in various systems, but in many regions humans need freshwater where it is least abundant. Because the human need for freshwater is often some distance from available sources, people have learned how to store water in artificial lakes and even in underground reservoirs and then transport it for use where and when it is needed.

Reservoirs

Man-made reservoirs are artificial lakes or ponds designed to store freshwater for future use. They are usually created by damming up large streams or rivers, often in valleys, canyons, or areas where the ground is depressed and is less permeable to water seepage. Many cities have dammed flowing water to create reservoirs to supply the needs of their citizens. Aqueducts were built by the Romans centuries ago to transport freshwater to cities. In the year 100 C.E., the city of Rome received ~85 million gallons of fresh spring water through its aqueducts each day. The Roman Pont du Gard aqueduct, with arches three stories high, had a channel on its top. It still stands near the city of Nimes, France.[31] New York City has an extensive system of man-made reservoir lakes and aqueducts to transport water over many miles into the city. It also has one of the best municipal water systems in the world, providing high-quality, fresh potable water to millions of citizens.

Ground Water

Ground water is what the word sounds like—water found below the surface of the land. It fills the small cracks as well as large and small voids in bedrock, and it is absorbed like a sponge by some soils and humus. Although ground water comprises only a small percentage of all Earth's water, there is about 20 times more ground water below the Earth's surface than there is surface freshwater located on continents and islands.[32] Also, as previously mentioned, there is about 5 times the amount of water in the lower mantle than in all the oceans, but of course, this water is not available to us.

- *Porosity* is the open spaces in the soils and rocks where ground water is found. In other words, porosity indicates the amount of water that soil or rocks can hold. Mud has the greatest porosity, at about 90% or more, primarily due to the way the clay particles attract water. According to its structure, clay can absorb a great deal of water or it can be almost impermeable. Sand and gravel can have a porosity of 40% or more due to the spaces between particles. Sandstone has a porosity of 10 to 30%, shales have a porosity of less than 10%, and igneous rocks have very little or no porosity, unless fractured. The amount of water that can be contained in these types of rocks depends on the extent of consolidation of their rock particles, gravels, and sands. If they are unconsolidated they have great porosity; if they have metamorphosed into rocks, they have very low porosity.
- *Permeability* is the capacity of any substance to permit passage of water. The permeability of soils, gravel, and rocks is determined by the size of the pores and spaces in a particular substance that can transmit or aid the flowing of water. Many soils and some rocks of high porosity are also high in permeability. Compacted clay is impermeable while it maintains some porosity. Ground water generally flows faster through soils and rocks with high permeability. Rocks that *slow* the flow of ground water are known as "aquitards," whereas impermeable basement rocks that *hinder* the flow of ground water are known as "aquifuges."
- The *water table,* also referred to as the "ground water table," is the upper level of an area just below the surface where the soil and/or rocks are permanently saturated with water. The top surface of the water table is known as the *zone of aeration* where the water's pressure is the same

as the atmospheric pressure. The underlying region of the water table, known as the *zone of saturation,* is the area at which more water cannot be absorbed. There are regions in the United States (e.g., Florida) where rain is plentiful, resulting in a water table near the surface, but it is often below a layer of *hardpan,* which is a claylike layer of compacted soil or gravel cemented by relatively insoluble materials such as iron oxide or silica that make this layer relatively impervious. The water table is often just below this shallow hardpan, and if you dig a posthole in such an area, it might fill with water before you can install the post. The depth of the water table for many areas is dependent on seasonal rainfall. During a dry period the water table may be low, but during wet seasons rain *recharges* the ground water, raising the water table.

- *Aquifers* are large underground areas in soil and rocks that can store great quantities of water. Aquifers are both porous and permeable, which means that as water is pumped out of an aquifer, it can be replaced by rain and surface water—unless more is withdrawn than can be recharged. Sand, gravel, sandstone, and broken bedrock make excellent aquifers, while shale, clay, and crystalline rocks generally make poor aquifers.
- *Artesian aquifers,* sometimes called "artesian wells," are formed when an inclined layer of permeable rock is contained by over and under layers of impermeable rocks. The water held in the raised end of the artesian rock exerts pressure on the water located in the lower end of the sloping layer of permeable rock. Through gravity, the weight of water at the raised end of the aquifer creates enough pressure to force the water to the surface at the lower end, either as a spring or a free flowing artesian well.

Caverns and Sinkholes

Caverns and sinkholes are landforms created by underground acidic ground water. The atmosphere contains about 0.03% carbon dioxide (CO_2), while the soil contains about 10%. This high concentration of CO_2 plus sulfur compounds and organic matter dissolves in water, forming acidic solutions in ground water. These acids, although relatively weak, slowly react with the minerals in limestone and cause long-term erosion, resulting in cracks, crevasses, holes, and caverns in bedrock.[33]

- *Caves* are formed over many thousands of years as acid water seeps into cracks in carbonate bedrock, dissolving the rock. In time some cracks in the rocks enlarge to form larger openings known as caves and sinkholes. Caverns are usually larger than caves and often involve a system of cave chambers. Caves are found in many parts of the world where there is limestone bedrock. Two well-known caverns and caves in the United States are Carlsbad Caverns in New Mexico and Mammoth Cave in Kentucky. Although most caves are formed when water that contains acid dissolves limestone, caves can form in other ways. One example is the caves carved out by the ocean surf crashing against coastal bedrock, for example, the Sea Lion Cave located in Oregon. Another type of cave is formed when tubelike cavities are created in cooling lava flows. Examples of lava caves are found in the northwestern United States and Hawaii. It is also believed that lava caves exist on the moon. Still another type of cave is formed at the base of melting glacier ice, which can just as easily disappear with new snow and freezing cycles. One example of a large, long-lasting ice-like cave is located near the summit of Mt. Rainier in Washington state.

- *Stalactites* and *stalagmites* are formed in some enlarged caves when the water and dissolved calcite from the chemical reaction produces CO_2 in solution. The solution collects on the ceilings of caves where, as the CO_2 is removed from the calcite, it leaves behind the mineral to form icicle-like deposits. When these minerals drip from the ceiling of a cave, solid calcite icicle-like formations known as stalactites grow toward the cavern floor. When the drippings from these hanging mineral icicles collect on the floor of the cave, they form stalagmites that grow from the ground up, sometimes meeting the stalactites to form a column.
- *Sinkholes* form on Earth's surface when the roof of a cave collapses, creating a circular, often cone-shaped, depression ranging from a few feet to hundreds of feet in width and more than dozens of feet in depth. Where you find one, you are likely to find others as they exist in extensive limestone areas. Streams often flow into sinkholes and drain out their bottoms into deeper caverns. Central Kentucky has about 60,000 sinkholes, and the southern part of the state of Indiana has more than 300,000 sinkholes.[34] Florida, which has an abundance of limestone bedrock, is also well known for its sinkholes. An ancient but famous one, located outside Gainesville, Florida, is called the Devil's Millhopper. This sinkhole is cone-shaped with steep sides. Sharks' teeth have been found at the bottom near the limestone bedrock, which indicates that this area was once covered with seawater. In May 1981, a large sinkhole formed in Winter Park, Florida, swallowing up part of a four-lane highway, a home, a swimming pool, and six automobiles when it formed a hole ~122 m (~400 ft) wide and ~38 m (~125 ft) deep. Another sinkhole, called the December Giant, formed in Alabama. It was 140 m (450 ft) wide and 50 m (165 ft) deep.[35] The collapse of the Florida sinkhole was due to an extended drought, while the one in Alabama resulted from extensive pumping of ground water. Both resulted from the dropping of their respective water tables, which removed the water support for the roofs of deeper caverns, thus collapsing the surface land areas.

Karst Topography

Karst topography is a unique type of landform that occurs in areas where limestone exposed near the surface of the Earth is subjected to flowing water. About 20% of surface land area of the Earth, and ~15% in the United States, is exposed limestone that is subjected to karst landform development. There are a great variety of karst formations, ranging from plains with slight depressions, to streams in valleys that just seem to disappear into the ground, to natural rock bridges and holes, to towering monoliths of insoluble rock. The term "karst" comes from a German word adapted from the Croatian word that describes this type of landscape. Areas of relatively heavy rainfall and above average temperatures promote a chemical reaction with limestone that erodes its surface and widens cracks and openings in the limestone rock formations. This reaction also forms caves and sinkholes, but karsts are mostly on the surface and appear as hollowed-out holes in rocks or, as found in China, *tower karsts,* which rise as monolithic peaks above the surface. Karst topography is found in many parts of the world, including Canada and the United States (Kentucky, Tennessee, Indiana, and Florida).[36–38]

Hot Springs

Hot springs are created when ground water is heated by (a) the heat in the Earth, which increases 30°C per kilometer of depth in the crust; at just a few miles deep the temperature can reach near the boiling point; (b) the heat from recent volcanic activity where the hot magma or hot igneous rock is still near the surface of the Earth (e.g., the hot geysers and hot springs of Yellowstone National Park); or (c) chemical reactions in the crust that produce heat; one example is the reaction of the sulfide chemicals in iron pyrite (FeS_2, fool's gold) with acidic water to produce hydrogen sulfide (rotten egg smell) and heat.

The main difference between a hot spring and a geyser is that a spring merely bubbles at the surface, whereas geysers, because of steam built up in cracks in rocks, are released explosively as eruptions spouting hot water and steam many feet into the air (e.g., Old Faithful at Yellowstone National Park).

Steam from geothermal energy can be used to produce electricity, or the hot water may be used to heat homes in areas that have an adequate supply of hot ground water. Several western states in the United States and several Northern European countries use this source of heat to produce electricity. Currently, about 2500 megawatts of electricity are being produced by geothermal power plants in several states, but this is just a drop in the bucket for what is needed to replace fossil fuels. The potential of geothermal energy is much greater than is currently used.[39] Someday it may be technically feasible to drill to depths where great quantities of water could be pumped down to hot spots and then extracted as steam to produce electricity.

Wetlands

Wetlands not only store great quantities of water but are also an important part of biological life in the biosphere. Wetlands also assist in the purification of polluted water. "Wetland" is a generic term for swamps, bogs, marshes, sloughs, mud flats, and flood plains. The term has a specific meaning for the U.S. Environmental Protection Agency (EPA) and some environmentalists in the United States. Federal, as well as some state, regulators consider any land area that is wet or soggy, even for part of the year, or dry every other year or so but having depressions of standing water in some years, as wetlands. This narrow definition has caused many problems with farms, private landowners, and even homes that have soggy yards part of the year, since, technically, they are wetlands and are subject to governmental regulations. Other nations do not have the same narrow definition of wetlands. Even so, the protection of some authentic wetlands is important, particularly on coastal saltwater tidal and inlet areas. For instance, about two-thirds of the Atlantic fish and shellfish species depends on coastal wetlands and estuaries for part of their life cycle. Wetlands are also breeding grounds for many birds and other life forms. In the past, large swamp areas in the United States were drained to produce farms and grazing land. Some years ago the Everglades area in Florida was partially drained to develop farmland, but recently a program to halt the drainage and reflood parts of this massive wetland has been instituted to not only restore the swampland but also help replenish Florida's receding supply of fresh ground water. Wetland restoration in large swamp areas is important for water purification, flood control, and replenishing ground water, as well as providing habitat for wildlife.[40]

The Hydrological Cycle

The hydrological cycle is a complex global system involving water circulation between and among the components of the hydrosphere within the larger system of the biosphere. It is an energy-driven system, mainly by radiation (heat) energy from the sun. However, the hydrological system also involves gravity and, to a lesser degree, geothermal energy generated by internal heat from the Earth. The United States currently has several satellites that monitor the atmosphere, humidity, water, temperature, and so forth, and return data to Earth that assist scientists in understanding how the water cycle between the Earth and the atmosphere and its surface affect life and our environment. See Figure 7.3 for a depiction of the interactions of water between the Earth and the atmosphere.

Water is unique in both its chemical and physical structure. Even the ancient Greek philosophers recognized it as one of the four essential substances of the Earth (water, air, fire, and earth). Over the centuries water has been recognized as an essential ingredient for life and has even been incorporated into religious theologies. As is well known, water is composed of two atoms of hydrogen combined with one atom of oxygen ($^{H}\text{-}o\text{-}^{H}$). The two hydrogen atoms more or less sit on the oxygen at a 105° angle (a structure for which the water molecule is sometimes called the Mickey Mouse molecule) where each of the two H^+ ions share an outer electron with one O^{--} ion, joining oxygen's six outer orbital electrons. This fulfills the required two electrons for the hydrogen atom's inner shell and the 8 electrons to fill the oxygen atom's outer

Figure 7.3
The hydrological cycle (or water cycle) is the constant circulation of water in the atmosphere, land, and sea. More than 97% of Earth's water is salty, about 1.8% is frozen, and the rest is freshwater on land and water vapor in the atmosphere. Precipitation of freshwater from water vapor in clouds is deposited on land and sea. In time, water on the land flows into the oceans, where it is again evaporated, forming water vapor that again condenses and is again deposited onto the Earth's surface to repeat the cycle.

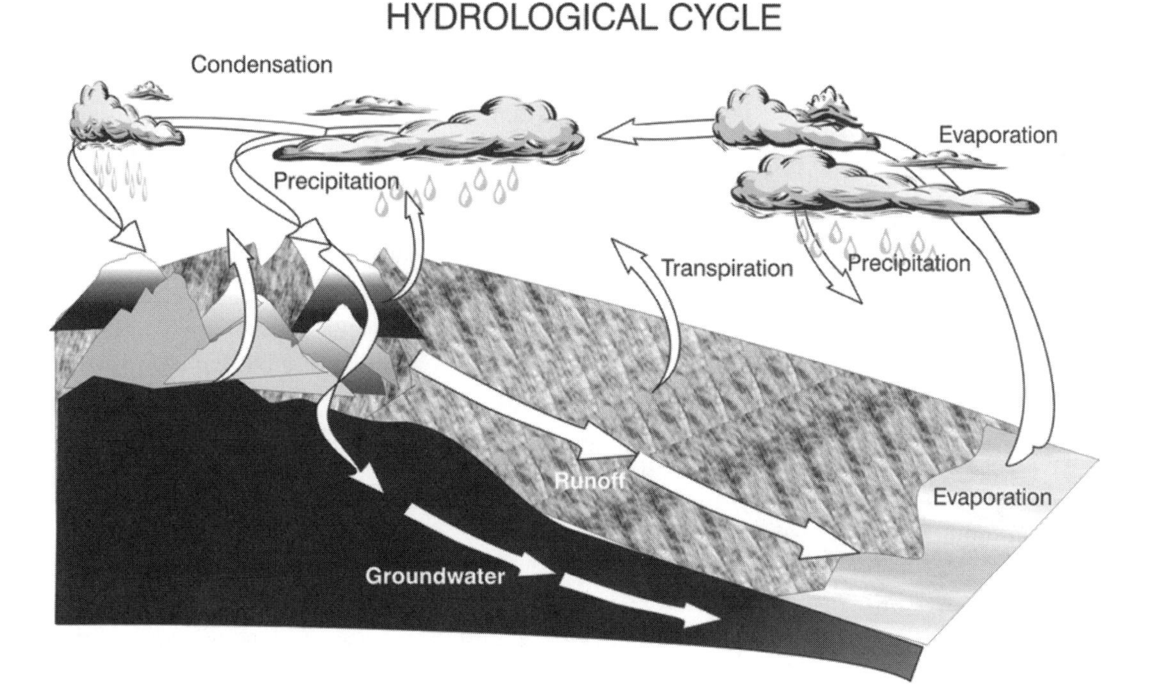

shell and results in the formation of a molecule of water. This means the water molecule (H_2O) has a slight imbalance of electrical charge on its outer surface. This polar nature of the water molecule is one factor that makes water chemically and physically important. This factor not only permits individual water molecules to cling together to form a liquid but also contributes to water's reputation as the universal solvent, because of its ability to dissolve so many different substances. Water is also one of the few substances whose physical state can be changed easily from a liquid to a solid or to a gas by relatively limited changes in temperature (and to some degree pressure).

The following physical attributes of water are some of those involved in various phases of the hydrological cycle.

- *Evaporation* is the change in the physical state of water from a liquid to a gas (water vapor). Water from the land and oceans evaporates to form water vapor. In addition, plant transpiration and animal respiration contribute water vapor to the atmosphere. Molecules of water vapor mix freely with the air molecules in the atmosphere. This is purely a physical, mechanical process driven by temperature and, to some extent, by pressure differences. Evaporation is a major component of the hydrological water cycle.
- *Condensation* is the reverse of evaporation. When the molecules of water suspended in the air as water vapor are cooled to a certain critical point, they condense from the gaseous state to liquid water. When the humidity is high in a warm room, and the outside temperature is low, water vapor will condense on a cold glass window, forming droplets. For this to occur in the atmosphere as rain, tiny particulate matter, such as salt, dust, soot, or pollution, must also be suspended in the atmosphere so that the water vapor has something on which to condense (analogous to the cold glass window). These microscopic particles act as condensation nuclei onto which the water vapor molecules can adhere and form droplets. If such particulate matter were not available in the atmosphere, we would have large amounts of water suspended in the atmosphere that would not condense into rain, snow, hail, or sleet. In other words, the relative humidity would be exceedingly high, which would affect not only living organisms on Earth but also geological features.
- *Humidity,* or "relative humidity," is not an absolute measurement of the amount of water vapor in the atmosphere. The amount of water vapor mixed with air is dependent on the temperature, as well as the shape of the surface over which the measurement is made. "Specific humidity" is the mass of water vapor per unit mass of moist air. In other words, it is a measure of how much water by weight can be compared to the weight of moist air. Because humidity is relative to the degree of saturation of the air above a surface where the measurement is made, the relative humidity may be more than 100%, but not by much more before the water vapor in clouds will condense as precipitation.
- *Saturation,* vapor pressure, and condensation of water vapor in the atmosphere are often incorrectly related to the temperature of the air in which water molecules are suspended. First of all, air is a mixture of gases with great spaces between atmospheric gaseous molecules. Therefore, there is much room remaining for molecules of water from evaporation to join this mix. The key point is that the water molecules are mixed with the air

molecules. The air does not hold the water vapor, nor is water vapor dissolved in air—humid air is a mixture of air molecules and vaporized water molecules. Thus, condensation leading to precipitation of water vapor is not dependent on the temperature of the air molecules but instead on the temperature of the water molecules as determined by the temperature of their surroundings at particular altitudes. (Even the *Encyclopedia Britannica* (CD 2001) entry for humidity confuses this concept.) More water vapor can mix with warm surface air than with cooler air at higher altitudes. Rain droplets are not formed when warm air molecules rise and then cool but instead when the water molecules condense onto cooler microscopic particulate matter that act as nuclei for the condensation process. It is the temperature at various altitudes that determines the temperature of these particulate nuclei that results in the condensation of water droplets to form precipitation. During the condensation process the air and water molecules are still a mixture. As nascent droplets continue to cool, and more moisture is condensed out of the water vapor onto the developing ice particles or droplets, they reach a point where their size is adequate to form rain, snow, or hail and fall to Earth as precipitation. Even though the temperature of the air and clouds of condensed moisture is usually about the same, it is the cooling of the water vapor molecules in clouds that causes precipitation.[41]

In summary, the hydrological cycle is global and involves a number of processes, subcycles, and systems. Hydrology is the science of the history and properties of water on the Earth and its interactions with the environment within an endless circulation between the oceans, land, and atmosphere. An interesting way of looking at the components of the hydrological cycle is to compare how long water resides in each of the cycle's components (see Table 7.2).

It is also interesting to compare the figures in Table 7.2 with the components of the hydrological cycle depicted in Figure 7.1. The exchange of water between and among these environmental components totals millions of cubic meters of water every day—it just takes longer to be processed and recycled by some than by others. The end result is that water is used over and over again in a variety of ways. Even some of the molecules of the water Aristotle drank are still circulating in the hydrological cycle, and no doubt even today people are drinking some of the same molecules of water that Aristotle drank.

Chapters 8 and 9 address the atmosphere, climate, and weather, which are also part of the hydrological cycle, as they contribute to the complex global systems that operate within the Earth's outer spheres.

Table 7.2
Residence Times for Water in the Different Components of the Hydrological Cycle

Component of Water Cycle	Residence Time
Oceans and seas	~2000 years
Lakes and reservoirs	~10 years
Swamps	1–10 weeks
River channels	Several weeks
Soil water	2 weeks–1 year
Groundwater	2 weeks–10,000 years
Ice caps and glaciers	10–100 years
Atmospheric water	10 days
Biosphere water	1 week

Sources: Adapted from R. A. Freeze and J. A. Cherry, *Groundwater* (New York: Prentice Hall, 1979); W. G. Ernst, ed., *Earth Systems: Processes and Issues* (New York: Cambridge University Press, 2000), 103.

Issues Concerning Water

The main issues concerning water are related to the availability, conservation, use, and misuse of freshwater. Although the quality of both freshwater and seawater has improved significantly over the past several decades, pollution is still a major concern. Before we examine issues related to these areas, let's look at some of the myths related to our drinking water.

Myths about Water

1. *We have less water than we did 100 years ago.* Wrong. The amount of water on the Earth today is essentially the same as it was billions of years ago. Because the total amount of freshwater may vary at different places on Earth (but not the total amount of *all* the water on the Earth), and because we often consume more from natural underground storage areas than can be replenished, the impression is that there is less water on Earth than in past ages. Since the human demand for freshwater has increased since 1900, and will continue to increase, and the supply of freshwater does not significantly increase through the water cycle, the myth that there is less total water then in past ages persists. During the past century consumption of freshwater increased twice as much as global population; thus, about a third of the world's population lives in countries with moderate to high water stress. Water stress is defined as water consumption that is more than 10% of the renewable freshwater supply. About 20% of the world's population lacks access to safe drinking water, while 50% lacks adequate sanitation systems.[42] These figures also have been reduced in the last few years, even though populations have increased. The pressures on limited natural freshwater supplies increase with the steady increase in global population and the demands to grow crops to feed these multitudes. However, as Bjorn Lomborg, in his book *The Skeptical Environmentalist,* states, "the data do not support [this] view of a mammoth problem. Our wells are not drying up; we are not facing insurmountable shortages. Rather the water challenges emphasize that we need to manage water more carefully, price it realistically, and accept a movement away from self-reliance in food production in arid parts of the world." Lomborg notes that this general conclusion is also expressed in the *Comprehensive Assessment of the Freshwater Resources of the World,* a report produced by the United Nations (UN). This UN report also claimed that the problem, rather than water stress, was "largely as a result of poor water allocation, wasteful use of resources and lack of adequate management action." It seems the main problem with management of water resources is the continued mining of ground water, estimated to be at 160 km^3 annually and which will increase at the present rate of mining to 600 km^3 annually by the year 2025. Two possible solutions, in addition to better management, are to dam and store more runoff freshwater and to use water for agricultural purposes more efficiently.[43]
2. *Once water is used, it is gone.* Wrong. According to the physical law of conservation of mass, water is essentially a renewable resource. Water is composed of the same H_2O molecules when it is pure as when it is used and discarded. Also, water is not fragile, nor is it destroyed when used. It can be treated and recycled over and over again without damage to the water molecules. While water molecules are broken down by chemical processes in living organisms, they are also reformed by plants and animals. Even so, water molecules are not basically altered before or after use and move freely throughout the hydrosphere.
3. *New water is better than old, used water.* Wrong. Almost all water has been used over and over again by plants and animals for the past several hundred million years. So-called pristine water may contain harmful bacteria and should be treated. Due to the processes involved in the hydrological cycle,

water is continually circulated throughout the biosphere. It is entirely possible that a few water molecules you drank today were the same molecules used by plants and animal many centuries ago.

4. *There are more pollutants in drinking water than there were 25–50 years ago.* Not necessarily so. This depends on the country and its history of public health and implementation of sewage and water treatment systems. In developed nations, the quality of drinking water has improved greatly over the past 50 years. Although about 50% of the drinking water in the United States comes from ground water, most of the world depends on streams, rivers, and lakes for their drinking water.
5. *Bottled water is safer to drink than tap water.* Not always. The quality of tap water is regulated by federal, state, and local governmental agencies, whereas there is no regulation to assure the quality of bottled water. Also, bottled water costs about 1000 times more than tap water per unit, and boutique bottled water costs even more per gallon than does gasoline.
6. *Lead found in your drinking water is the fault of the treatment plant.* Wrong. First, in the United States there is very little lead found in drinking water. Second, what lead is found is the result of water standing in pipes in older home water systems in which lead solder was used to seal pipe connections.
7. *Home water purification systems in the United States provide potable drinking water that is purer than municipal systems.* Not necessarily so. Home water treatment systems, such as reverse osmosis, may improve the taste of some tap water, but these small systems cannot *kill* all the microorganisms found in untreated water. Even so, reverse osmosis will filter out most particulate matter and larger microorganisms. Most municipal water systems not only kill bacteria but also try to improve (not always successfully) the clarity and taste of tap water.[44] On the other hand, a home water distillation system that uses heat to evaporate untreated water and condense it into pure water produces very pure drinking water, with little or no taste.

Freshwater Issues

Since the beginning of civilization humans have settled close to water. They moved if there was too little of it (drought) or if there was too much of it (floods). All people, no matter where on Earth, require a daily supply of freshwater for most of life's activities, including drinking, cooking, washing, industry, energy, transportation, rituals, and entertainment; if they can't get it, they will go to war to acquire a supply.[45]

Less than 3% of all the water on Earth is freshwater, and 2.15% of that is tied up in ice caps and glaciers, which leaves humans dependent on the remaining 0.65%, of which 0.62% is ground water. Much of that remaining freshwater supply is not where humans may need it most. Thus, all types of systems have been devised to transport water from where it exists to where it is needed, including canals, aqueducts, pipelines, diverting streams and rivers, and pumping from ground water and aquifers.[46]

Most ground water is found in aquifers that are bodies of porous and permeable rock where water flows some distance below the Earth's surface.[47] It would take 150 years to replenish all the ground water to a depth of 750 m in the United States if it was all removed.[48] This naturally stored freshwater is pumped by wells to provide both irrigation and drinking water. About 38% of county and city dwellers in the United States, and about 97% of rural households who do not depend on a city water company, secure their drinking water from ground water (wells and aquifers). The city of San Antonio, Texas, which pumps its drinking water from the great Edwards artesian aquifer, is the only city to be supplied completely by an aquifer, but it may soon have to find other sources of freshwater because its aquifer is being drained faster than it can be recharged by nature. One possible solution may be to drain excess river water during floods into aquifers. Other proposed solutions to recharge the Edwards aquifer include harvesting rainwater, cloud seeding, building recharge dams, using recycled water (treated sewage), diverting water from the Colorado River into the aquifer, desalination, and conservation. This latter solution is now being enforced by the city.[49]

The Ogallala Aquifer under the central plains states in the United States is being rapidly drained for irrigation to the extent that agriculture and jobs may soon be at risk.[50] Florida has extracted ground water from its aquifers to the extent that saltwater is encroaching into the underground supply.

Per person consumption of freshwater (for all uses including industrial and agricultural) has increased from about 1000 L/day to ~2000 L/day over the past century. At least 50% of this increase has been in agriculture. On the average, each person takes in more than 16,000 gallons of water in his or her lifetime, but most of it is, sooner or later, returned to the hydrological cycle. About 80 to 90% of the freshwater used for cooling electric power generating plants is returned to the hydrological cycle, and about 30–70% of irrigation water flows back to the rivers or percolates into the aquifers. Therefore, much of the water used for personal, business, industrial, and agricultural purposes is recycled.[51]

It is estimated that in the United States ~69% of the freshwater consumed is used for agriculture, ~23% for industrial purposes, and just 8% for home use.[52] The U.S. Department of Agriculture estimates that 90% of the total freshwater used in the United States is consumed by agriculture and industry, while municipalities, retail commerce, and households account for only 10% of the freshwater used. The percentage of water used for agriculture is even higher for areas such as the Rio Grande Valley of Texas and the Imperial Valley of California, where irrigation is extensively used.

The public is constantly encouraged to conserve water in their homes, which is a worthwhile but not very effective effort in actually reducing the use of freshwater. For instance, if all homes reduced their water consumption by 10%, the savings would be 10% of 10%. In other words, only 1% of all water used would be conserved. But if agriculture and industries reduced their consumption by 10%, the savings would be 10% of 90%, or 9% of all freshwater used, which would be a huge conservation effort.

Some suggested methods for conserving and reducing consumption of water are given in the following lists.

Home Water Conservation Ideas

1. Check toilets and pipes for leaks—a leaking toilet can waste up to 200 gallons of water a day. You can do this by checking the water meter when all water is turned off. If the dials move, you have a leak. Or place colored vegetable dye in the toilet tank and wait to see if any appears in the toilet's bowl.
2. Reduce the level of water in the toilet tank (place a brick or plastic bottle of water in it).
3. Run your dish and clothes washers only when you have a full load.
4. Replace all faucets with new flow-constricting, aerating fixtures. This reduces the flow of water up to 50%.
5. Reduce time for showers. Turn water off in sinks when brushing teeth or washing dishes, and so forth.
6. Reduce lawn watering, and use a bucket for car washing.[53–54]
7. Conserve freshwater through indirect methods: drive less—it takes 6000 gallons of water to produce 1 gallon of gasoline; eat less meat—5200 gallons of water are

required to produce 1 pound of beef, while only 23 gallons are required to produce 1 pound of tomatoes; and recycle—in general, it takes less water to produce goods from recycled materials than from raw material.[55]

Agricultural Water Conservation Methods

1. Plant drought resistant crops where possible, for example, Xeriscaping for both farming and home gardens.
2. Use limited freshwater for raising crops instead of animals. Feeding plants to humans instead of animals not only reduces the amount of water needed but also increases the total food supply available for humans by a factor of about ten.
3. Change the types of plant irrigation. Eliminate flood, row, and sprinkler types of irrigation that lead to excessive runoff, seepage, and evaporation. In 1995, approximately 32.7 million acres of irrigated land used flood irrigation, 24.9 million acres used spray irrigation, and only 1.7 million acres used drip irrigation. Using drip or subirrigation systems in areas where topography makes such irrigation possible is desirable.[56–57] Although drip systems are more expensive to install, in the long run they greatly reduce evaporation and conserve freshwater.
4. Use a regional irrigation water delivery system that has enclosed irrigation pipes instead of open canals and ditches to greatly reduce water loss by ground seepage and evaporation. Some areas in the southwestern United States use both open and closed irrigation systems in conjunction with reservoirs.
5. Use reclaimed water as a potential source for increased irrigation. In 1995, more than 92,600 square miles of land were irrigated in the United States (about 59,250,000 acres). During that year, 49,000 million gallons per day of *ground water* and about 84,000 million gallons per day of *surface water* were used for irrigation, but only a small amount (718 million gallons/day) of *reclaimed water* (treated sewage) was used for agriculture irrigation.[58] One reason it is not used more extensively is that reclaimed water is produced in urban areas but is needed for irrigation in rural regions.

One of problems in arid parts of the world is the continued attempt at self-sufficiency in food production that requires extensively irrigated crops. It would be more efficient (but most likely politically impractical) for countries with adequate water and large, often excess, agricultural production to provide food that would supplement reduced food production in those countries with severe water problems. Another method of increasing the supply of **potable** freshwater is to reduce the pollution of the supplies.

Pollution: Freshwater

Contrary to statements by some environmentalists, there does not seem to be an explosive global water crisis in the twenty-first century. In 1997, the UN produced the report *Comprehensive Assessment of the Freshwater Resources of the World,* which questions this gloomy assessment. It concluded that data do not support such an assessment and that wells and rivers are not running dry. Instead, the problem is poor allocation, wasteful use, and lack of adequate management of existing freshwater resources. The World Water Council stated in the global report *World Water Vision* that "[t]here is a water crisis today. But the crisis is not about having too little water to satisfy our needs. It is a crisis of managing water so badly that billions of people—and the environment—suffer badly."[59]

Even if there is adequate freshwater globally, about 20% of the world's population lacks safe drinking water and about 50% lacks adequate, safe sewage sanitation. For example, in most of the rivers on the Asian continent the fecal coliform (intestinal bacteria) count is 50 times higher than the standard set by World Health Organization (WHO) guidelines, and only 2% of the sewage in Latin America is treated. WHO estimates that fresh drinking water is polluted for about 1.2 billion people, worldwide, which contributes to the deaths of about 15 million children under the age of five each year. Sewage pollutes underground water in most underdeveloped countries.[60] The Worldwatch Institute's report on progress toward a sustainable society, *State of the World 2002,* states that "[t]he damage from deforestation borne by developing countries is especially disturbing when linked to wasteful consumption habits [of freshwater]."[61]

About a half century ago many rivers and lakes in the United States, as well as in northern industrial areas of Europe, were severely polluted. The Cuyahoga River in Ohio was overwhelmed with volatile pollutants to the extent that in 1969 the river actually burned for several hours. Some of the Great Lakes in the United States and rivers in Europe had fewer fish than 50 years before, and those fish that were still in the rivers and lakes were toxic. Swimming was banned in many rivers, including the Hudson and East Rivers in New York State, and many beaches. However, by the late 1960s the quality of water in the United States began to improve for a number of reasons, including new government pollution control regulations and efforts by private industries and agriculture. The *National Water Quality Inventory* conducted in 1973 found that water pollution levels had greatly decreased, and *Resources for the Future* stated that 93 to 96% of the nation's waters were fishable in 1972.[62] Bjorn Lomborg, in *The Skeptical Environmentalist,* states that over the past 40–50 years the quality of water in rivers and lakes in both North America and Europe (and to a lesser extent in some other parts of the world) has greatly improved. This includes vast improvement in the percentage of oxygen in the water for the following rivers: the Hudson River and the New York Harbor (from ~18% to ~60% in dissolved oxygen between ~1940 and ~1995), the Thames River in Great Britain (~36% to ~65% between ~1950 and ~1975), and the Rhine River in Germany (~21% to ~74% between ~1970 to ~1995). The increase in oxygen in rivers since World War II, especially in European rivers, is attributed to biological treatment of domestic and industrial wastewater that is recharged into the hydrological cycle. This higher percentage of oxygen has improved the quality of the rivers to the extent that they again thrive with aquatic life.[63] What happened?

Most of the water pollution originated from what are called "point sources," which means that the pollutant was released at the site of its origin—usually an industrial site, but also from antiquated and inefficient municipal sewage disposal systems, farm erosion, and animal feedlot runoff. Regulations designed to reduce or eliminate pollution also played a part in improving the quality of water. The *Federal Pollution Control Act Amendments* of 1972 required factories to install pollution control technologies and provided grants up to 75% of costs to municipalities for upgrading old sewage disposal plants and building new ones. These efforts, plus improved farming techniques and an awareness of the general public, soon improved the quality of the U.S. rivers and lakes. Fishing and swimming have returned, more than 75% of the population in the United States is served by wastewater treatment facilities, and the safety of drinking water in the United States is of the highest quality. Yet, even with these positive improvements over the past decades, more can be done, particularly in developing countries where poor water management and inadequate pollution controls still exist.[64]

Saltwater Issues

Coastal areas and the seas have been important to human beings since the beginning of human existence. Not only did life most likely originate in seawater, but humans tended to settle near both freshwater and seawater. Cultures developed on different continents partially due to travel by sea, which provided a means for commerce and trade to develop. The oceans still provide a water highway for human activities. The exploitation of the seas has both benefits and hazards. One hazard of current concern is pollution.

Pollution: Seawater

The story of pollution of the oceans is similar to that of freshwater. Some countries still send raw sewage and garbage out to sea, and ships of all kinds at one time dumped their wastes into the open sea, fouling coastal areas. Much of this type of pollution is being controlled by international agreements that require ocean-going vessels to dispose of their wastes once they dock. The open oceans are now relatively pollution free, while coastal areas are less so.

Oil tankers and some other types of vessels still, from time to time, dump ballast seawater both close to the shore and in open seas. This ballast may contain oil, toxic wastes, sewage, or foreign organisms. For instance, some years ago the freshwater zebra mussel, which is not native to the United States, was brought to Lake Erie by foreign ships. It has no natural enemies in its new environment and thus thrives. Though relatively small, these exotic bivalves rapidly reproduce, clogging intake water pipes for power plants and municipal water systems.

Damaged and wrecked oil tankers, from time-to-time, leak millions of gallons of oil near coastal areas. However, accidental oil spills by supertankers, such as the *Exxon Valdez* in Prince William Sound in Alaska, are not the only source of oil in the oceans. Other sources include offshore drilling and drainage from natural oil seeps on the ocean floor. The major sources of oil in lakes and rivers, as well as coastal beaches, are from barges and motors of fishing and pleasure crafts.

Both international cooperation and new technologies have reduced the ecological damage from massive oil spills near coastal areas. Three new technologies to remove spilled oil are Inipol, Eco Spheres, and Elastol. (1) Nature provides several natural biomicroorganisms that consume oil in seawater, but they take some time to break down crude oil. Inipol is a genetically altered natural microorganism that eats spilled oil, thus reducing damage to the environment. This man-made microorganism is related to the first genetically altered life forms to receive a United States patent and was used after the 1989 *Exxon Valdez* oil spill. It speeded up by two or three times the period required for natural microorganisms to digest the spilled oil. (2) Eco Spheres (Heller's Beads) are tiny glass beads coated with titanium dioxide that, upon becoming coated with oil, break down oil's molecular structure into carbon dioxide and water. The floating residue is then skimmed off the surface of the water. This technology was not used in the *Exxon Valdez* oil spill. Had it been used, it might have cleaned it up in a few weeks. (3) Elastol is a chemical that congeals oil into a gooey, filmlike skin that can be pumped off the surface area while returning purified oil to a tanker.

It is interesting to note that the coastal areas of the *Valdez* oil spill that were left uncleaned recovered in just 18 months, whereas the treated beaches required three to four years to recover. According to the U.S. National Oceanic and Atmospheric Administration (NOAA), the original cleanup did more harm than good. NOAA estimated that 20% of the original oil evaporated, 50% was broken down by chemicals and natural activity, 12% is lying in lumps on the bottom of Prince William Sound, and about 3% is still on the beaches as nontoxic tar balls. It appears that the pressure-washing of the coastal areas killed much of the marine life.[65]

Since the 1990 Oil Pollution Act, damage from ocean oil pollution has been greatly reduced. The act included increased financial liability for owners of supertankers, established an Oil Spill Liability Fund, required all tankers arriving in U.S. ports to have double hulls, provided for unlimited liability if the ship's captain is intoxicated, and made alcohol and drug testing of crews mandatory. Reduction in ocean pollution, including oil spills, is most evident in the United States, but unfortunately, not all companies outside the United States comply with these regulations.[66]

Many environmental activists and the media overreact to major oil spills in the oceans. Oil experts claimed that the clean up procedures for the *Valdez* accident would be more harmful than merely allowing nature to take its course, but this did not harmonize with the public view that the cleanup had to be undertaken, regardless of costs, for betterment of the animals.[67]

Oil spills in the ocean are not nearly as long lasting or damaging as some groups would have the public believe. A study by the National Academy of Sciences found that offshore drilling accounts for only 2% of the oil drained into the world's oceans, with a small percentage of the balance from tanker accidents. By far, most of the oil found in the world's oceans comes directly from natural seepage from the ocean seabed. Researchers from Texas A&M University discovered thriving communities of mussels and tube worms feeding on natural oil and gas seeps on the floor of the Gulf of Mexico, which they claim is proof that oil is a natural part of the marine ecosystem.[68] Jack Hilton reported in the May 15, 1989, issue of the *New York Times* that Prince William Sound contained about 262 trillion gallons of water, while the *Exxon Valdez* oil spill amounted to 10 million gallons, which he stated was equal to about a teaspoon of oil spilled into a swimming pool. No one should claim that man-made oil spills do not cause some ecological damage in the short term, but recovery is much faster and more complete than many environmental activists claimed. In 1993, NOAA reported that "the Sound has made a remarkable ecological comeback because of human efforts and nature's own cleansing process."[69] Today, there is little evidence of damage where the *Valdez* spill occurred—the Earth has great power to heal itself.

In conclusion, environmentally things are better—but not necessarily good. Although over the past 40 years the use and treatment of the hydrosphere's water resources have greatly improved, particularly in the developed countries, there is still much that can be done globally to improve our water resources.

Notes

1. "Spacecraft Finds Water beneath Ganymede," *USA Today,* December 18, 2000.

2. *Earth Science: Origin of Water on Earth,* Science-Week Focus Report 18, December, 1998, accessed 2001, http://scienceweek.com/swfro65.htm.

3. Greg Clark, *Extraterrestrial Water: Water From the Heavens,* accessed 2001, http://www.space.com/science/astronomy/meteorite_water.html.

4. William Steigerwald, Goddard Space Flight Center, Greenbelt, Md., *Dying Comet's Kin May Have Nourished Life on Earth,* accessed 2000, http://www.gsfc.nasa.gov/news-release/release/2001/01-46.htm.

5. *A Taste for Comet Water,* accessed 2000, http://www.solarviews.com/eng/linear2.htm.

6. *Water in Orion,* accessed 2000, http://cfa-www.harvard.edu/cfa.hotimage/worion.html.

7. James C. G. Walker, *Scientific American*: Ask the Experts, accessed 2001, http://www.sciam.com/askexpert/environment/environment13.html.

8. *Encyclopedia Britannica,* CD-ROM 2001, s.v. "Hydrosphere."

9. Stanley Chernicoff, *Geology: An Introduction to Physical Geology* (New York: Houghton Mifflin, 1999), 390.

10. Patricia Barnes-Svarney, ed., *The New York Public Library Science Desk Reference* (New York: Macmillan, 1995), 397.

11. Graham R. Thompson and Jonathan Turk, *Earth Science and the Environment* (New York: Harcourt Brace, 1999), 314.

12. Howard Herzog, Baldur Eliasson, and Olav Kaarstad, "Capturing Greenhouse Gases," *Scientific American,* February 2000, 72–79.

13. Thompson and Turk, *Earth Science,* 315.

14. Kenneth W. Hamblin and Eric H. Christiansen, *Earth's Dynamic Systems* (Upper Saddle River, N.J.: Prentice Hall, 1998), 212.

15. Isaac Asimov, *Asimov's Chronology of Science and Discovery* (New York: Harper and Row, 1989), 297–98.

16. Hamblin and Christiansen, *Earth's Dynamic Systems,* 212–18.

17. Thompson and Turk, *Earth Science,* 318–21.

18. *Encyclopedia Britannica,* CD-ROM 2001, s.v. "Boundary Ecosystems between Waters: Estuaries."

19. U.S. Geological Survey, *Water Science for Schools: Glaciers and Icecaps,* accessed 2001, http://www.ga.water.usgs.gov/edu/earthglacier.html.

20. Bjorn Lomborg, *The Skeptical Environmentalist: Measuring the Real State of the World* (New York: Cambridge University Press, 2001), 150.

21. *Encyclopedia Britannica,* CD-ROM 20001, s.v. "Environmental Works: Surface Water Sources."

22. Thompson and Turk, *Earth Science,* 212.

23. Thompson and Turk, *Earth Science,* 212–26.

24. Hamblin and Christiansen, *Earth's Dynamic Systems,* 277–306.

25. U.S. Geological Survey, *Water Science for Schools: Earth's Water: Lakes and Reservoirs,* accessed 2001, http://www.ga.water.usga.gov/edu/earthlakes.html.

26. Marguerite Holloway, "The Killing Lakes," *Scientific American,* July 2000, 92–100.

27. Hamblin and Christiansen, *Earth's Dynamic Systems,* 275.

28. Thompson and Turk, *Earth Science,* 226–31.

29. *Encyclopedia Britannica,* CD-ROM 2001, s.v. "Hydrosphere: Lake Water."

30. Hamblin and Christiansen, *Earth's Dynamic Systems,* 374.

31. Jacqueline L. Harris, *Science in Ancient Rome* (New York: Grolier, 1998), 14.

32. *Microsoft Encarta,* CD-ROM 1994, s.v. "Groundwater."

33. Hamblin and Christiansen, *Earth's Dynamic Systems,* 233–34.

34. Chernicoff, *Geology,* 458–62.

35. Chernicoff, *Geology,* 461.

36. Thompson and Turk, *Earth Science,* 236.

37. Hamblin and Christiansen, *Earth's Dynamic Systems,* 326–30.

38. Chernicoff, *Geology,* 458.

39. Thompson and Turk, *Earth Science,* 237.

40. Thompson and Turk, *Earth Science,* 238–39.

41. Steven M. Babin, *Water Vapor Myths: A Brief Tutorial,* from Bohren and Albrecht, Penn State University 1998, accessed 2001, http://www.fermi.jhuapl.edu/people/babin/vapor/index.html.

42. *Geo-2000: Freshwater: State of the Environment,* accessed 2001, http://www.grid2.cr.usgs.gov/geo2000/english/0046.htm.

43. Lomborg, *The Skeptical Environmentalist,* 157.

44. There are several internet references for these myths, including http://www.1.stpaul.gov/depts/water/pages/myths.htm (accessed 2001); http://www.awwa.org/bluethumb/mythsrealities.htm (accessed 2001); and http://www.hashville.org/ws/drinking1.htm (accessed 2001).

45. Mikhail Gorbachev, "Out of Water: The Distant Alarm Comes Closer," *Civilization,* October/November 2000, 82–83.

46. Lomborg, *The Skeptical Environmentalist,* 149.

47. *Scientific American Science Desk Reference* (New York: John Wiley and Sons, 1999), 289–90.

48. Lomborg, *The Skeptical Environmentalist,* 149.

49. *Aquifer: The Edwards Homepage,* accessed 2001, http://www.edwardsaquifer.net.

50. Albert Gore, *Earth in the Balance* (New York: Plume, 1993), 110–11.

51. Lomborg, *The Skeptical Environmentalist,* 150–51.

52. Lomborg, *The Skeptical Environmentalist,* 154.

53. Texas Natural Resource Conservation Commission Media Relations, *Texans Can Be Water Wise by Making Simple Changes at Home,* accessed 2001, http://www.tnrcc.state.tx.us/water/wu/drought/nycu.html.

54. *10 Tips for Water Conservation,* accessed March 4, 2001, http://cleanoceanaction.org/COATips/COASTips/H2Oconserve.html.

55. Lori Dombroski, "The Grain Dilemma: Feeding Livestock or Feeding the World," in *Everybody Eats,* Fort Collins Food Co-op, September/October 1995), 4.

56. U.S. Geological Survey, *Water Science for Schools,* accessed 2001, http://www.ga.water.usgs.gov/edu/qausage.html.

57. *Encyclopedia Britannica,* CD-ROM 2001, s.v. "Irrigation and Drainage: Water Application."

58. U.S. Geological Survey, *Water Science for Schools,* accessed 2001, http://www.ga.water.usgs.gov/edu/qausage.html.

59. As cited in Lomborg, *The Skeptical Environmentalist,* 157.

60. *Geo-2000: Freshwater: State of the Environment,* accessed 2001, http://www.grid2.cr.usgs.gov/geo2000/english/0046.htm.

61. Christopher Flavin, Harry French, and Gary Gardner, *State of the World 2002.* Worldwatch Institute (New York: W.W. Norton, 2002), 9.

62. Joseph L. Bast, Peter J. Hill, and Richard C. Rue, *Eco-Sanity: A Common-Sense Guide to Environmentalism* (New York: Madison Books, 1994), 15–18.

63. Lomborg, *The Skeptical Environmentalist,* 202–5.

64. Bast, Hill, and Rue, *Eco-Sanity,* 16–17.

65. Lomborg, *The Skeptical Environmentalist,* 193.

66. Bast, Hill, and Rue, *Eco-Sanity,* 151–54.

67. Lomborg, *The Skeptical Environmentalist,* 193.

68. Bast, Hill, and Rue, *Eco-Sanity,* 150.

69. Bast, Hill, and Rue, *Eco-Sanity,* 151.

8

Atmosphere: Sphere of Air

The term "atmosphere" is a combination of the Greek *atmos,* which means "vapor," or "air," plus the Latin *sphaera,* which means "sphere," or "globe." Although most of the planets in the solar system have atmospheres composed of various combinations of chemical elements and compounds at varying temperatures and pressures, the planet Earth's atmosphere is unique in the solar system because of its somewhat symbiotic relationship to the other spheres in the Earth's biosphere. The Earth's atmosphere, one of the three major spheres comprising the biosphere (hydrosphere, soil/crust, and atmosphere), consists of several different, fluid, gaseous subspheres, strata, or envelopes. The fluid nature of the atmosphere means there is some movement of atoms, molecules, and ions between layers, and thus these layers are not well differentiated except at the highest altitudes. Some examples of these subsphere transition zones are the *tropopause, stratopause, mesopause,* and *thermopause.* More than 99% of the Earth's total atmospheric mass is located 40 kilometers (km) (25 miles [mi]) from the Earth's surface. The major factors used to distinguish one stratum (layer) from another are related to properties of the air's chemistry, as well as the temperature and density/pressure at various altitudes.[1] The fluid nature of the atmosphere's different layers is composed of molecules of different gases, ions (charged particles), water vapor, and minute particulate matter that influence the varying degrees of density/pressure and temperature of Earth's atmospheric environment. This gaseous mixture results in a complicated system of moving air and water that affects the Earth's hydrosphere, ocean currents, climate, weather, constantly changing temperatures, and amount of water vapor in the atmosphere. All of these factors relate to Earth's geological systems.[2] The processes of condensation and **sublimation** within the atmosphere result in cloud and fog formations consisting of water vapor that forms droplets of ice crystals that can precipitate as rain, hail, sleet, snow, dew, or frost. Clouds also contain minute particulate matter from meteoric or surface origins, including natural and man-made pollution. Without the atmosphere the Earth would be a lifeless sphere of rock. (For more on weather and clouds, see chapter 9.)

Origin of the Atmosphere

As with the origin of the Earth's geosphere, its hydrosphere, the biosphere, and life, there are several theories for the origin of the Earth's atmosphere.

Outgassing Theory

The outgassing of hydrogen (H_2) and helium (He), the most common gases of the early universe, occurred during the formation of the Earth. These light gases escaped from the Earth's surface because Earth's gravity was not strong enough to hold them close to the surface. (This is still true today.) The early Earth experienced continuous episodes of volcanism and plate tectonics in which a number of gases were expelled into the early atmosphere. In addition to H_2, there was water (H_2O), carbon monoxide (CO), carbon dioxide (CO_2), hydrochloric acid (HCl), methane (CH_4), ammonia (NH_3), nitrogen (N_2), and various sulfur (S and SO_2) gases. The outgases consisted of ~68% water, ~13% carbon dioxide, and ~8% nitrogen, with traces of the other gases. There was no free oxygen, meaning that the Earth's original atmosphere was **reducing**, not oxidizing, and thus could not support life that required oxygen for metabolism. Evidence for a lack of oxygen in the early atmosphere is based on several facts. First, minerals such as urananite and pyrite oxidize readily in today's oxygen-rich atmosphere, but they are still found in Precambrian stabilized sediments. During this same period there were many iron-rich minerals and rocks, but no oxidized iron compounds (rust) as there are today. Second, some simple, early forms of life were anaerobic, meaning the metabolism of these types of ancient bacteria required no oxygen. Several species of anaerobic archaebacteria still exist (e.g., botulism) but are killed when exposed to the present atmosphere. Third, when adequate oxygen accumulated in the atmosphere, between 2.5 and 3.5 billion years ago, the essential ingredients such as amino acids, DNA, and proteins developed in that oxygen-rich atmosphere.

Accretion Theory

The accretion theory is based on the compaction of matter on the Earth and other planets of the solar system when they were bombarded by a multitude of meteors, asteroids, and comets that deposited numerous chemicals and minerals on their surfaces during their formation. These impacts on Earth created high internal temperatures. When combined with internal radioactivity, the Earth was so hot that it required millions of years to cool down. At the same time, accretion covered the planet's surface with a blanket of debris that helped retain heat.[3] During the accretion phase, volatile substances such as the gases mentioned previously were deposited in and on the Earth, but oxygen was not. Not until ~3 to ~4 billion years ago, after the Earth cooled, did the primitive atmosphere lose much of its carbon dioxide through the weathering of minerals by the chemical reaction $CaSiO_3 + CO_2 \leftrightarrow CaCO_3 + SiO_2$ at the Earth's surface.[4]

There are several theories explaining the origin of oxygen on the Earth. One involves the *thermal dissociation* of H_2O (water) that occurred when the Earth was cooling down but was hot enough for the hydrogen to be separated from the oxygen in water. The hydrogen escaped from the Earth's surface by the process of *hydrodynamic escape,* in which hydrogen's thermal energy (kinetic molecular motion) is adequate for the molecules to escape Earth's gravity.[5] This theory for an oxygen-rich atmosphere is based on the idea that O_2 molecules seldom achieve adequate thermal energy for hydrodynamic escape to occur and thus were left behind as the source of early oxygen. Another producer of oxygen was early *photochemical dissociation* (photolysis), in which water molecules are broken down by ultraviolet light and the H_2 is again lost by hydrodynamic escape while the O_2 remains in the atmosphere. This process accounts for about 1 or 2% of the current oxygen in the atmosphere. By far the most important source of oxygen, both historically and currently, is by the process of *photosynthesis* that takes place in green plants in the presence of sunlight: CO_2 + H_2O + sunlight → living matter + O_2. Early cyanobacteria produced only small amounts of ancient oxygen, but these were enough to assist higher plants to evolve to the point where they supply most of the oxygen in the atmosphere.[6–7]

Other theories suggest that volatiles from volcanoes were insufficient to form an atmosphere and that much of the Earth's atmosphere was formed when the solar system's comets and meteors, which contain ice and rock, frozen carbon dioxide, ammonia, methane, and other gases, bombarded the Earth's surface. Organic molecules such as methane are thought to be common in the distant, myriad asteroids found in the Kuiper belt beyond the planet Pluto. The Kuiper belt, located in the outer reaches of the solar system, is similar to but much larger than the asteroid belt between Mars and Jupiter. It contains Kuiper belt objects (KBOs) ranging from 50 km to 1,200 km in size, with more than 100,000 KBOs greater than 100 km in size. According to S. Alan Stern of NASA's New Horizons mission to Pluto, billions of years ago some of these KBOs could have added gases and chemicals necessary to support life on Earth.[8]

As the Earth cooled, the water vapor condensed from this mixture, forming primordial oceans. Earth's heavier elements were produced from both material that formed the sun and solar system and some heavier elements that were generated inside the sun.[9] Recently, Robert M. Hazen of the Carnegie Institution of Washington's Geophysics Laboratory theorized that inorganic minerals were also required for carbon-based life to form. His theory is not accepted by all geologists, but he stated that ancient carbon-based gases on Earth, such as carbon dioxide (CO_2), carbon monoxide (CO), and methane (CH_4), had only a single carbon atom. However, complex building blocks of living organisms require complex molecules with dozens of carbon atoms in their molecular chains and rings. His theory asserted that some mechanism was necessary to protect short carbon chains from the harshness of ultraviolet radiation so that they could, in time, combine to form longer chains required for self-replicating molecules, namely, life. His candidates as protectors of early, prelife carbon-based molecules were different types of early minerals (pumice, clays, and crystals) that supplied (1) tiny compartments as shelters to protect carbon-based molecules; (2) scaffolding on which the molecules could assemble and grow; (3) shelter, providing support for early life functions; (4) metallic ions required for organic chemical reactions resulting in self-replication; and (5) dissolved substances that became part of the living matter.[10] The use of minerals as a means to assist long-chained carbon-based molecules to get a start, or some similar theory, would be required for early life to advance. Brooks Hanson and others suggested that methanorganic bacteria created a drop in the levels of carbon about 2.5 to 3 billion years ago. They claimed this is evidence that paved the way for an oxygen-rich atmosphere.[11]

Svante August Arrhenius (1859–1927), the Swedish electrochemist who received the 1903 Nobel Prize in chemistry, later in his life proposed a theory called **panspermia.**[12] Panspermia states that life came to Earth as bacteria spores or other simple forms from outside the solar system and were transported by radiation pressures or comets and meteors to Earth.[13] Although not all scientists accept the panspermia theory, it recently has been revived. Theories of spontaneous synthesis of life in space do not address the problem of survival of long-chain carbon molecules. Long exposure to radiation in space and on the Earth's surface would make their survival problematic. These theories require some mode of protection for the microorganisms to survive to become complex, carbon-based molecules once they reached Earth. Without protection as proposed by Hazen, they could not survive to form the long carbon chains necessary to self-replicate and evolve to the stage of photosynthetic plant life that later developed.

Chemistry of the Atmosphere

The gases of the original atmosphere that were brought to the Earth and the gases from volcanic activity changed over eons of time as they interacted with Earth's geology. These nascent gases evolved as they reacted with Earth's minerals, rocks, water, and, in time, living organisms. With the development of photosynthesis, plants became capable of converting the sun's energy, CO_2, and H_2O into carbohydrates and oxygen. In time, the level of atmospheric reactive oxygen reached a chemical/physical balance at a concentration of about 21%, as compared with about 79% of nonreactive gases. The chemical composition of air, expressed as percentage as found in dry air at sea level today, is given in Table 8.1.

An article by David Catling, Kevin J. Zahnle, and Christopher McKay in the August 3, 2001 issue of *Science* magazine describes some new ideas about the chemistry of the ancient atmosphere. There seem to be two theories. One idea is that the concentration of oxygen increased in the ancient atmosphere due to photosynthetic plants. The other theory is that plants provided a biogenic source of methane (CH_4) that combined with carbon dioxide (CO_2) in a reaction that caused a net gain of oxygen in the Earth's atmosphere. That reaction is described as follows:

$$CO_2 + 2H_2O \rightarrow CH_4 + 2O_2 \rightarrow CO_2 + O_2 + 4H \uparrow \text{(into space)}$$

Much of the methane was broken down by this reaction. This allowed the lighter hydrogen gas to escape into space (hydrodynamic escape), resulting in an increase in the concentration of oxygen to the present level of about 21%. It is also theorized that much of the ancient methane was the result of methane hydrate deposits (methane ice) within the seafloor slowly leaking massive quantities of methane into the atmosphere that replaced carbon dioxide. Ancient methane acted as a greenhouse gas, and, even today, methane is 60 times more powerful as a greenhouse gas than is carbon dioxide.[14] In essence, the ancient Earth's atmosphere changed from one that promoted chemical *reduction* reactions to one leading to chemical *oxidation* reactions. This explains why the present-day surface of the Earth is irreversibly oxidized.[15]

The chemical composition of the atmosphere not only gives clues to its origin but also tells us much about the interactions of the gases that make up air with both the geology and the biology of the Earth. The present nature of soil/crust, minerals/rocks, lakes/oceans, and life would not be the same as it is today if the gases, particularly oxygen, did not evolve as they did in the past or

Table 8.1
Chemical Composition of the Air at Sea Level

Components of Air (Gases)	Percentage of Dry Air (by volume)
Nitrogen (N_2), not very reactive, 780,900 ppm*	78.09
Oxygen (O_2), most reactive, 209,500 ppm	20.94
Argon (AR), most abundant noble gas, 9,300 ppm	0.93
Carbon dioxide (CO_2), 300 ppm, % varies	0.03
Neon (Ne), an inert noble gas, 18 ppm	0.001
Helium (He), inert noble gas lighter than air, 5 to 5.2 ppm	0.0005
Methane (CH_4), organic compound molecule, 1.5 to 2 ppm	0.00015
Krypton (Kr), an inert noble gas, 1 ppm	0.00011
Hydrogen (H_2), lightest gas, 0.5 ppm	0.00005
Oxides of Nitrogen (N_2, NO_2), by solar radiation	0.00005
Ozone (O_3), varies with altitude, 0.4 ppm, avg. %	0.00004
Carbon monoxide (CO), poisonous, % varies	0.00003
Xenon (Xe), a rare inert noble gas, 0.08 ppm	0.000009
Water vapor (H_2O), varies greatly with temperature, avg. %	$<$1.0
All other gases	$<$0.000001

Sources: Patricia Barnes-Svarney, ed., *The New York Public Library Science Desk Reference* (New York: Macmillan, 1995), 423; Kenneth W. Hamblin and Eric H. Christiansen, *Earth's Dynamic Systems* (Upper Saddle River, N.J.: Prentice Hall, 1998), 201; Graham R. Thompson and Jonathan Turk, *Earth Science and the Environment* (New York: Harcourt Brace, 1999), 347; *Encyclopedia Britannica,* CD-ROM 2001, s.v. "Average Composition of the Atmosphere."
*ppm = parts per million by volume.

continue to slowly evolve as they are in the present. Without the envelope of air containing a high percentage of carbon dioxide, methane, and water vapor to form a so-called greenhouse blanket that retains some of the heat from solar radiation and sends it back to the surface, the Earth would be a giant ice ball of frozen rock and water. The percentages of gases that comprise the mixture we know as air have been changing since the beginning of time and, no doubt, will continue to change, but only slightly. There now appears to be a stable balance maintained by a series of natural cycles interacting between and among the mixture of gases that makes up the atmosphere, the geosphere, and the biosphere. The percentage (by volume or weight) of some of the atmosphere's gases may change, but not by much. These changes are due mostly to natural events such as volcanic activity and other natural chemical, physical, and biological phenomena. Also, the very slow alteration of the Earth's orientation to the sun and the sun's variation of output of radiation affects Earth's climate and the composition of the atmosphere. Living organisms are part of the balance equation, as are the natural physical forces driving the geological changes of the Earth. Humans and other living organisms also have an effect on and alter the percentages of some atmospheric gases. All living things interact with each other, as well as with their environments (i.e., ecological systems). In other words, living organisms cannot be separated from their environments, and all living (and dead) things have an impact on the atmosphere's chemical systems. Note also that basically this balance of systems is dependent on the energy of the sun, but this balance is not the same as thermodynamic equilibrium. Likewise, all living organisms, sooner or later, reach a chemical equilibrium with their environment—which is death and decay. Living plants and animals, including humans, produce carbon dioxide, oxygen, and water vapor through respiration and transpiration. In addition to carbon dioxide, methane is released into the atmosphere through digestive elimination, death, and decay. Human social activities also alter the ratio of some gases in the atmosphere, but not nearly as much as is often attributed to human activities, particularly when compared to natural activities—that is, unless life and activities of living plants and animals (including humans) are not considered natural. The atmosphere is a giant, efficient sink that has assisted the Earth in maintaining a self-correcting balance of its chemical, physical, and biological cycles over the past ages. Note also that slight increases or decreases in the atmosphere's temperature and chemical composition are somewhat periodical and cyclical, and these have been normal since the beginning of time. The critical balance may fluctuate, but for the most part it is maintained by a series of cycles involving the three main spheres of the biosphere (atmosphere, hydrosphere, lithosphere), as well as to a lesser extent all the spheres of the Earth's geosphere. A short discussion of a few of these cycles follows.

Carbon Cycle

The carbon cycle involves the atmosphere, the hydrosphere, and soil/rocks, as well as the biology/ecology of living organisms. The global carbon cycle involves the exchange of carbon dioxide (CO_2) between and among the atmosphere, the hydrosphere, and the crust/lithosphere of the biosphere by a series of chemical and physiological activities.

The metabolic processes of microorganisms produce most of the organic CO_2, as well as perform most of the recycling of carbon in the global carbon cycle. About half of the global fixation of carbon is accomplished by biomass (mainly land plants). Inorganic CO_2 is mainly produced by volcanic action and is used in the chemical weathering of silicate rocks. Carbon dioxide is also dissolved in the oceans and is essential in the formation of carbonate rocks. The rate and amount of exchange of carbon in the cycle is dependent on the concentration of CO_2 in the water, or in the air above the water, or on volcanic and human activities.

If there is excess CO_2 in the atmosphere, it is diffused into the water, where it becomes concentrated and used by marine life to form carbon compounds in the form of carbonates and bicarbonates (shells and skeletons) that, in time, precipitate to form seabed sediment. Some carbon also remains in or on the surface of the Earth in the form of organic matter that, over thousands of years, becomes carboniferous peat, coal, oil, or natural gas. Both organic and inorganic processes are responsible for the great reduction of the atmospheric CO_2 that was present in high concentrations in the primitive atmosphere.

The carbon cycle also involves the ecology of living organisms. In fact, carbon is considered the basis for all organic matter, not only by definition but also by the fact that all organic matter contains complex compounds of carbon. Note that both so-called organically grown vegetables and so-called nonorganically grown vegetables are all *technically* organic, and are also all natural, because they both contain the essence of living matter, that is, carbon. The main difference is that the organically grown vegetables use less fertilizers and pesticides and thus cost more—they are no more nutritious than vegetables produced by other methods. Animal manure is often used as a natural fertilizer for organic gardening, and it also contains the same elements (nitrogen, phosphorous, etc.) as do chemical fertilizers—and the plants don't know the difference.

The microbial carbon cycle is related to both the production and the decomposition of organic matter. The cycle might be thought of as layers: (1) the aerobic zone is the uppermost region of soil and surface atmosphere, where oxygen-loving organisms exist; this is the zone where photosynthesis, respiration, and the nitrogen cycle take place; (2) the next deeper zone is the anaerobic zone, where sulfate reduction reactions dominate and denitrification occurs; (3) the next deeper zone is the anaerobic zone where the sulfate reactions cease and fermentation and methane production occur; and (4) the deepest zone in the carbon cycle is the burial zone, where organic matter forms sediments.[16]

Herbivore animals consume the carbon-rich plant food, converting it into carbon dioxide, water, tissue, and energy by processes of metabolism, including respiration, transpiration, digestion, and elimination. In addition to animals returning carbon to the cycle by consuming plants, carbon is incorporated in animal tissue that, on death, decomposes and releases carbon dioxide (along with other elements/compounds) back into the atmosphere and soil to continue the carbon cycle. Earth is home for a gigantic pool of carbon in the form of both organic and inorganic compounds. The total amount of carbon is estimated to be about 50,000 metric gigatons (1 metric gigaton = 10^9 metric tons). The percentage distribution of carbon is as follows: ocean carbons, 71%; fossil carbons, 22%; Earth's ecosystems, 3%; and carbon found in the atmosphere and used in photosynthesis, less than 1%.

Not only is the consumption and release of CO_2 from the weathering of rocks greater than that involving living organisms, but its cycle is much longer. Millions of years are required for CO_2 to be incorporated with minerals that are then buried as igneous and metamorphic rocks for eons. The following are two typical mineral reactions that consume carbon dioxide.

1. CO_2 gas tied up as calcium carbonate:

$$CO_2 + CaSiO_3 \rightarrow CaCO_3 + SiO_2$$

2. CO_2 buried as magnesium carbonate:

$$CO_2 + MgSiO_3 \rightarrow MgCO_3 + SiO_2$$

The other half of the carbon dioxide/mineral cycle is the slow *release* of CO_2 from igneous and metamorphic sources (where it was stored), resulting from lava and gases released at the mid-oceanic ridge and the subduction zones of tectonic plates. This reaction is: $CaCO_3 + SiO_2 \rightarrow CaSiO_3 + CO_2\uparrow$, and is somewhat the opposite of the two reactions listed above, which buried carbon dioxide gas. This reaction is partly responsible for the continued supply of carbon dioxide gas in the atmosphere that is available for photosynthesis, as well as for maintaining the Earth's greenhouse effect.[17–19] The greenhouse effect and global warming are addressed in chapter 9, but we do know that a tremendous amount of formerly stored CO_2 is released by burning fossil fuels. Scientists know that not all of the man-made CO_2 stays in the atmosphere, but they are not sure just what happens to it. It is assumed that the bulk of excess atmospheric CO_2 is recycled into the biomass sink of both earth and ocean plants and is then stored in plant materials, the soil, seawater, and ocean sediments.

Nitrogen Cycle

The nitrogen cycle is not only complicated but also is essential for all living organisms, which are, as in the carbon cycle, the major participants in this cycle. Nitrogen is the most abundant gas in the atmosphere, comprising about 79% of the Earth's atmosphere as N_2 diatomic molecules. Even so, it is not a very reactive gas due to the nature of its atomic structure, which consists of very strong covalent bonds that are difficult to break. This factor makes it difficult for nitrogen to form usable compound molecules. Only the reaction of some bacteria, lightning, light/radiation, volcanic action, and some chemical reactions can force nitrogen to combine with a few other elements. Nitrogen compounds are found in living tissues and are essential for the formation of amino acids/proteins, nucleic acids (RNA and DNA), and other nitrogen compounds essential for living organisms. The N_2 first must be converted by bacterial processes into forms of nitrogen compounds that can be used by plants. Note that animals cannot use nitrogen directly from the atmosphere but must secure the required nitrogen compounds from eating plants and other animals. The processes that move nitrogen through its cycle are (a) nitrogen fixation, (b) assimilation and biosynthesis, (c) decay and decomposition, (d) ammonification, and (e) nitrification.

Nitrogen fixation is a major component of the nitrogen cycle essential for plant growth. Atmospheric nitrogen is converted to biological nitrogen compounds by the actions of bacteria and algae. Legume plants such as alfalfa and soybeans develop nodules on their roots that contain specialized bacteria that take nitrogen from the air and convert it into ammonia (NH_3), which is soluble and thus dissolves to become NH_4^+. The fixed ammonia is converted into nitrogen compounds (nitrate, NO_3 and nitrite, NO_2) that provide the chemicals necessary to form plant tissue, carbohydrates, proteins, and so forth, which provide food for animals.

All living things produce waste and sooner or later die and decompose, resulting in the slow breakdown of the proteins and so forth into ammonia compounds and, finally, into nitrites. At this stage denitrifying bacteria in the soil reduce the nitrates and nitrites to nitrous oxide (N_2O) and gaseous nitrogen (N_2), which is released into the atmosphere to complete the cycle. The development of agriculture and subsequent use of fertilizers to increase plant growth has greatly increased not only food production but also pollution. The use of fertilizers containing nitrogen and the burning of fossil fuels (acid rain) has increased soil acidity, which in turn affects the chemical activity of roots and reduces nitrogen in the cycle.[20–24] (For more on acid rain, see chapter 9.)

Sulfur Cycle

As pointed out, organisms are mostly responsible over the short term for the carbon and nitrogen cycles. This is also the case for the sulfur cycle, which takes place near the surface of the crust. In this case, the major players are prokaryotes, primitive single bacteria cells that lack nuclei and other cell components. (Eukaryote cells are more advanced and contain nuclei and membranes containing organelles. Metabolically eukaryotes require oxygen.) Prokaryotes are archaic cells that do not use oxygen but instead get their energy by assimilating sulfur. They are responsible for the cycling of sulfur compounds in the soil, water, and air. Most of the prokaryote reduction and oxidation reactions of sulfur compounds takes place in salt marshes or buried in coastal marine sediments where the oxygen supply is limited.

Both microorganisms and multicellular organisms have been discovered living on the deep ocean floor where light does not penetrate. For some time it was a puzzle as to how these organisms secured and oxidized food, but they live near sulfur vents on the ocean floor and use sulfides originating in the vents to provide electrons for the fixation of carbon. The bacteria/microorganisms live off hosts that are mouthless, gutless tubeworms.[25]

There are links between the carbon, nitrogen, and sulfur cycles that constitute a global biochemical cycle. The evolution from simple cells that use sulfur for their energy to more metabolically complex oxygen-consuming organisms is recorded in geological and geochemical records. Rocks formed by lithification and lamination contain layers of fossil microbial organisms. The biological relationships among these three cycles confirm their importance to the geology of the Earth.

Oxygen Cycle

Oxygen gas (O_2) is the second most abundant element in the atmosphere (21% by volume). It is the most abundant element found in the Earth's crust (being reactive, it combines with many other elements), and it comprises about 89% of the oceans as water (H_2O).

The oxygen cycle involves the exchange of oxygen between the **biotic** and **abiotic** aspects of the environment. This cycle is closely related to, and interacts with, both the carbon cycle and the water (hydrological) cycle.

The original atmosphere of the Earth contained much more carbon dioxide, methane, and ammonia than oxygen. It is generally assumed that after primitive life was established, photosynthetic plants converted CO_2 to O_2 and that the ratio of gases in the atmosphere changed over many millions of years to the percentages that now exist. During photosynthesis, O_2 is produced by splitting oxygen from water and returning the gas to the atmosphere. The cycle is continued as atmospheric oxygen is used by both plants and animals in the process of respiration in which it is used in the oxidation of food to produce energy, with the resulting waste carbon dioxide gas being returned to the atmosphere, making it again available for the carbon and water cycles.[26–28]

Water Cycle

We addressed the water cycle as the hydrological cycle in chapter 7 (see Figure 7.3). A few important aspects of this cycle are as follows:

- Water is constantly changing from one phase to another (gas, liquid, solid), as well as from one place to another geographically. The amount of water recycled on Earth is essentially the same as the amount existing many billions of years ago.
- In essence, water progresses through the following cycle: (1) water evaporates from lakes, rivers, soils, and the oceans to become water vapor (gaseous phase) in the atmosphere; (2) water vapor is also added to the atmosphere by transpiration of plants and respiration, perspiration, and elimination by animals; (3) water vapor is transported by clouds that have formed over water to land areas, where it condenses from a gas to a liquid or possibly a solid (ice) and falls to land as precipitation (rain, snow, hail, sleet, etc.); great quantities of water are also encapsuled as sea ice; (4) once on land as a liquid water runs off the surface into streams, rivers, and finally into oceans; some of it percolates through the soil and rocks to form ground water, where it is

returned, in time, to the oceans, where the cycle is continued.

Nutrient Cycle

The scientific term for the nutrient cycle is the "biogeochemical cycle." This cycle traces the movement of chemical elements between living organisms (biotic, organic) and their nonliving (abiotic, inorganic) environments. The biogeochemical cycle might be thought of as the ultimate group of cycles. It includes the carbon cycle, the nitrogen cycle, the oxygen cycle, the water cycle, the rock cycle, the phosphorus and sulfur cycles, and a few other minor cycles. It represents the cycling between the chemicals of metabolism of living organisms and the geological or physical chemicals of the environment, including the atmosphere. It is also simply the cyclic conversion of inorganic compounds to organic compounds by living organisms and back again to inorganic compounds once the organisms die. Solar energy is the driver of all these cycles.[29–30]

Physical Structure of the Atmosphere

All the planets, with the possible exception of Mercury, have atmospheres. Some are very tenuous, some extremely dense, and most have layers of gases differentiated by temperature, pressure, or chemistry. No other planet has an atmosphere consisting of the unique combination of gases as are found on Earth. The Earth's atmosphere extends hundreds of miles into space and is divided into concentric subspheres, or envelopes (layers), by temperature differences and pressure gradients (see Figure 8.1).

Both horizontal and vertical motion of the air occur within the envelopes (spheres) of air; also, there are no sharp boundaries between them. Therefore, there is no sharp change in temperature, pressure, or chemical composition between layers.

Figure 8.1
The Earth's atmosphere is unique among the planets in the solar system. It is composed of a number of different gases that are maintained by circulation with the oceans, rocks, soil, and living organisms. The atmosphere is divided into subspheres or layers according to the pressures and temperatures of the layers' altitudes. (Image not to scale.)

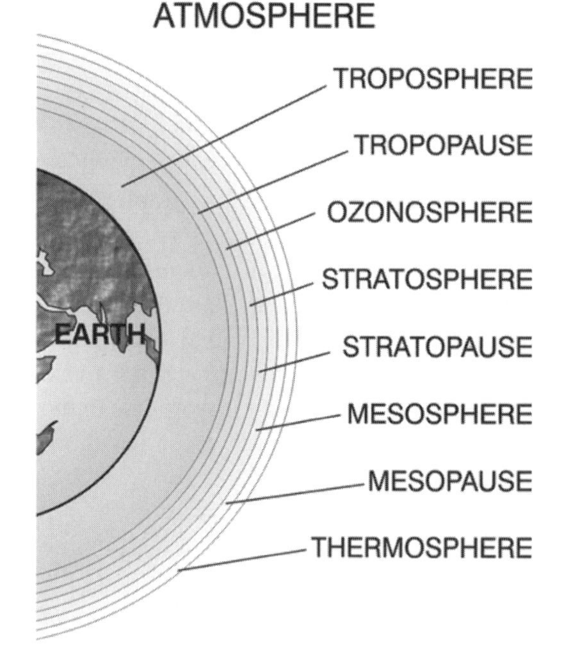

The sun's radiation, mostly as light and heat, penetrates the mixture of gases comprising the different stratalike atmospheric envelopes. Light from the sun, at times, shows properties of *waves,* while in other situations it acts as *particles*. Sir Isaac Newton (1642–1727) arrived at the concept that white light consists of waves. He used a prism to demonstrate that white light is composed of a variety of colored light, and when he used a second prism, each color joined again to reform white light. This led other scientists to develop the concept of diffraction and the wave nature of light. The particulate nature of light was confirmed by Albert Einstein's theory of the photoelectric effect, which demonstrated that light also behaves as particles, which he called "photons." John Tyndall (1820–1893) is known for the Tyndall effect, which states that light is scattered when it passes through gas molecules and particles, thus making the light beam visible. The Tyndall effect explains why the sky appears blue (when it is not obscured by clouds). The blue light of the sun's electromagnetic spectrum is scattered more than are the other wavelengths of visible light when it encounters air molecules in the atmosphere. At the same time, the red wavelengths of light make the sun appear yellow when high in the sky. When sunsets occur, the sunlight is traveling a longer path through the Earth's atmosphere, and these longer wavelengths of light (orange/red) produce the colored sunset.[31–32]

These atmospheric gases also prevent a great deal of heat (infrared radiation) from escaping from the Earth back into space, so it is not unexpected to find temperature differences on the surface of the Earth, as well as in the ascending subspheres of the atmosphere. The average global surface temperature for the Earth is about 15°C (59°F), with extreme temperatures ranging from about minus 90°C to a high of about 65°C (−130°F to 120°F).[33] Temperature differences above the Earth's surface are one of the factors determining the nature and separation of the different layers of the atmosphere. These gaseous layers, or envelopes, are named in the following list in order from the Earth's surface to the outer reaches of the atmosphere (note that all temperatures given are approximate).

1. The troposphere and tropopause (−60° to +20°C)
2. The stratosphere and stratopause (including the ozone layer) (−65° to +5°C)
3. The mesosphere and the mesopause (D-layer) (–80° to +5°C)
4. The ionosphere (E-layer) (−80°C)
5. The thermosphere and exosphere (F-layer) (−80° to −20°C)
6. The magnetosphere

Let's examine some of the characteristics of each of these subspheres or layers of the atmosphere (see also Figure 8.2).

Troposphere

The troposphere contains ~75 to ~80% of all the gases in the atmosphere. It is the layer of air closest to the surface of the Earth and is essential for processes important to both life and geology. The troposphere is also where air is most turbulent and where weather and climate occur.

- The troposphere extends to ~11 km (~7 mi) above the Earth's polar regions and ~17 km (~10–11 mi) at the equator.
- Because it is closest to the Earth, the troposphere is where the air pressure is greatest (gravity pulls all the air above the troposphere toward the Earth's surface; thus, air is densest at the surface). It is the layer where the movement of air known as wind, and water vapor, that is, humidity, occurs. These factors, along with surface temperatures and pressures, are referred to as "weather," and this is the sphere where most all of the weather takes place. This is the region of both horizontal and vertical air movement, including the lower level of the jet stream. (The geological and meteorological aspects of weather are discussed in chapter 9.)
- The troposphere is the layer that contains almost all the clouds, water vapor, and precipitation.

Figure 8.2
The Earth's atmosphere is not consistent in its chemical and physical composition. Great variations of temperature and pressure exist throughout the atmosphere, and the ratio of different gases is not the same throughout all its levels. These and other factors change with altitude. The various layers are not divided by abrupt distinctive separations but instead by regions where atmospheric attributes change gradually.

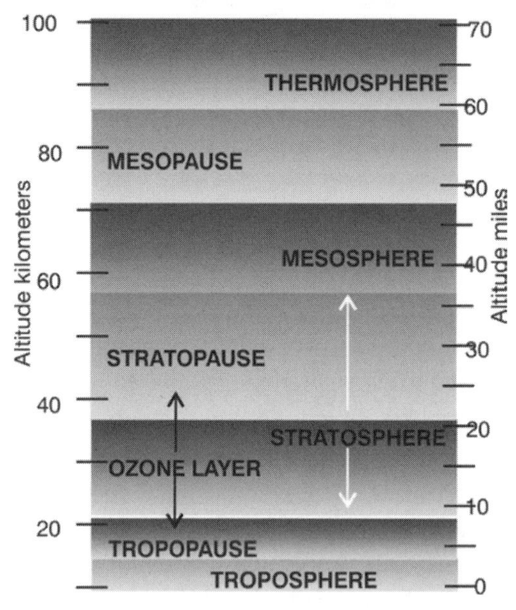

- The temperature ranges in the troposphere from about 18°C (65°F) at the surface to about −60°C (−76°F) near its upper region.
- Temperatures in the troposphere decrease with altitude at an average rate of 2°C per 305 meters (m) (3.6°F per 1000 feet [ft]). Although the rate of cooling varies from place to place on the Earth, it is quite obvious that as you go up a high mountain, the temperature decreases. The same is true the higher up one goes in the troposphere.
- The *tropopause,* the top region in the troposphere, is the transition boundary region where the troposphere thins and the stratosphere begins. It varies in altitude from 8 km (5 mi) to 18 km (11 mi) and is distinguished by a change in the air's temperature from −60°C at the top of the troposphere to above ~15°C at the top of the next higher layer, the stratosphere. This drastic temperature change in the tropopause is an indication that you have crossed from one layer to another, that is, from the troposphere to the stratosphere.[34–36]

Stratosphere

In addition to the essential processes of the troposphere, several important processes occur in the stratosphere that make life possible on Earth. In some ways both the chemistry and physics of this envelope of thin air are quite different from the Earth-hugging troposphere. For instance, the air is so diluted (thin) and the pressure so low that humans must wear pressurized suits and be supplied with oxygen to survive—unless, of course they are in a pressurized high-altitude aircraft.

- The stratosphere is located just above the tropopause and extends from ~11 km (~7 mi) to ~50 km (~30 mi).
- The air in the stratosphere does not rise and fall as it does in the troposphere, nor are there ordinary clouds or precipitation. In other words, there is no Earth-type weather at this altitude. (For more on weather and climate, see chapter 9.) However, in the lower stratosphere there are horizontal winds, known as the jet stream, that reach ~300 km/hour (~186.5 miles per hour [mph]). The jet stream usually flows from west to east, giving aircraft flying in this direction and at this altitude a strong push.

- Unlike the temperatures in the troposphere that gradually *decrease* with altitude, the temperatures in the stratosphere gradually *increase.* The temperature is about −60°C to −65°C (−76°F to −85°F) at the lower level of the stratosphere, gradually rises to above 0°C (32°F) at its central region, and then rises again at the upper region. The temperature in this upper region, just below the mesosphere, is close to what is found on the Earth's surface (~59°F) and is the result of sunlight radiation reacting with oxygen at this level to form ozone.
- Near the upper region of the stratosphere are areas where chemical activity results in temperatures as high as 65°C (149°F). These narrow bands or layers located within different regions of the stratosphere are known as the chemosphere, the ozonosphere, and the mesosphere and are distinguished by their origin and content.
- Of the most interest to humans, at least in recent times, is a relatively thin region in the lower part of the stratosphere known as the ozonosphere, where the sparse oxygen (O_2) of the stratosphere absorbs energetic ultraviolet rays from the sun. This radiant energy breaks up the diatomic O_2 molecular oxygen to form O as free oxygen atoms, sometimes known as nascent oxygen. The next part of the reaction occurs when the free atomic O recombines with diatomic molecular O_2 to form a heavier molecule known as ozone (O_3, or triatomic molecular oxygen). This reaction releases heat that increases the temperature at this zone. The thin ozone layer has the ability to absorb much of the sun's ultraviolet radiation that is harmful for most living organisms. (For more on ozone, see chapter 9.)
- The ozone layer starts at the lower regions of the stratosphere but does not continue into its higher altitudes. Therefore, at ~56 km (~35 mi) above the Earth the temperatures again decline steeply with altitude. This boundary area of changing temperatures is known as the stratopause, which leads to the next subsphere of the atmosphere, the mesosphere.[37–39]

Mesosphere

The mesosphere, also known as the "middle sphere," exists at ~48 km (~30 mi) to ~88 km (~55 mi) above the Earth. It is a region of the atmosphere that does not have well-defined boundaries, but it is an important region that couples the middle and upper envelopes of the atmosphere.

- The most distinguishing characteristics of the mesosphere are the rapid decrease in temperatures and its very cold, thin air. At the boundary between the stratopause and the mesosphere the temperatures range from about 0°C to about 10°C (32°F to about 50°F) at its base, and then the temperature rapidly decreases with altitude in the mesosphere to about −90°C to about −110°C (−130°F to about −166°F).
- The mesosphere is so cold that thin ice-clouds form that contribute to beautiful sunsets. Even though the air in the mesosphere is very thin, it is the region where most meteors burn up as their energy of motion produces friction with the sparse air molecules as they approach Earth.
- *Mesopause.* The mesopause is located at the upper boundary of the mesosphere where decreasing temperatures are stabilized at the lowest temperatures found in the atmosphere. At the upper boundary of the mesopause the temperatures again begin to rise in the next layers, the ionosphere and the thermosphere.[40–41]

Ionosphere

The ionosphere is really part of the next much larger sphere, the thermosphere, which extends to near the outer edge of the atmosphere. The ionosphere is a region that overlaps other layers, located just above the mesopause and extending well into the thermosphere. The ionosphere is important for several reasons and is usually described by its unique, mirrorlike ability to reflect radio waves.

- The ionosphere is found at ~50 km (~31 mi) at the base of the mesopause and extends into the thermosphere to more than 400 km (250 mi) above the Earth's surface.
- The ionosphere is a region of slightly rising temperatures and is composed of electrified particles and ions. These ions are produced by the action of solar radiation on rarefied gases. A negative ion is an atom of a gas that has attained an extra electron, thus becoming a particle with a negative charge. A positive ion is an atom that has had one of its electrons stripped from it, leaving the ion with a positive charge but also producing a free electron. These charged particles (ions) plus free electrons exist as far out as gases are found in the atmosphere. In addition, several forms of radiation exist in the outer atmosphere.
- The ionosphere is divided into several layers to which scientists have assigned letters (D, E_1, E_2, F_1, F_2, and G) for designation. These layers act as a mirror ceiling that reflects radio and television waves back to Earth. A radio signal is first sent from Earth into the ionosphere, where it bounces back to Earth, which also acts like a mirror that sends the signal back to the ionosphere, where it is again reflected back to the Earth at a different location, and so on across the surface of the Earth. Because the Earth's surface is curved and these waves travel in straight lines, bouncing them back and forth is the only way AM and short-wave radio signals can be transmitted around the Earth. Otherwise, as with certain FM and TV wavelengths, the straight-line signals could only be sent 40 or 50 miles.
- The D and E layers of the ionosphere are activated by sunlight; therefore, they are only active during daylight hours. The E layers reflect the AM band of radio waves, whereas the F layers reflect the AM and FM bands of radio waves, as well as TV signals. At night, the two F layers combine and the signals are not absorbed as much as in the daylight hours, thus increasing the degree of reflection and the quality of nighttime reception.
- In the Northern Hemisphere, during nighttime of the winter months, the sun is still shining on the ionosphere. The ionized gases become charged to the extent that they glow, producing what is known as the "aurora borealis," or northern lights. This same phenomenon is called the "aurora australis" in the Southern Hemisphere (see Magnetosphere).
- Some scientists believe the ionosphere extends beyond the thermosphere to the exosphere, which is the very edge and extent of the atmosphere. At this altitude the Van Allen belts are considered part of the exosphere (see Magnetosphere).[42–44]

Thermosphere

The thermosphere is defined by its great temperature changes; thus, it encompasses other regions (layers) of the upper atmosphere. The thermosphere generally is considered to include the upper region of the mesosphere/mesopause and the ionosphere. At times, the exosphere is also included as the outer part of the thermosphere. The lower region of the thermosphere is where artificial satellites, particularly communication satellites, orbit the Earth.

- The thermosphere extends to ~700 to ~800 km (~435 to ~500 mi) from the surface of the Earth and consists of only 0.01% of the Earth's atmosphere.
- The chemistry and physics of the upper region of the mesosphere (55 to 80 km [35 to 50 mi]) and the lower region of the thermosphere (80 to 180 km [50 to 112 mi]) are being studied by scientists using rockets to place unmanned satellites with sensing instruments into space that return data by telemetry. (The lower strata of the atmosphere are studied by instrument packages sent aloft by balloons.) These sensing instruments include spectrometers, photometers, resonance fluorescence detectors, thermometers, and pressure gauges that measure the concentrations of O, O_2, O_3, N_2, H_2, OH^-, CO_2, H_2O vapor, CH_4, Cl_2 and ions, as well as this region's temperatures and densities. Molecular nitrogen and oxygen are the main gases found in this very thin atmosphere, but the trace constituents (less than one part per million) are of great interest because not much is known about them and how they affect the chemical dynamics of the upper atmosphere. Scientists are trying to determine how these chemicals interact so that they can devise improved computer models for predicting how the upper atmosphere affects Earth's climate.[45]
- In 1998, an important atmospheric satellite named TIMED (thermosphere, ionosphere, mesosphere energetics, and dynamics) was launched by NASA in a high-inclination, low-Earth orbit. Its instruments investigated, remotely sensed, and radioed back to Earth data concerning the physical and chemical processes acting in the outermost layers of the Earth's atmosphere. This region is collectively referred to as the MLTI (mesosphere, lower thermosphere, and ionosphere). This is a region where energetic solar radiation is absorbed, energy input from the aurora maximizes, intense electrical currents flow, and upwardly propagating waves and tides break (including gravity and planetary waves). Scientists are also investigating chemical and physical changes in the MLTI to establish a baseline for future investigations.[46]
- The upper thermosphere (exosphere) is the first region that receives solar radiation, which makes it very hot because the thin air at this altitude is heated faster than it can cool off. The density of air decreases with altitude in the thermosphere, whereas the temperature increases. The sun's short-wave radiation is absorbed by the ionization process of the air molecules at this region, which accounts for the thermosphere's high temperatures.[47] At the region of the outer layer of the thermosphere the temperatures can reach and exceed 1475°C (2687°F).[48]

Exosphere

The exosphere is the very outer limit of the atmosphere where gas molecules and ions are found. It is the transition region of highly separated air particles and the vacuum of space.

- The exosphere begins at the upper reaches of the thermosphere at ~700 km (~435 mi) and extends at least to ~65,000 km (~40,000 mi).
- The atmosphere at this altitude is attenuated to the extent that the distance that air molecules travel without hitting each other is equal to the radius of the Earth.

Even so, at this great distance from Earth gas molecules and ions still have some kinetic energy (i.e., heat) and a few can be knocked by other particles with enough thermal energy (hydrodynamic escape) to attain a velocity great enough to escape the influence of Earth's gravity. The velocity required of a mass to escape from a planet is related to that planet's mass and, thus, its gravity. To escape from Earth's gravity, a rocket, gas particle, or anything with mass must attain a velocity of 11.2 km/second (sec), or about 7 mi/sec, or about 25,000 mph.[49]

- Beyond ~9,700 km (~6,000 mi), the sparseness of matter forms a better vacuum than can be produced on Earth with vacuum pumps.
- Earth's gravity and magnetic field still affect gas molecules and ions with high electrical charges as far out as ~65,000 km (~40,000 mi), which is considered to be the extent of the exosphere before the vacuum between the Earth and the sun, also known as the region of solar wind.[50]

Magnetosphere

The magnetosphere is the outermost region that shields the Earth from harmful radiation by magnetically trapping electrically charged particles from the sun. There are few air molecules in this region.

- Both the sun and Earth act as giant electromagnets with magnetic fields composed of lines of force extending far beyond their immediate atmospheres. In 1600, William Gilbert (1544–1603), an English physician, was the first to establish the cause and effect relationship that explained why a compass needle aligned itself with the Earth's North and South Poles. He used a large spherical lodestone to demonstrate that a compass needle dipped (declination) toward the north end of the lodestone as well as toward Earth's North Pole, which proved that Earth was a magnet. He also coined the term "magnetic pole" and demonstrated that the Earth acted like a giant lodestone or bar magnet and had magnetic poles more-or-less congruent with the Earth's geographic poles.[51]
- As previously mentioned, at the fringes of the atmosphere gas molecules are heated and bombarded by radiation that strips electrons from the atoms to form positive ions and free electrons. Since these particles now have an electrical charge, they are attracted to and guided by the lines of force of Earth's magnetic field. Note that at lower altitudes the molecules of oxygen and nitrogen that we breathe are essentially electrically neutral; since they have no charge, they are not captured by the Earth's magnetic field. In contrast, the high-altitude charged particles (ions and free electrons) behave somewhat like beads strung on a curved wire that rise near the south magnetic pole, arc high over the equator, and enter the Earth again near the north magnetic pole. These guided charged particles can be accelerated and drift as they enter the Earth's atmosphere, causing electrical disturbances as well as the aurora borealis.[52]
- The Van Allen radiation belts are two very high intensity, high energy, doughnut-shaped bands of charged particles that have been captured by the Earth's magnetic field and are concentrated over the equator. James Van Allen (1914–) used NASA's *Explorer I,* and later other satellites, to study cosmic radiation. He found radiation so strong at one area that it jammed the satellite's Geiger counter. These early satellites from both the former Soviet Union (U.S.S.R.) and the United States discovered that the Earth's magnetic field captured high-speed

charged particles that encircle the Earth. But it was Van Allen who discovered two odd-shaped belts of charged particles in the Earth's magnetic field concentrated over the equator. One belt is located at ~1000–5000 km (~600–3000 mi) above the Earth's surface and consists mainly of positively charged protons and electrons. The second belt, located at an altitude of ~15000–25000 km (~9,000–15,000 mi) above Earth's equator, consists of mostly electrons. It was later determined that these charged particles in the upper atmosphere were electrons and protons (ions) that resulted from collisions with cosmic rays from space and from solar-wind radiation from the sun.[53] Since the Earth's magnetic field attracts and traps these charged particles, they follow the path of the field that leads to the polar regions. Scientists continue to study these high radiation belts to see if or how they may affect the Earth. Unmanned artificial satellites provide a more efficient and cost-effective platform for conducting research in the outer reaches of the atmosphere (including planetary research) than do manned spacecraft.[54]

Solar Activity: Sunspots, Storms, and Flares

The sun, although not physically part of the Earth, is an important player in the Earth's geology, particularly the biosphere. The unique atmosphere of the Earth would not exist without solar energy, and, thus, neither would life as we know it. The following list presents some important facts and processes related to solar energy and geology.

- The sun is a huge nuclear fusion reactor slowly converting hydrogen into helium while releasing enormous amounts of energy in all directions into space. The sun's intense gravity as a result of its great mass keeps this fusion reaction from becoming an exploding hydrogen bomb.
- **Sunspots** are cooler, darker regions on the sun's surface caused by temporary distortions of its magnetic field. Sunspots spawn tremendous eruptions or flares of electrified gas into the sun's atmosphere, where they are hurled out into space in all directions—including toward the Earth.[55]
- The internal nuclear reaction of the sun produces bulges on its surface that flare outward for hundreds of thousands of miles, resulting in star-size magnetic storms in space. If Earth is in the path of these storms, known as "coronal mass ejections," they can damage satellites and power transformers and disrupt communications as they bend and follow the line of force of the Earth's magnetic field. These captured charged particles can reach speeds of 1 million mph.
- The sun's nuclear reactor produces radiation across the electromagnetic spectrum, giving rise to very short, high-frequency, high-energy, cosmiclike gamma radiation and x-rays, as well as longer infrared (heat) radiation and very long, relatively low frequency, electric currents and radio-type radiation.
- Ultraviolet radiation begins at the short wavelength end of visible light, and infrared (heat) radiation begins at the long wavelength end of the light spectrum, whereas visible light is near the middle of the electromagnetic spectrum. Human eyes can only detect a small segment of the electromagnetic spectrum (namely,. visible light), but instruments have been developed that can see or detect all wavelengths and frequencies of the spectrum. James Clerk Maxwell (1831–1879) developed the theory of electromagnetism by combining several concepts of magnetism and electricity as a single equation. He concluded that although

radiation may be of different wavelengths and frequencies, all categories of electromagnetic radiation travel at the same speed as light in a vacuum, namely, 186,000 mi/sec.[56]

- The sun goes through an eleven-year maximum and minimum cycle of intense internal solar activity. The high point, or maximum intensity period, of a cycle may last a few years. The most recent maximum period started in the year 2000 and continued to about the year 2003.
- During one of the sun's previous 11.1 year cycles in 1989, 6 million Canadians were left in the dark when a huge solar storm caused a massive outage across the Hydro-Quebec power grid.[57]
- This cyclic activity produces solar flares, or huge tongues of solar radiation, on the surface of the sun that send out in all directions a stream of highly charged ions and particles known as solar wind. This solar wind is not exactly similar to wind on the surface of the Earth, but it does create a weak pressure on objects in its path. NASA has considered using the pressure of solar wind as a means of propulsion to push large, kitelike satellites, accelerating them into deep space. As solar wind approaches Earth, it distorts the flow of charged particles in the magnetosphere and Van Allen belts. This causes the patterns of charged particles to bulge out from the Earth on the side opposite the sun.
- For July 5, 2000, the solar wind was 447 km/sec (1.61 million km per hour, or about 1 million mph) with a solar density of 6.3 protons per cubic centimeter. On this date the geomagnetic conditions were quiet to unsettled with a 15% chance of a low-level, mid-latitude solar storm in the next 24 hours and a 1% chance for a severe solar storm. In March 2001, the biggest sunspot cluster seen in the past 10 years developed on the upper right quarter of the side of the sun visible from Earth. NASA scientists said the most powerful flare that erupted at this time rated a classification in the most potent category. It triggered a brief blackout on some high-frequency radio channels and low-frequency navigational signals.
- Because solar geomagnetic storms are one of nature's natural hazards, referred to as "celestial hurricanes," the Space Environment Center (SEC) of the National Oceanic and Atmospheric Administration (NOAA) launched the ACE (Advanced Composition Explorer) satellite on August 25, 1997 to give the public at least a one-hour warning of severe geomagnetic storms that can cause communication problems and electric utility blackouts over large regions. Later, NASA launched a sun-orbiting satellite that can detect solar activity and flares on the back side of the sun; thus, as the satellite becomes aligned with Earth, it can signal a warning two weeks in advance of the actual solar storm. Scientists also use instruments to listen to the boiling sun's insides and its vibrations to determine what is occurring on both its inside and far side.[58]

NASA's *Pioneer 10* was the first spacecraft to travel to and beyond the outer planets in the solar system. It passed Mars and sent back a close-up image of Jupiter as it traversed the hazards of space. *Pioneer 10,* launched in 1972, remained active, but aged, for more than 25 years. It weighed only 570 pounds but was able to collect and send back valuable data on solar wind at the far edge of the solar system and cosmic radiation from interstellar space. Since other, more advanced artificial satellites were launched to explore the solar system and deep space,

Pioneer 10 was turned off in 1997. However, it will keep on speeding into deep outer space for eons to come until destroyed by radiation or hits by micro-meteors and asteroids.[59]

Summary

The Earth's atmosphere is unique, and we still have much to learn about its chemistry and physical structure as it becomes less and less dense with altitude. Exactly what are the causes and effects of the reactions of its atoms, molecules, ions, charged particles, and so forth, and their interactions with Earth? Today we use unmanned spacecraft as vehicles to support instruments, and we use scientific instruments to extend our powers of observation, to study the changes at different strata or layers to assess how the conditions of this fluid composed of many different gases may affect the future of Earth's environment.

Issues

Social, economic, and political issues related to the atmosphere are addressed in chapter 9.

Notes

1. Atmospheric Chemistry Data and Resources, *Atmospheric Structure,* accessed 2001, http://www.xtreme.gsfc.nasa.gov.
2. Kenneth W. Hamblin and Eric H. Christiansen, *Earth's Dynamic Systems* (Upper Saddle River, N.J.: Prentice-Hall, 1998), 198–99.
3. *Encyclopedia Britannica,* CD-ROM 2001, s.v. "Earth: Shock-Heating of the Earth During Accretion."
4. *Volcanic Gases and the Origin of the Atmosphere,* accessed 2001, http://www.volcano.und.nodak.edu/vwdocs/Gases/origin.html.
5. S. Alan Stern, "Journey to the Farthest Planet," *Scientific American,* May 2002, 61.
6. *Origin of the Earth's Atmosphere,* accessed 2001, http://www.us1.eiu.edu/~cfjps/1400/atmos_origin.html.
7. Rick Behl, *Earth Systems and Global Change: The Atmosphere and Hydrosphere,* GEOL 300i, California State University Long Beach, accessed 2001, http://www.seis.natsci.csulb.edu/rbehl/300i-L06.htm.
8. Stern, "Journey to the Farthest Planet," 61.
9. Graham R. Thompson and Jonathan Turk, *Earth Science and the Environment* (New York: Harcourt Brace, 1999), 176–77.
10. Robert M. Hazen, "Life's Rocky Start," *Scientific American,* April 2001, 77–85.
11. Brooks Hanson, "Atmospheric Science," *Geology* 29 (2001): 1003, as reported in *Science,* Highlights of Recent Literature—Editor's Choice, November 16, 2001.
12. *Encyclopedia Britannica,* CD-ROM 2001, s.v. "The Origin of Life: Hypotheses of Origins."
13. *Evolution of Life on Earth,* accessed 2001, http://www.scibridge,sdsu.edu/coursemats/introsci/evolution_of_life.
14. Sarah Simpson, "Triggering a Snowball: Did Methane Addiction Set Off Earth's Greatest Ice Ages?" *Scientific American,* September 2001, 20.
15. David C. Catling et al., "Biogenic Methane, Hydrogen Escape, and the Irreversible Oxidation of Early Earth," *Science,* August 3, 2001, 839–43.
16. W. G. Ernst, ed., *Earth Systems, Processes and Issues* (New York: Cambridge University Press, 2000), 287.
17. M. Battle et. al., "Global Carbon Sinks and Their Variability Inferred from Atmospheric O2, O3, and C," *Science,* March 31, 2000, 2467–70.
18. *Microsoft Encarta,* CD-ROM 1994, s.v. "Climate: Carbon Cycle," by Robert Leo Smith.
19. *Earth's Cycles,* accessed 2001, http://www.rainbow.ideo.
20. *The Nitrogen Cycle,* accessed 2001, http://www.sturgeon.ab.ca.
21. *Nitrogen Cycle: CHM 110,* accessed 2001, http://www.elmhurst.edu.
22. *The Nitrogen Cycle,* accessed 2001, http://www.clab.cecil.ccmd.us.
23. Peter M. Vitousek et al., *Human Alteration of the Global Nitrogen Cycle: Causes and Consequences,* accessed 2001, http://www.esa.sdsc.edu.
24. Ernst, *Earth Systems,* 288–91.
25. Ernst, *Earth Systems,* 287–88, 291.
26. *Oxygen Cycle,* accessed 2001, http://www.xrefer.com.
27. *The Atmospheric Oxygen Cycle,* accessed 2001, http://www.agu.org.

28. *Encyclopedia Britannica.* CD-ROM 2001, s.v. "Oxygen Cycle."

29. *Biogeochemical Cycle: Nutrient Cycle,* accessed 2001, http://www.xrefer.com.

30. *Nutrient Cycling: High School Biology,* accessed 2001, http://www.homeworkhelp.com.

31. James Trefil, *1001 Things Everyone Should Know About Science* (New York: Doubleday, 1992), 111–13.

32. *The Handy Science Answer Book* (New York: Carnegie Library of Pittsburgh, 1997), 38.

33. Hamblin and Christiansen, *Earth's Dynamic Systems,* 201.

34. Edward J. Denecke Jr., *Let's Review: Earth Science* (Hauppauge, N.Y.: Barron's Educational Series, 1995), 9.

35. Patricia Barnes-Svarney, ed., *The New York Public Library Science Desk Reference* (New York: Macmillan, 1995), 420–21.

36. *Peel the Planet: Atmosphere,* accessed 2001, http://www.discovery.com.

37. Thompson and Turk, *Earth Science,* 355–56.

38. *Columbia Encyclopedia,* 6th ed., s.v. "Atmosphere," accessed 2001, http://www.bartleby.com.

39. Barnes-Svarney, ed., *The New York Public Library Science Desk Reference,* 420–22.

40. James E. Bobick and Margery Peffer, *Science and Technology Desk Reference,* 2d ed. (New York: Gale, 1996), 497.

41. Thompson and Turk, *Earth Science,* 356.

42. John Tomikel, *Basic Earth Science: Earth Processes and Environments* (New York: Allegheny Press, 1981), 43–45.

43. *Peel the Planet: Atmosphere,* accessed 2001, http://www.discovery.com.

44. *Scientific American Desk Reference* (New York: John Wiley and Sons, 1999), 266–67.

45. *Atmospheric Chemistry in the Mesosphere and Thermosphere,* accessed 2001, http://www.sprl.umich.edu/SPRL/research/atmo_chem.html.

46. TIMED: *Science at the Crossroads of the Earth's Atmosphere,* accessed 2001, http://www.nascom.nasa.gov/solar_connections/TIMED.

47. *Atmosphere: Layers of the Earth's Atmosphere,* accessed 2001, http://www.bartleby.com.

48. Bobick and Peffer, *Science and Technology Desk Reference,* 497.

49. Michael Upshall, ed., *Hutchinson Dictionary of Physics* (Oxford, England: Helicon, 1993), 55.

50. *Columbia Encyclopedia,* 6th ed., s.v. "Atmosphere," accessed 2001, http://www.bartleby.com.

51. John Daintith, Sarah Mitchell, Elizabeth Tootill, and Derek Gjertsen, *Biographical Encyclopedia of Scientists,* 2d ed., vol. 2 (Bristol, England, and Philadelphia: Institute of Physics Publishing, 1994), 346.

52. David P. Stern, *The Magnetosphere: The Earth Is a Huge Magnet, and Its Magnetic Influence Extends Far into Space,* accessed 2001, http://www.spof.gsfc.nasa.gov/Education.

53. Daintith et al., *Biographical Encyclopedia of Scientists,* vol. 2, 900–901.

54. *The Handy Science Answer Book* (New York: Carnegie Library of Pittsburgh, 1997), 72.

55. Joseph B. Verrengia, "Solar Eruptions to Disrupt Signals: Scientists Say Intense Storms Rage on the Sun," *Valley Morning Sun,* Arlington, Tex., March 3, 2001.

56. Robert E. Krebs, *Scientific Laws, Principles, and Theories: A Reference Guide* (Westport, Conn.: Greenwood Press, 2001), 223.

57. John Yaukey, "Solar Storms Could Wreak Maximum Havoc," *USA Today,* April 4, 2000.

58. John S. McNeil, "Peeking behind the Sun: New Ways to Predict Stormy Solar Weather," *U.S. News & World Report,* June 5, 2000.

59. Robert L. Park, *Voodoo Science: The Road From Foolishness to Fraud* (New York: Oxford Press, 2000), 88–89.

9

Climate and Weather

"Climate" is the term used for global, long-term atmospheric conditions, while "weather" is the term used for local, short-term atmospheric conditions near the surface of the Earth. In other words, climate may be thought of as the average conditions, as well as cyclic variations in the atmosphere. Weather is considered the day-to-day atmospheric conditions, including occasional extreme atmospheric events (e.g., hurricanes, tornadoes, etc.). Climate relates to the lower and middle layers of the atmosphere that cover large regional zones or the entire globe. Weather relates more to changing, day-to-day atmospheric conditions that occur near the surface of the Earth over a local geographic area. Therefore, atmospheric conditions that produce weather are much more predictable than are regional or global climates over longer periods of time. Air, water, and heat are the major physical and chemical components of the atmosphere that relate to climate, and they are also the major agents responsible for geological changes on and near the surface of the Earth.[1]

From the time of early human existence, survival depended on developing some understanding of and ability to respond to both the Earth's weather and climate. Humans, as well as many animals, learned how to not only protect themselves from the daily and seasonal elements but also read clues as to what they might expect of their weather and climate in both the near and more distant future. Long-term climatic conditions determined the migration and expansion of civilizations over many thousands of years, and short-term weather conditions accounted for how people lived and worked on a daily basis. Using instruments, such as thermometers, to record the conditions of local weather dates back less than 150 years to the middle nineteenth century.[2] Since then, modern weather instruments have provided accurate information about our local weather and global climates. Also, scientists have developed methods that can estimate the general climatic conditions of the Earth many thousands, even millions, of years before human civilizations developed.

History of Climate (Paleoclimatology)

The science of studying past climates is called paleoclimatology. The word is a combination of the Greek word *paleo,* meaning "ancient," the Latin word *clima,* meaning "region of the Earth" (or climate), and *ology,* meaning a "branch of learning." Paleoclimatologists use natural or proxy environmental artifacts to infer past climate conditions many eons before recorded history. Scientists are also interested in the geology, hydrology and atmospheric and other Earth processes that caused and altered past climates.[3] Records indicate that over the past few million years, major glaciation cycles or ice ages occurred about every 100,000 years. There were also shorter and less severe cycles. The most recent ice age glacier was at a maximum ~20,000 years ago. It subsided ~10,000 to ~11,000 years ago. Some meteorologists and geologists believe that this most recent ice age has almost reached the bottom of its cycle. There were also warm periods, some going back hundreds of millions of years to the age of the dinosaurs. Not to be confused with these historic, drastic climate changes are a series of more recent alterations in regional climates. The Medieval warm period lasted from ~1000 to ~1300 C.E. and was followed by the Little Ice Age of ~1300 to ~1850 C.E.[4–5]

Paleoclimatologists have learned from environmental and proxy clues that there were ice ages over the past millions of years with glacial cycles. These cycles were partly influenced by significant changes in temperatures caused by changes in the sun's radiation level. (The sun was ~20–30% fainter than it is today, yet the oceans did not freeze.)[6] Several other causes of ancient climate changes were the movements of continents resulting in volcanoes, variations in ocean temperatures and currents (El Niños), alterations in the percentage of carbon and oxygen in the atmosphere, and vast regions of changing plant growth resulting in the formation of oil and coal. Many other chemical and physical factors also contributed to the cycles of warm and cold climates of the Earth.[7]

Proxy Climate Data

Paleoclimatologists gather environmental and proxy clues from many sources to estimate past climatic conditions on the Earth. They use the information about past climates gained from environmental proxies to assist them in predicting future climatic changes.

Historical Records

Recorded historical data found in old writings of farmers and in public records is the first source of information on past climates and weather. This provides a start, but it does not go back far enough in time to answer many questions that can be useful in predicting future weather and climate. Herodotus, the Greek historian who visited Egypt in the fifth century B.C.E., was curious about the annual flooding of the Nile. During his visits he developed several explanations for this flooding phenomenon. More important, his observations provide a record of the climate for that region from more than several thousand years ago.[8]

Tree Rings (Dendrochronology)

Tree rings are indicative of yearly weather conditions. When many rings are considered, they reveal proxy chronologies of the climate during a particular period of the tree's growth. Scientists have gathered data from more than 300 tree-ring chronologies from North America, Europe, Siberia, and several countries in the Southern Hemisphere. Using advanced statistical methods, they compared ancient proxy temperatures indicated by tree rings with the surface temperatures measured by modern twentieth-century instruments. They noted natural changes in temperatures, including a global cooling in 1816 that followed the eruption of the Tambora volcano in Indonesia. Their conclusions were that solar radiation and volcanic emissions forced ancient global climate changes and that climate can change much more rapidly than they formerly believed. Not surprisingly, they also concluded that, in the twentieth century, both humans and natural phenomena have affected climate. Scientists not only count growth rings in trees, but they measure their width, their patterns of branching, and their chemical isotopic composition. Since trees normally produce one ring each year, and some species can live for hundreds, or even thousands, of years, they can provide evidence of weather conditions and climate as far back as the trees are old.[9] Using isotope carbon-14 (^{14}C) dating techniques on recovered ancient logs buried in glacial till, paleoclimatologists can estimate past climates from the chemical composition of the preserved wood.[10]

Fossil Pollen

Fossil pollen grains provide some proxy clues to ancient climatic conditions. Most plants, including ancient plants, produce distinctly shaped pollen grains. Some ancient pollen is preserved in the layers of sediment at the bottoms of ponds, lakes, and the oceans, where, in time, it becomes fossilized. Scientists identify specific plants by their unique pollen. Then, by determining the layer of sediment in which the pollen was found, they infer the climate at the period of time the plants thrived. Pollen grains have been found in ancient burial chambers, but these date back only a few thousand years. Pollen from spruce trees was found in an 11,000-year-old Minnesota bog. Scientists contend that spruce trees were the most abundant species at the time the bog formed. Comparing this date with the cold Canadian climate where spruce trees are currently plentiful, they determined that the climate ~11,000 years ago was much colder than in more recent times. Also, after examining pollen found in the Minnesota swamps, scientists believe that other species of pine trees replaced the spruce, indicating that the climate in that region had become warmer over the last 11,000 years.[11]

Coral Skeletons

The skeletons of sea coral are another natural proxy for determining ancient climates. Sea corals extract minerals and oxygen from seawater that are then incorporated into their skeletal structure. The main mineral in coral is calcium carbonate ($CaCO_3$). It is a very stable, widely dispersed mineral that contains the isotopes of oxygen, as well as traces of metals and other minerals, which can be used to date the coral. Note that the regular oxygen atom is ^{16}O with 8 protons and 8 neutrons. However, there are two other flavors of heavy oxygen, ^{17}O and ^{18}O (both with 8 protons but with 9 and 10 neutrons, respectively). The ^{18}O isotope of oxygen makes up only 8 parts to 10,000 parts of regular oxygen. This same ratio of isotopes exists in regular water (H_2O) molecules found in glacial ice that can also be used to determine ancient climates.[12] Hydrogen atoms also come in three varieties of isotopes: the ^{1}H atom with a single proton and one neutron; ^{2}H (deuterium) with two neutrons, one of the heavy versions; and ^{3}H (tritium) with three neutrons plus their single protons, the other heavy version. These two isotopes of hydrogen are components of heavy water molecules. This percentage of oxygen changes as the uptake of oxygen is incorporated into the calcium carbonate that forms the coral. Thus, scientists can measure the differences in isotopic composition, which enables them to determine the sea surface temperature at the time coral was growing (or glacial ice was formed). Thus, they can make assumptions as to what the global climate was like at the time the coral skeletons formed. These proxy factors are also used to determine the temperatures of the seawater at the time of coral growth, hence providing a record of the temperatures used in reconstructing the climate at the time the coral was living.

Ice Cores

Deep cores of ice collected in northern Greenland and the Antarctic polar cap contain ice from snowfall over many thousands of years. Since the ice does not melt at these locations, even in summer, new snow just adds another layer on top of the ice beneath it. Scientists sink a hollow drill rod as deep as several thousand feet into the many layers of old ice. When the drill is brought up, an approximately 3 inch (in.) diameter core of ice, divided into sections each several feet long, is removed from inside the drill shaft. These ice cores are examined, measured, recorded, and stored in subfreezing warehouses for future study. Thousands of cores, totaling up to 2 miles (mi) in length, from areas in Greenland and Antarctica are currently being stored and examined for air bubbles and water that contain ancient atmospheric oxygen and hydrogen isotopes, dust and ash particles from past volcanoes, and so forth. These proxy clues can be used to interpret the climate of those regions thousands of years ago. A recent core taken from the ice in Vostok Station, Antarctica, is about 2000 meters (m) (6,436 feet [ft]) long and contains deposits from ~160,000 years ago. Scientists measure trapped molecules of gases, including greenhouse gases, to determine the ratios of their isotopes. This enables them to estimate past climate conditions and possible environmental conditions as well. They also use lasers to measure dust that forms annual layers in ice over many past seasons (similar to tree rings). The isotopes provide evidence of long episodes of intense cold, interrupted every 90,000 years or so by short intervals of warmer climates that last about 10,000 years. There seems to be an approximately 100,000-year cycle of major ice ages. Ice cores also reveal very rapid past changes in global temperature and climate, as well as natural causes for these rapid changes.[13–15]

Ocean Floor

The bottoms of oceans and lakebeds accumulate layers of sediments deposited from land areas. These sediments, over time, accumulate materials that become fossilized in layers that record past climates. Each year approximately 10 million tons of sediment collect on lake and ocean floors. Scientists use core-type drills to penetrate deep into these sediments. After the cores are extracted and stored, the preserved fossils and minerals and solutions in the cores are examined. Subsequently, estimations of climates for the past hundred to hundreds to millions of years can be made.[16]

Caves

The formations of caves and their bottoms provide excellent information about past weather patterns and climates but only for limited periods of time. Speleotherms are stalagmites, stalactites, and flowstones that are found in old, deep sections of many caves located on almost all continents. Dr. Karin Holmgren, a Swedish Earth scientist, examined a cave named Lobatse II in Botswana, on the African continent, an arid region where few proxy clues exist that hint of past climates. This cave provided her with a wealth of information. After working her way deep into the cave through piles of bat guano to get speleotherm samples, she compared stable oxygen and carbon atoms with isotopes of both elements (^{16}O, ^{17}O, ^{18}O, ^{12}C, and ^{14}C) to determine the ratios that indicated the rainfall, vegetation, and other factors in the area between 20,000 and 30,000 years ago.[17]

Rocks

Rocks and fossils found imbedded in rocks provide abundant evidence of ancient climates but a limited amount of historical data on global temperatures and rainfall. Nevertheless, geologists can estimate the chemical and physical properties of ancient ecosystems, and thus the climate at that period of time, by comparing fossils from the Cambrian period (~570–505 mya, and younger) with modern relatives of ancient organisms. For earlier periods (Proterozoic era, ~3,800–2,500 mya) before life was abundant, fossils do not provide organic proxies that can be used to determine ancient climates. However, sedimentary rocks from glacial **tillite** debris, carbonate rock participates (e.g., limestone), and iron deposits give clues to the chemistry and temperatures of very old global atmospheres.[18]

Climate (Climatology)

Climate may be thought of as statistically averaging the variations of local weather over relatively long periods of time. In addition, extreme events that occurred over long periods of time are also considered in the averaging of data. Climatology not only describes climate but also encompasses the consequences of climate change. It is directly related to a wide range of other sciences, including astronomy (solar system), oceanography, geography, geology, geophysics, and biology. It is also peripherally related to agriculture, engineering, economics, medicine, sociology, politics, and statistics.[19]

Weather influences our daily life activities for short periods of time. If it rains, we may use an umbrella. If it snows, we may stay indoors, or if it turns sunny and warm, we may go for a swim. In contrast, climate influences many aspects of peoples' lives over longer periods of time, not just the day-to-day activities. One example of how climates influence our lives is the migration over the past century from the colder, northern states, and in particular the northeastern regions, to the much warmer climates of the southern and southwestern United States.

Factors That Determine Climate

There are five spheres or layers of the Earth that interact in a feedback system that affects the Earth's climate. These spheres can be thought of as elements in a process diagram, arranged in a circle with two-way arrows between each of them: lithosphere (crust) ↔ biosphere (region of life) ↔ cryosphere (frozen water and land) ↔ hydrosphere (water) ↔ atmosphere (air) ↔ lithosphere (crust).[20]

Climates are also influenced by latitude, land and water distribution, altitude, land barriers, ocean currents, air movement, and high- and low-pressure storm areas. Let's take a look at some of the factors that determine geographic and global climates.

Latitude

One of the most important factors is a region's latitude on the face of the Earth. Latitude is the angular distance of a point on the Earth's surface measured from the equator (zero latitude) north or south to the poles, which are at 90° N or 90° S. What this means is that the climate is partially determined by the location of the geographic region in regard to its position between the equator and the poles. Climate is also partially determined by the tilt of the Earth on its axis by 23.5 degrees to its ecliptic (celestial equator). Thus, the closer to the equator, the more direct is the radiation received from the sun and the more even the distribution of night and day, which determines the length of time the Earth is exposed to the sun's radiation. As one moves north, the angle of exposure to the sun's radiation increases, resulting in less direct rays striking the Earth's surface and less total time that the area is exposed to sunlight. Only the equatorial region between 23.5° N (tropic of Cancer) and 23.5° S (tropic of Capricorn) receives vertical rays of the sun during the equinoxes. During the spring, the vernal equinox occurs on approximately March 21, whereas during the fall, autumnal equinox occurs on approximately September 23. During these equinoxes the day and night are both 12 hours long for virtually every portion of the Earth. Both the angle at which the sun's rays strike the Earth's surface and the extent of daylight hours—and thus the climate—are determined by a region's latitude.[21] Contrary to a common belief, the sun is not closer to the Earth during the summer months. Instead, the converse is true. Since the Earth's orbit is slightly elliptical, there is a six-month difference in its distance to the sun. Between January 1 and 4 the Earth is about 91,400,000 mi (147,000,000 kilometers [km]) from the sun, whereas between June 2 and 6, it is approximately 94,500,000 mi (152,200,000 km), a difference of ~3 million mi.[22] Actually, this relatively small difference is not what causes climate differences, but, as mentioned, the Earth's tilt and latitude are major factors that determine a region's climate.

Surface Land-Water Distribution

The distribution of land and water over the surface of the Earth is also a major factor that influences climates. Land areas heat up and cool off much faster than do large bodies of water. One reason is that the sun heats ocean water to a depth of 3 to 10 m (9.8 to 32.8 ft), while at the same latitude, the sun heats land areas to only about 5 to 10 centimeters (cm) (2 to 4 in.). This factor is one reason that landmasses have many more climatic extremes than do oceans. Air temperatures over land areas are warmer in the summer and cooler in winter than they are over bodies of water located at the same latitudes. Water has a much higher **latent heat** capacity than does continental crust. Oceans maintain an average temperature with less hot and cold extremes than do land areas. Although water has a greater capacity for absorbing heat, much of this heat energy is used in evaporation of water from its surface. (The energy is not lost but is later conserved by condensation and precipitation of the water vapor in the atmosphere.) These factors are also responsible for the moderating influence of temperatures for coastal land areas. Coastal cities may have an annual average temperature that is 20°F or cooler than inland cities located at the same latitude.[23–24]

Altitude and Air Pressure

Altitude, air pressure, and wind are related factors that affect regional, as well as global, climates. Altitude is how high up one goes from the surface of the Earth. Thus, the higher one goes, the more spread out the gas molecules are that comprise the atmosphere. When molecules are dispersed, the molecular motion responsible for kinetic energy (heat) decreases. In addition, the types of gas molecules found at different altitudes absorb the sun's radiation at different rates, which also affects temperatures at different altitudes. Altitude is also one of the major controllers of climate.[25]

Air pressure may be thought of as the weight of air pushing down on the surface of the Earth from its highest altitude. Even at high altitudes where gas molecules are spread far apart (very low density), they are still under the influence of Earth's gravity and, thus, exhibit weight. You might also think of air pressure as the attraction of the air to the surface of the Earth and water pressure as the attraction, or push, of water on the seafloor or lakebed. Both are measured by the extent (degree) of force exerted by gravity, that is, by weight. Air pressure is measured at sea level as the weight of a 1 square in. column of air extending upward as high as air goes, approximately 50 mi or more in altitude. The weight of this 1 square in., 50^+ mi tall column of air is referred to as *one atmosphere* and is equal to about 14.696 pounds of force exerted on each square inch of surface (see Figure 9.1).

This force is exerted on everything on Earth, including people. Humans have between 15 and 20 square feet of skin surface area.[26] Thus, the average human has about 25,000 square in. of skin area. Therefore, air above and around us is pushing on our skin with a total force of ~36,750 pounds. We are not crushed by this air pressure because there is an equal force inside our bodies that pushes outward, thus the forces are balanced and we are oblivious to the tremendous weight of the air above and around us. As the number of air molecules around you decreases, the air pressure decreases. This causes your ears to pop as you go up a mountain, as the air inside and outside your ears tends to equalize the pressure. Also, if you attain an altitude many miles above the Earth and are not protected, the air pressure is so much less than at ground level that your body bloats from the inside out. The special pressure suits worn by astronauts are designed to protect them from the lack of air pressure in space. The reverse is true for water pressure. As a human dives deeper and deeper into a lake or ocean, the pressure difference can become so great as to collapse the person's lungs, but since our bodies are mostly water, our body tissues are compressed but not crushed.

Figure 9.1
Air pressure and temperature are related, as they both are the result of gas molecules colliding with each other. The higher the temperature, the faster air molecules move, creating more space between them and, thus, less weight (pressure). Because air pressure is related to weight, it can be measured and recorded by barometers as atmospheric pressure.

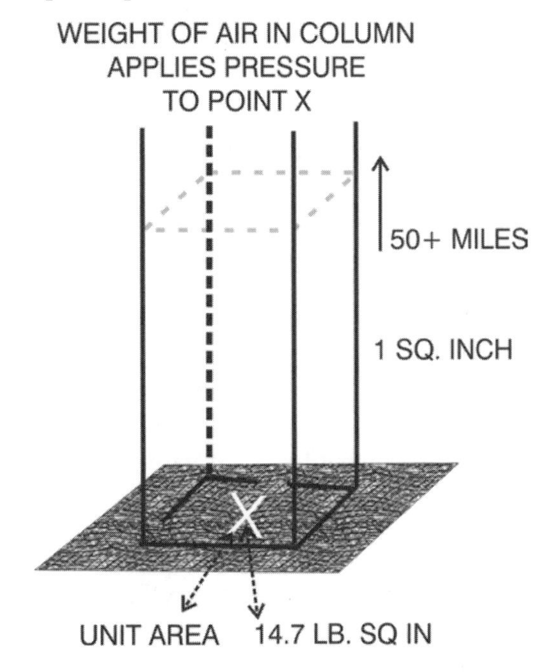

An everyday example of how air pressure pushes is the ordinary drinking straw. Your mouth forms a low-pressure area in the straw as you drink, but the air pressure above the liquid in the glass is responsible for pushing down on the surface of your drink and thus forcing the liquid up the straw into your mouth.

Altitude (and air pressure) also control temperature and precipitation, which are basic factors giving rise to adjacent high- and low-pressure belts of wind. Wind may be defined as both the local movement of air and the movement of huge air convection cells that form because of unequal heating on the Earth's surface that results in global pressure belts that drive wind. Local winds are important factors related to local weather conditions, but winds influence regional and global climates as well. At the equator, water is heated and evaporated and water vapor then rises as moist, less dense, low-pressure air, resulting in a belt of wind outward from the equator. These winds are known as easterlies, or trade winds, as they move away from the equator toward the poles. The air at the polar regions is cold, dense, and dry, resulting in high-density, high-pressure atmospheric conditions. When the cold polar air moves south, it is warmed and thus rises, creating subpolar low-pressure belts at about 60° latitude for both the northern and southern hemispheres. These different warm and cool air masses (low- and high-pressure areas) give rise to planetary wind belts that carry moist air over landmasses, causing condensation in the atmosphere for water vapor to cool, resulting in precipitation. Global wind belts modify the cold polar temperatures, just as the hot equatorial temperatures greatly influence Earth's regional and global climate.[27–28]

Topographical Factors

Topographical relief refers to the differences in the heights of natural land barriers, such as mountains, that have an effect on both climate and weather. The atmosphere near the crest of high mountain ranges exhibits lower atmospheric pressure and temperatures that affect precipitation, and thus climatic conditions. The sides of mountains facing the prevailing winds produce what is known as the *orographic effect.* In other words, the sides of the mountains facing the winds (windward side) force the warm, moisture-laden air upward. As the moisture-laden air rises, it cools, as does the water vapor, which condenses on minute particulate matter as clouds and precipitation. This results in a greater amount of rain and snow falling on the windward side of the mountain than on the leeward side, which, because deprived of rain and is often desertlike. For example, (1) the Alps in southern Europe protect the Mediterranean coast from severe weather by extracting moisture and by blocking some of the northern colder, windy, wetter weather from the Italian and Grecian beaches. The Mediterranean coast has an entirely different climate than the landmasses located on the northern side of the Alps. Also, (2) the Himalayas, located at the southern border of Tibet (China) and the northern region of India, remove much of the water from the atmosphere on the mountains' north slopes, leaving the northern portion of India with a dearth of precipitation. And (3) in the United States the Cascade Mountain Range located on the western edges of Oregon and Washington states forces the winds from the west that normally move east up to a higher altitude. By the time the air reaches the top of the mountains, it has cooled, forming clouds and, consequently, losing much of its moisture. The cooler, dryer air proceeds down the eastern side (leeward) of the Cascade Range where it is warmed, but it no longer can hold much water vapor because it has become denser at lower altitudes.[29–30] Many people consider Washington state to be a wet state because Seattle has many rainy days. Actually, desertlike conditions exist for the majority of the state that is located east of the Cascades. The central regions of Washington and Oregon receive only about 10 in. of precipitation a year, including snowfalls. Desertlike conditions in these areas necessitate extensive irrigation for farming. Washington is one of the few states exhibiting a wide spectrum of climates: a humid, rainy, mild-winter climate (west coast and Seattle); a dry, semi-arid, desertlike climate (central and southern regions); a steppe-like climate (eastern Washington); a temperate rainforest (far northwestern Olympic Peninsula); and a cold-winter climate (northern Washington forest region bordering Canada).

Ocean Currents

Ocean currents have a major effect on climates by transferring heat from the equatorial zone to the polar regions. These currents are responsible for cooling the ocean and landmasses near the equator while at the same time warming the oceans and landmasses located closer to the poles. For instance, the city of London, England, and the province of Newfoundland in northeastern Canada are both located at about the same northern latitude (51°). However, London has a temperate climate where it seldom snows, while Newfoundland has a polar climate. (The reason for this discrepancy is explained in the section on oceans in chapter 7.) The Gulf Stream current carries water warmed near the equator northward past the east coast of the United States until it meets the North Atlantic Drift, which transports the warmed water that bathes the shores of England, northern Europe, and Scandinavia. At the same time, the Labrador Current of the North Atlantic Ocean flows southward, transporting frigid water from the North Pole toward the equator to complete the cycle. The cold water from the North Pole is saltier and is thus denser, so it sinks as it flows all the way to Antarctica, cooling the coast of South America. Ocean currents might be thought of as rivers of warm or cool seawater that greatly influence the climates of coastal land areas by moderating their temperatures.

Global Air Movements

Large masses of air that move on a global basis are closely related to ocean currents. Warm air from the tropics circulates in currents northward, as do the ocean currents. As previously mentioned, warm air contains more water molecules, as vapor from ocean evaporation, than does cold air. As air (wind) moves over cooler landmasses, the water vapor is condensed and precipitated as rain and snow. Winds also drive surface circulation of the oceans that form a pattern related to the circulation of the prevailing winds. This wind movement affects shallow ocean water, which also contributes to moderating coastal climates. In the 1950s, an interesting experiment was conducted to demonstrate the global effect of temperature on wind/ocean currents. A shallow, round pan of water with a heating element wrapped around its outer edge was placed on a turntable. Colored dye was placed in the water at the edges of the pan. The water and dye at the outer edge of the pan represented water at the equator. As the water warmed, currents of dye in the water rose at the pan's edges, moved across the surface, and sank near the cooler center of the pan. The currents of dye then returned along the bottom of the pan back to the edge, to be rewarmed and continue the cyclic current. This circulation in the pan demonstrated what is known as a Hadley cell, named after George Hadley (1685–1768), who in 1735 developed the theory that global winds were generated by temperature differences at the Earth's equator and poles. His theory also stated that the airflow toward the equator was deflected by the Earth's rotation from west to east, an observation made a hundred years before Gaspard-Gustave de Coriolis (1792–1843) arrived at the idea of the Coriolis force, in 1835. The Coriolis force is the inertial force that acts on a rotating surface at right angles to its direction of motion causing air or water to follow a curved path rather than a straight line[31] (see Figure 7.2.) In the pan experiment, when scientists slowly rotated the pan on the turntable, the cell-like current flow of the dye moved from the edge to the center of the pan. The cell pattern was retained but deflected. When scientists speeded up the turntable, the cell structure was destroyed, but halfway from the edge to the center there was a pattern of whirls and eddies similar to the ocean currents and storms observed at the mid-latitudes of the Earth.[32]

Today, climate models are constructed by computers that consider the sun as the main source of energy that drives many of the factors that determine Earth's climates. One of the problems with computer models, even very advanced ones, in predicting future climates and weather is that there are so many variables to consider, along with many unknowns. These problems are associated with the study of complex systems that involve an understanding of **chaos theory**. Chaos theory, in essence, states that it is almost impossible to tell from the beginning of an event (and no two events have exactly the same beginning conditions) exactly how that event will progress as innumerable (and possibly unmeasurable) physical factors both act on and respond to the event's progress in space and time. As an example, can any person or computer predict with any accuracy the exact shape a cloud will take 5 or 10 minutes after it begins to look like a particular figure?

Climate Zones

The fact that there are different climates for large geographic regions has been known for thousands of years. People naturally described areas, or zones, where they lived or visited according to the distinctions in temperature, precipitation, wind, cloudiness, topography, vegetation, animal species, and so forth. It was not until 1900 that Wladimir Peter Köppen (1846–1940) developed a mathematical system of climate classification consisting of five broad climate types that provided some systematic approach to the study of climates for different geographic regions. Köppen's original system was based mainly on the latitude of geographic areas. He assigned the first five uppercase letters of the alphabet to his basic system for latitudes, namely, (A) represented the equatorial zone, and (B), (C), (D), and (E) were zones related to latitudes progressing from the equator toward both poles of the Earth. Today these letters have meaning beyond just latitude designations, representing climate zones such as A: hot, moist, tropical rainy climates; B: dry, but with wide range of temperatures; C: mid-latitude, moist, mild temperate climates: D: cold, snow, moist forest climates; and E: frozen tundra, polar zones (see Table 9.1).[33]

Köppen later defined three patterns of precipitation to add to his system, using lowercase letters to indicate (f): no dry period; (s): dry summers only; and (w): dry winters only. Later, four geographic subzones were added to the system. These geographic zones are represented by uppercase letters, but they are only used as subclassifications to the original five classes: (S): steppe; (W): desert; (T): tundra; and (F): perpetual frost. Still later, Köppen added six additional temperature classifications, which really complicate the system. These multiple classes based on latitude, temperature, precipitation, vegetation, and so forth provided climatologists a multitude of theoretical climate zones, but Köppen believed that only eleven classes were necessary to determine the climatic zones of Earth.[34] Köppen's system is not the only one used by scientists, but because it provides a means of classifying great varieties of climates, it is widely used. See the following examples for a typical Köppen classification system. Also, a number of internet Web sites can be accessed to print out examples of charts using Köppen's classification system.[35]

Table 9.1
Köppen Climate Chart

A Tropical humid	Af	Tropical rain forest	No dry season
	Am	Tropical monsoon	Short dry season; heavy monsoonal rains in other months
	As	Tropical wet/dry High sun	Summer dry season Dry savanna
	Aw	Tropical wet/dry Low sun	Winter dry season Dry savanna
B Deserts	BWh	Subtropical desert	Low-latitude desert, arid, hot
	BSh	Subtropical steppe/dry	Semi-arid, hot low-latitude
	Bwk	Mid-latitude desert	Mid-latitude arid, cool
	BSk	Mid-latitude steppe	Mid-latitude semi-arid, cool
C Mild Mid-Latitude	Csa	Mediterranean	Warm with dry, hot summer
	Csb	Mediterranean	Cool with dry, cool summer
	Cfa	Humid subtropical	Mild with no dry season, hot summer
	Cwa	Humid subtropical	Mild with dry winter, hot summer
	Cfb	Marine west coast	Mild, rain all year, warm summer
	Cfc	Marine west coast	Mild, no dry season, cool summer, heavy cloud cover
D Severe Mid-Latitude	Dfa	Humid continental	Humid with severe winter, no dry season, hot summer
	Dfb	Humid continental	Humid with severe winter no dry season, warm summer
	Dwa	Humid continental	Humid with severe dry winter, hot summer
	Dwb	Humid continental	Humid with severe, dry winter, warm summer
	Dfc	Subarctic	Severe winter, no dry season, cool summer
	Dfd	Subarctic	Severe very cold winter, no dry season, cool summer
	Dwc	Subarctic	Severe dry winter, cool summer
	Dwd	Subarctic	Severe very cold and dry winter, cool summer
E Polar	ET	Tundra	Polar tundra, no true summer
	EF	Ice cap	Permanent ice

Note that the various climate zones in the biosphere are related to multiple causes, including geological factors, the hydrosphere (mainly ocean currents), the atmosphere (wind and humidity patterns), and the latitudes of the zones (i.e., orientation to the sun). The following sections provide descriptions of a selected few of the major climate zones (after Köppen).

Tropical (A) Climate Zones

As the name implies, tropical climates are warm-to-hot with abundant rainfall as the result of the very large atmospheric low-pressure zone located near the equator. The tropical climate classification (A) varies from the rainforest (Af), to tropical monsoon (Am), to tropical wet-dry (Aw) climates.

(Af): Tropical Rainforests

Tropical rainforests consist mainly of very tall trees that branch out only at their tops, forming a canopy that blocks sunlight from the forest floor. Therefore, very little sunlight reaches ground level (about 0.1%), resulting in few ground plants. The excessive moisture keeps rainforest soil soggy and everything else wet or at least damp. Tropical rainforests are found in the Amazon Basin of South America, the Congo Basin of Africa, the Indonesian/Malaysian areas of Asia, the western coastal areas of India and Southeast Asia, the equatorial regions of South America, and the eastern coastal areas of Central America. The following list provides some statistics for typical rainforest climate zones:

- Latitude range: 0° to 15° north and south of the equator
- Average annual precipitation: 200 cm/year to 500 cm/year (79 in./year to 197 in./year)
- Average annual temperature: 23°C to 28°C (73.5°F to 82.5°F)

There are really three levels of vegetation in tropical rainforests. The **kapok** trees are ~200 ft tall and grow about 12 in. per year. The middle layer trees range from ~100 to ~130 ft high and form the main canopy that blocks the sun from ground plants, resulting in sparse undergrowth. Seventy percent of all the vegetation in tropical rainforests are trees with slender, wet trunks that have few branches. Soil is thin but wet, which prevents tree roots from penetrating deeply into the ground, resulting in roots that spread out near the surface for 10–15 ft. Whatever falls to the ground from the tree canopy is quickly eaten, or it decays. Thick layers of dead plant material build up on the soil but decay rapidly, resulting in very acidic and infertile soil. Rainforests are excellent examples of great biodiversity and stable ecosystems that have been in balance for many centuries. Some products from the tropical rainforests, besides timber, are rubber, cacao (chocolate beans), vanilla, figs, Brazil nuts, and many plants used for medicines.[36] When rainforests are cleared (usually by burning) and farmed, the thin soil is leached, depleting in a few years whatever nutrients previously existed. Fields are then abandoned, resulting, in time, in the establishment of secondary forests. In the meantime, new forest patches are destroyed for more farmland, as population growth requires more food. The aftermath is the elimination of huge tracks of rainforests that are major sinks for atmospheric carbon dioxide and the home of yet-to-be-identified species of plants and animals.

(Am): Tropical Monsoon

Tropical monsoon climate zones are also hot but, in contrast to tropical rainforests, are only seasonally wet and humid. While the total rainfall in monsoon zones is adequate to support rainforests, unlike in rainforests, the rain is seasonal, which makes it more ideal for farming. Two well-known (Am) tropical zones are located in northern and central India and southeastern Asia. In India, for example, low-pressure areas in the summer bring warm, moist air from the Indian Ocean across the continent, resulting in torrents of rain. The high-pressure areas that form in the winter produce masses of cooler and very dry air. In northern regions of India, the beginning of the monsoon season, usually about July, is evoked as a period of rejoicing after a difficult dry season by walking and playing in the rain. Northeastern Brazil is located in an (Am) climate zone with a distinctive, short, one-month dry season and less than a 5°C (41°F) annual variation of temperatures. This one dry month in Brazil receives less than 6 mm (2.3 in.) of precipitation, while the other months receive much more rain. The following list provides some statistics for (Am) climate zones:

- Latitude range: 5° to 30° north and south of the equator
- Average annual precipitation: 150 cm/year to 400 cm/year (60 in./year to 158 in./year)
- Average annual temperature: 20°C to 30°C (68°F to 86°F)

As expected, vegetation changes as climate zones shift from a constant wet area to one that is only seasonally wet. Forests are less dense, while ground cover is denser as more sunlight reaches the forest floor. Deciduous trees have adapted by maintaining leaves during dry seasons while sinking their roots deep into the soil to obtain moisture.

(Aw): Tropical Wet and Dry

Tropical wet and dry climates are located moving from the equator toward both the North and South Poles. Compared to the (Am) zones, the dry periods are much longer and the year-round water surplus found in the monsoon zones no longer exists. This results in a **xerophytic** type of vegetation adapted to dryer climates. These dry periods become longer and longer and the water deficit greater and greater as you move farther and farther from the equator. Trees with thick bark and thorns are more scattered, while grasses with dense, water-absorbing roots are dominant. This prevents much of the moisture from penetrating deep into the soil. The massive root systems of grasses permit them to retain enough moisture to survive long dry periods. Tree roots grow about 10 times deeper into the ground than the height of the tree itself. This describes what is known as savanna climate and vegetation found in the semi-arid grasslands of central Africa. The following list provides some statistics for savanna-like (Aw) climate zones:

- Latitude range: 5° to 25° north and south of the equator
- Average annual precipitation: 100 cm/year to 180 cm/year (39 in./year to 71 in./year)
- Average annual temperature: 20°C to 30°C (68°F to 86°F)

Desert (B) Climate Zones

There are several classifications for (B) climate zones. Two of the most common are semi-arid (BS) and arid (BW). Both of these classifications can be further defined by temperatures and precipitation as (BSh) and (BWh), where h stands for hot.

(BS): Grasslands, Steppes, and Taiga

Grasslands, steppes, and taiga types of vegetation are found in the (BS) and (BSh) semi-arid climate zones. The precipitation in these zones is not very great, yet there is considerably more annual rainfall than in the desert zones. Grasslands of the steppes receive about 10 in. of rain per year, but never as much as 30 in. Even so there is an annual water deficit in semi-arid regions. Steppes (BSh) are located toward the polar side of low-latitude deserts and receive most of their precipitation during winter months.

(BW): Low-Latitude Deserts

Low-latitude deserts (near the equator at 18 to 28°) are considered the (BW) and (BWh) climate zones. These zones are located next to the equatorial border of the subtropical, high-pressure, trade winds regions. Their temperatures vary greatly, with very hot summers during daylight hours followed by cold nights. Examples of great deserts are the Sahara, Thar, Sonoran, Kalahari, and Great Australian, and Death Valley in the United States. Xerophytic vegetation has adapted to these harsh climates by developing root systems that can absorb and store water. This type of vegetation may form compact ground growth, with plants having either no leaves or with thick leaves or needles. Many xerophytic plants also have green stalks or trunks where photosynthesis can take place. Some of the desert's deciduous evergreen shrubs and grasses are dormant during dry periods but grow rapidly in wet periods. There are many varieties of desert plants and cacti that can take in moisture from dew through their leaves and stems. The following list provides some statistics for desert climate zones:

- Latitude: varies by continent; arid regions are generally nearer the equator (15° to 28°), while semi-arid zones are further from the equator (about 15° to 35°)
- Average annual precipitation: semi-arid (BSh) = 25 cm to 50 cm (10 in. to 20 in.); arid (BWh) = less than 25 cm (10 in.)
- Average annual temperature: varies greatly from summer to winter as well as from day to night; the (BSh) and (BWh) climate zones cannot easily be distinguished by their temperatures

The following are examples of average maximum and minimum temperatures (°F) for a few cities classified as having (B) climates.

Reno, Nevada: maximum = 98, minimum = −1, variation = 99
Yuma, Arizona: maximum = 113, minimum = 31, variation = 89
Phoenix, Arizona: maximum = 122, minimum = 17, variation = 105
Kazalinsk, Russia: maximum = 103, minimum = −21, variation = 123
Cairo, Egypt: maximum = 117, minimum = 34, variation = 83
Damascus, Syria: maximum = 113, minimum = 21, variation = 92[37]

Note that annual precipitation for the listed cities ranges from 1.1 in. (Cairo) to 7.66 in, (Phoenix).

Mid-Latitude (C) Climate Zones

The (C) climate classification includes the middle latitude regions of the Earth where most of the population lives. It is a zone of climate transition from the (A) class to the (D) class. There are numerous subclassifications representing variations in seasonal precipitation and temperatures for (C) climates, as well as for all other climate zones. Some examples for the (C) zones are humid subtropical, (Cfa); Mediterranean, or dry subtropical, (Csa); marine west coast, (Cfb); and mid-latitude monsoon, (Cw). Note that precipitation levels are classed as (f), (w), or (s), while severity of winter is classed as (a), (b), or (c). It might be easier to consider the C zone as divided into two general regions. The southern region, which is more subtropical than the other (C) zones, experiences hot, humid summers and mild, short winters, with rainfall influenced by the subtropical high-pressure belt. The more northern region of the (C) zone may have mild to hot summers but longer and cooler winters. There is a transition in agricultural crops and natural vegetation from the southern part of the (C) zone to the northern part of the zone, located farther from the equator. Just a few examples of the many subclasses for (C) zone climates are given in the following sections.

(Cfa): Humid Subtropical, Mid-Latitude

Humid subtropical, mid-latitude climates have hot, long, and humid summers, with cool, short, winters that are only slightly less humid. Cyclonic storms (frontal low-pressure atmospheric areas) provide adequate rain and snow during the winter season. The average winter monthly temperatures seldom go below 7°C (44.6°F). Atmospheric convection currents that produce thunderstorms provide precipitation during the summer months. For both North America and Asia, huge polar air masses dip down toward the equator far enough to appear as cold fronts, bringing more extreme winter cold spells to the (Cfa) zones. Both conifer trees (with needles) and deciduous (broadleaf) trees grow in the (Cfa) zone. For the most part, however, the trees, are considered mid-latitude deciduous and are found in the eastern United States, southeastern Canada, northwest Europe, east Asia, and the southern part of South America. There are fewer species of trees in the (C) zone than in the (A) tropical zone, but they are of the variety valuable for lumber. Houston, Texas, and New Orleans, Louisiana, are examples of cities in the lower, more humid part of the (Cfa) zone, while most of the southeastern and lower Midwestern states are in the more northern, less humid (Cfa) zone. The following list provides some statistics for the (Cfa) zone:

- Latitude ranges: 15° to 40° north and south from the equator
- Average annual precipitation: 60 cm/year to 250 cm/year (24 in./year to 98 in./year)
- Average annual temperature: 7°C to 32°C (45°F to 90°F)

(Csa): Mediterranean or Dry Subtropical

Mediterranean or dry subtropical climate zones are found on the west coasts of continents at the middle latitudes. Examples are the west coast of California, the coastal region of central Chile, and the west coasts of Africa and Australia, as well as the southern European countries with coastlines on the Mediterranean Sea. Mediterranean regions comprise the largest geographic areas for (Csa) zones, which have long, hot, summers, and mild, foggy, rainy winters. The hot summers are the result of an atmospheric high-pressure cell that develops in the subtropics. The wet winters are the result of cyclonic storms (low-pressure areas) that are pushed eastward by the westerly winds that bring the rain to the west coast regions of several continents. An excellent example is the San Francisco area located on the west coast of the United States. The following list provides some statistics for (Csa) climate zones:

- Latitude range: 30° to 40° north and south of the equator
- Average annual precipitation: 35 cm/year to 75 cm/year (24 in./year to 30 in./year)
- Average annual temperature: 5°C to 30°C (41°F to 86°F)

The limited rainfall in summer (with a surplus in winter) determines the types of vegetation in this zone. Plants consist of evergreen trees and drought-resistant leathery foliage. Woody shrubs and trees range from 1.5 to 10 ft in height.

(Cfb): Marine West Coast

Marine west coast climate zones are an extension of the Mediterranean climate zones, both north and south of the equator about 65° latitude. Therefore, they are both cooler and wetter than (Csa) climate zones. These (Cfb) zones are located in the Pacific Northwest of the United States, most of central Europe, all of the British Isles, and the southwest coast of Chile from the coast to the mountains. These zones are naturally forested areas with giant fir trees that support the timber industry. Generations of settlers have reduced the size of the forests while creating much larger, productive agricultural industries in these zones. The prevailing westerlies constantly bring moisture from the warm, offshore ocean currents that results in a humid, mild climate. Some examples are the west coast of northern California; Portland, Oregon; Seattle, Washington; and Anchorage, Alaska. The following list provides some statistics for the (Cfb) climate zone:

- Latitude range: 40° to 65° north and south of the equator
- Average annual precipitation: 50 cm/year to 255 cm/year (20 in./year to 100 in./year)
- Average annual temperature: 0°C to 25°C (32°F to 77°F)

The wettest areas are those where mountains block the warm, moist ocean air moving from west to east. These areas become temperate rainforests where annual rainfall can be as much as ~100 in. per year. This part of the (Cfb) zone receives steady, heavy rainfall. The lower regions bordering the Mediterranean zones are also rainy, but differences in the total annual rainfall is striking. For instance, both Portland, Oregon, and Seattle-Tacoma, Washington, receive only ~37 in. of rain annually (which is less than the following cities: Houston, Texas, ~58 in.; Little Rock, Arkansas, ~42 in.; Atlanta, Georgia, ~51 in.; and New Orleans, Louisiana, ~62 in.). However, the Pacific rainforest in the Olympic Peninsula located in the (Cfb) zone receives ~100 in., about three times the annual rainfall as does Seattle. The reason people think Seattle is rainy is that their annual allotment is spread over most of the year as light rain or drizzle, while other temperate zones may receive their greater allotment of rain as infrequent downpours, often resulting in flooding. Because Seattle and other northwest coastal areas are considered rainy, it is difficult to comprehend that most central and eastern parts of the states of Oregon and Washington are classed as semi-arid, with ~9 or ~10 in. of annual precipitation.

Moist, Severe Winter (D) Climate Zones

The (D) climate zones are also referred to as "humid mid-latitude climate with severe winters." The (D) climates are found just poleward from the (C) climate zones. Their mid-continental, mid-latitude location results in hot summers but cold to severe winters, which is the typical climate of the northern Great Plains of the United States, as well as northern Europe and Asia. Forests persist in the more moist zones, while grasslands predominate in the dryer regions. There are a number of subclassifications for the (D) zones that relate to both precipitation [(s), (w), (f)] and temperature [(a), (b), (c), (d)].

(Dfa) and (Dfb): Boreal Forest

Boreal forest climate zones encompass large landmasses that makes these zones more prominent in the Northern Hemisphere than the Southern, since the only large landmass in the Southern Hemisphere is Antarctica, which is in the (E) polar climate zone. Although boreal forests may occur in other northern climate zones (e.g., the northern parts of the (C) zone), they generally exist in regions with cool summers and cold winters as found in the (D) zone. The trees are conifer evergreens that have adapted to severe climates as found in the geographic taiga forest regions of subarctic Siberia, Eurasia, and North America. Note that the taiga biome is characterized by conifer trees that have adapted to fluctuations of low precipitation and cold temperatures. They have leaves with thick cuticles and are conical shaped, permitting snow to slide off and not break branches. Think of the shape of a typical Christmas tree. The species of trees found in the taiga include spruce, larch, fir, birch, and pine. The following list provides some statistics for the (D) climate zones:

- Latitude range: 40° to 65° north and south of the poles
- Average annual precipitation: 50 cm/year to 150 cm/year (20 in./year to 60 in./year)
- Average annual temperature: −15°C to 25°C (5°F to 77°F)

The southern regions of the (D) climate zone are too dry for trees to survive; therefore, grasslands predominate. These vast grasslands are where millions of bison roamed North America and mammoths grazed in northern Asia.

Polar (E) Climate Zones

There are two basic types of climates at the poles, *tundra* (ET) and *frost* (EF). Some geologists also add a classification of (H) to the polar climate zone to represent the highland mountain climates at or above 1500 m (4800 ft). These (H) zones generally have the same temperature and precipitation as does the (E) zone.

(ET): Tundra Climate Zone

The tundra climate zone gets its name from its typical biome, which is an extension of the grasslands of the subarctic zones. The southern boundary is also the northern limit of the boreal forest zones. The (ET) is a zone of cold winters and cool, short summers with little precipitation. Much of the tundra consists of gravel as a result of strong winds blowing away the thin soil. This thin soil, plus the cold climate, prevents the growth of trees. What soil exists supports lichens, mosses, sedges, and low-lying plants, as well as a variety of grasses. The top layers of the surface soil and gravel experience seasonal freezing and thawing cycles that create problems when constructing buildings and roads, as the surface is not stable. This is the land of Eskimos and the Lapp, where the grasslands that thrive in the summer sun attract moose, elk, and musk ox. It is the land of winters with cold, long nights and the midnight sun. The daylight periods in the summer are long, thus providing a growing season adequate for some crops. The following list provides some statistics for the ET tundra zone:

- Latitude range: 65° to 80° north and south of the equator
- Average annual precipitation: 10 cm/year to 35 cm/year (3.9 in./year to 13.8 in./year)
- Average annual temperature: −35°C to 10°C (−31°F to 50°F)

(EF): Frost Climate Zone

The frost climate zone is even farther toward the Earth's poles than is the tundra zone and thus experiences even colder weather and a 6-month cycle of dark winters and light summers. The temperatures, however, are rarely above 0°C (32°F). The frost zone includes the areas north of the Arctic Circle, all of the Antarctic, most of Greenland, and the very northern parts of Asia. Frost climate zones are also known as permafrost regions because the soil and upper crust is permanently frozen. Permafrost thickness varies from 10 cm (4 in.) to more than 1000 m (3281 ft) in depth.[38] Permanent snow, ice, and freezing temperatures characterize the frost zone. Since it is a zone with not much precipitation, even as snow, and is a cold desolate region, it is sometimes referred to as a frozen desert. The frost zone is seldom visited except for explorers and scientists.[39–44]

El Niño and La Niña

El Niño and La Niña are related opposites that are blamed for much of the severe weather around the world. Humans tend to ascribe the term "normal" to the average climate and weather conditions that exist for their geographic location. However, climatologists and meteorologists know that "normal" is not a viable term when used to describe climate and weather patterns. Climate and weather phenomena exhibited many changing patterns in the past, and these patterns will continue to change in the future. Conditions known as El Niño and La Niña are cyclic—they wax and wane and, depending on where you live, may affect you, but they are not responsible for global climate phenomena.

El Niño

The phenomena of El Niño (warm current of water from the south Pacific Ocean to the northwest coast of the United States) and La Niña (cold surface water) affect both regional and global climates, as well as more local weather conditions. Although meteorological records of both go back as far as the beginning of the twentieth century, their connection with the cyclic warming and cooling and variable rainfall, floods, and droughts for different geographic regions is not well understood by most people. El Niño, is the Spanish word for "boy," translated to mean the "Christ Child." It was named by fishermen who plied the Pacific Ocean off the west coast of Ecuador and Peru when they noticed that periodically the water became warmer about the time of Christmas and the New Year. It is referred to as ENSO (El Niño–southern oscillation), which is a change in the ocean and atmospheric system in the eastern portion of the Pacific Ocean. In essence, El Niño is a weather phenomenon attributed to the periodic lessening of the normal Pacific Ocean currents and trade winds. The tropical south Pacific trade winds, under normal conditions, push equatorial water westward, where it is exposed to long periods of solar heating. During El Niño, this easterly trade wind weakens, allowing the warmed water to now accumulate along the west coast of South America. This can have a devastating effect on the fishing areas, since cool water pulled from deeper levels in the ocean holds more oxygen and increased nutrients for fish, while warm surface water is depleted, resulting in fewer fish and less marine plant life. Warm-water fish, such as tuna, are now spotted off the northwest coast of the United States, while fish usually found off the coast of Seattle are now found off the Alaska coasts. El Niño appears about every 3 to 7 years and generally lasts only a few months. Only recently have scientists realized that El Niño can affect patterns in weather conditions worldwide that result in economic disasters. During the 1982–83 El Niño year, tropical rains shifted eastward and changed atmospheric wind patterns worldwide, shifting the jet stream and monsoon patterns and causing unusual weather for many regions. There was drought in southern Africa, India, Sri Lanka, Philippines, Indonesia, Australia, southern Peru, Bolivia, Mexico, and Central America. During that time there were also heavy rains and flooding in Bolivia, Ecuador, northern Peru, Cuba, and the Gulf states of the United States. Hurricanes devastated Tahiti and Hawaii. The 1982–83 El Niño also warmed the Pacific Ocean along the west coast of the United States as far north as Alaska, resulting in unusual local weather conditions that included flooding and coastal damage in California and wetter weather in the southern states. In 1997–98, a mass of warm water in the Pacific Ocean caused one of the strongest El Niño weather disturbances so far recorded.[45–46]

Some people claim that recent, strong El Niño events are a result of global warming or, at the very least, that there is a connection between the two. The reasons for a periodic cycle of warming and cooling of the southern Pacific Ocean are not completely understood. During El Niño years when the surface of the northern Pacific Ocean and the atmosphere become warmer, global temperatures are *not* affected. El Niño is not responsible for warmer than usual winter in the northeastern United States nor in other regions. Global warming may play a role in this, along with an increase in greenhouse gases (see the section on global warming in this chapter). The El Niño phenomenon is more localized to the eastern Pacific Ocean and the western coastal areas of South and North America. The cyclic nature of temperature and wind changes has existed long before the recent increase in CO_2 in the atmosphere attributed to human activities. Fossil evidence found in Peru indicates that there was little variation in climate until ~5000 years ago. This is also about the time El Niño began causing so much climatic disruption around the world and that some cultural changes are also attributed to these variations in climates. There is also some indication that most of the United States is about to have a repeat of the 1950–60 decade when, on the average, the winters were cooler and wetter, while at the same time the southwest and southern states experienced warmer weather and droughts. It is also predicted that, along with fewer El Niños that warm the western Pacific, there will be an equal and opposite cooling of the eastern tropical waters by La Niñas.[47]

La Niña

La Niña means "the little girl." It is also referred to as El Viejo, or the anti-El Niño, as well as a "cold event" or "cold episode." As with El Niño, La Niña is one of many factors that affect regional and global climate, as well as local weather—particularly during the winters of the northern latitudes. La Niña is characterized by unusually cold ocean temperatures in the equatorial Pacific, while just the opposite characterizes El Niño. La Niña is caused when equatorial winds in the eastern Pacific Ocean blow in a westward direction. This results in cool water being pulled up from deep in the ocean and cooling the atmosphere. Both El Niño and La Niña influence ocean and atmospheric temperatures, rainfall, and other weather conditions in the United States. El Niño winters are warmer than normal in the north central States and cooler in the south and southwest, but it is just the opposite for La Niña years, which occur only about half as often as does the El Niño cycle. La Niña follows El Niño by a year or two, resulting in cooler than usual winters in the north and northwest, and warmer in the south and southeastern parts of the United States.[48]

The influence of island ocean wakes may also result in some of the changes attributed to the El Niño and La Niña phenomena. An ocean wake is created south of an island when a wind-driven ocean current is flowing from north to south past the island. Satellites observed a 3000 km (1900 mi) wake in the ocean downwind from the Hawaiian Islands that creates a positive feedback that affects the sea surface temperature, clouds, and wind patterns. This effect may influence atmospheric transport of aerosols and trace gases (carbon dioxide, methane, etc.), as well as the distribution of plankton and other fishery resources.

Climate Summary

As previously mentioned, climate may be thought of as the long-term manifestations of weather conditions over a long period of time for geographic regions or the entire globe. Factors such as the inclination (latitude) of the region to the more or less direct rays of the sun, the energy output of the sun, temperature, pressure, ocean currents, winds, and precipitation all affect regional, as well as local, climates. Paleoclimatologists who use natural proxies (tree rings, fossils, ice and ocean sediment cores, caves, corals, and, more recently, isotopes of carbon, hydrogen, and other elements) in addition to historical records have established that the Earth's climate has changed, including some drastic changes, over time. Based on the geological time scale, climate changes can be very slow or relatively fast. Life has adapted to these changes through evolutionary processes, and no doubt will continue to do so for future climate changes.

Weather (Meteorology)

People have continued to predict weather conditions since prehistoric days but without any high degree of accuracy. Only in the past century have humans been able to make reasonable predictions for the future weather and climate of local, regional, and global areas but, again, with varying degrees of success. Local weather forecasters can predict with some accuracy what to expect weather-wise in their area for the next few days, but with only about a 50% record of accuracy beyond four or five days. With the assistance of weather satellites and new instruments and knowledge, meteorologists are getting better at short-term weather predictions, but long-term predictions are still not very accurate. A series of four satellites named *Aqua, Grace* (two satellites), and *Jason 1* are now tracking the atmospheric temperature, humidity, precipitation, soil moisture, sea surface temperature, Earth's gravity for movement of large masses (e.g., melting glaciers), and changes in topography (height of bodies of water and ice, and shifts in volumes). By observing the climate and weather cycles scientists hope to learn more about global warming.

Geological evidence provides indirect evidence of catastrophic ancient weather events that indicate geographic conditions were markedly different from those existing for present landmasses. Not only have continents moved to different climate zones (see chapter 2), but also there is evidence of tremendous rainfalls, flooding, glaciation, desertification, and weathering of landscapes, all indicative of weather patterns dissimilar to those that exist today. In an effort to determine what might be expected in the near future, scientists have produced records of temperatures for the northern hemisphere covering the last 600 years that can be compared to more recent temperature records. For instance, the temperatures for the years 1990, 1995, 1997, and 1998 (the hottest in recent history) were the warmest (average) since the end of the little ice age after ~1800 C.E. In addition to greenhouse gases, it is believed that fluctuations of solar energy and volcanic activity were mainly responsible for the minor increase in historic temperatures.[49] Debate currently exists regarding the extent to which human's roles have changed worldwide temperatures, weather phenomena, and global climate. This is an important debate because temperatures (heat) not only influence local weather but also, over long periods of time, are a major factor in determining climates.

Predicting Weather—Past

Humans have always been affected by weather on a daily basis, so it was natural for us to observe certain changes in the patterns of the behavior of insects and animals, as well as our own. Primitive people developed methods for reading weather patterns by observing many natural phenomena that acted as predictors of change, some of which had valid cause-and-effect relationships related to changes in weather conditions. By observing nature and using common sense, primitive people knew that rain or snow was imminent if birds roosted close to the ground, just as early hibernation of bears indicated a harsh winter.[50]

Even with some logical and accurate forecasting techniques, ancients primarily based their observations on mystical explanations. Myths and folklore about weather that are based on fear and superstition (as well as ignorance) have existed for thousands of years. One online Web page lists more than 300 "Folklore and Weather Myths."[51] The vast majority of weather myths have no validity as far as having insight into the future, including predicting the weather.

Weather Elements

Chapter 3 presented how different physical and chemical activity affected the weathering of the landscape and how atmospheric weather factors are related to the dynamic nature of the Earth's outer spheres. Now let's examine some of the more local and regional atmospheric phenomena that constitute what we call "weather." The physical elements that comprise weather are temperature (heat), atmospheric pressure (altitude and density of air), humidity (moisture), clouds (condensed water vapor in the atmosphere), and wind (movement of air).

Temperature

Temperature as related to climate, as well as weather conditions, is probably the most important factor that drives daily weather. Heat is associated with the kinetic energy from the motion of molecules. Therefore, one way to look at heat is to recognize that there is either more or less molecular motion, meaning that the terms "hot," "warm," "cool," or "cold" indicate the existence of a great deal of molecular motion, less motion, or very little molecular motion in matter. Heat is just more or less kinetic energy (molecular motion), usually measured on the Kelvin scale from absolute zero (−273.16°C or −459.69°F) where normal molecular motion ceases. Temperature, by definition, is the determination by an agreed-on scale of degrees of whether a body, the atmosphere, or even space has the ability to transfer heat to another body. Heat (i.e., kinetic energy of molecular motion) can only be transferred from a body that is hotter to a body that is cooler. Heat never flows from cold to hot.[52] Temperature as related to weather is the measure of the transfer of heat between the Earth's surface and atmosphere. It is usually determined empirically by reading assigned points on instruments, that is, thermometers. Even if we don't measure the extent of "hot," "warm," "cold," "cool," "humid," and "dry" with instruments, our senses provide subjective interpretations. Although these are relative and subjective terms, they are very useful in describing meteorological conditions, that is, the weather.

By far, the sun is the most important source of Earth's energy. (Note that a small amount of heat energy reaches Earth's surface from fission decay of radioactive elements in the Earth's crust, but this amount is insignificant.) Past, present, and future humans, animals, and plants, as well as petroleum/gas deposits, all owe their existence to the sun.

The energy that the Earth receives from the sun is referred to as "**insolation**," which is only about one two-billionths of the total energy output of the sun. Of this amount of solar radiation sent in Earth's direction, only about 40% is reflected back into space. Twenty percent is absorbed by the atmosphere, and only about 20% actually reaches the Earth's surface. In other words, most of the sun's energy is radiated in all directions and is lost in space. Therefore, the Earth receives a tiny, but important, proportion of the sun's total output of energy.

The main electromagnetic energy received on Earth's surface from the sun is in the visible light and infrared (heat) wavelength regions. The shorter wavelengths of solar radiation have greater effect on the upper levels of the atmosphere. The sun's heliosphere (magnetic field) acts as a shield that protects Earth from the full effect of high-energy, short-wavelength solar radiation. The magnetic activity of the sun produces ionized particles that **wax** and **wane** on an 11-year solar cycle (see chapter 8). The variations in the amount and strength of ionized particles is controlled partly by solar flares that result in changes in Earth's upper atmosphere. Thus, the solar cycle has a great periodic effect on both climate and weather. The fifth most massive solar flare activity on record was recently measured by solar satellites. This occurred on Easter Sunday, April 15, 2001.[53] Scientists are also investigating what happens to the Earth's weather when the sun's magnetic pole flips from north to south. The sun spins on its axis in only one direction, so that when these solar flips occur they result in the development of a strong magnetic field. This field affects both the sun's radiation and cosmic radiation in different ways, causing periodic changes in Earth's climate and weather systems.[54]

Pressure

Pressure (air pressure) is directly related to the density of air as affected by atmospheric temperature and altitude. (Note that pressure was defined and discussed in the climate section of this chapter.) As we mentioned, warm air is less dense than cool air; thus, it forms low-pressure air masses that tend to rise. Air in the upper atmosphere is cooler and thus denser, forming a high-pressure area that tends to descend. This creates movement of cooler air masses sliding under and replacing warmer air, causing a number of changing weather conditions. Alterations in pressure create cold fronts that produce air movements referred to as "wind" (see Figure 9.2).

The cooling of humid, warm air results in precipitation and can result in extreme changes in air pressure, forming violent storms, flooding, tornadoes, and tropical cyclones (hurricanes). A "pressure gradient" is the relative difference in air pressure and wind movement over a distance. For

Figure 9.2
A mass of air at a given altitude exhibits approximately the same temperature and pressure over a large region. Air masses of different temperatures collide with one another as they meet, forming a boundary called a "front." A cold front forms when a moving mass of cold air meets a warmer mass of air, forcing the warmer air to rise, where it cools, resulting in condensation of water vapor. A warm front forms when a moving mass of warm air collides with a slower moving or stationary mass of cooler air.

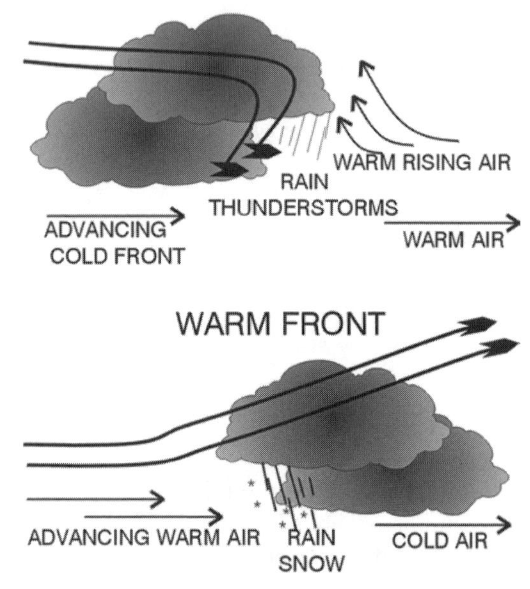

instance, the wind movement in a steep pressure gradient (e.g., a rapid and significant change from high to low pressure) is somewhat like a rock rolling down a hill. The steeper the hill, the greater the rock's movement. Thus, the steeper the pressure gradient, the faster the air will move (wind). Atmospheric weather maps use continuous lines called "isobars" to represent points of equal pressure. Meteorologists place isobar lines close together on weather maps to represent a steep pressure gradient at a well-defined boundary between high- and low-pressure air masses. Conversely, isobar lines are placed farther apart if the pressure difference (gradient) is less. Previously, we mentioned that standard air pressure was 14.7 pounds per square in. (29.92 in. of mercury) for dry air at the surface of the Earth. Meteorologists use a different system, called "millibars," when calculating isobars representing pressure gradients. There are several different units used to measure atmospheric pressure. For example, if atmospheric pressure is 14.7 lb/square in., it can also be recorded as 1 kilogram/cm^2, or as 29.92 in. of mercury (in a barometer), or as 1,013 millibars of pressure on a weather map. Since air temperatures and pressures are always changing, pressure gradients also change, often in just a few hours, as weather changes. This results in frequent corrections of isobar configurations on weather maps. If short-term isobars representing short-lived or transient weather changes are ignored, and only more steady or permanent isobars are used for a region's map, it is possible to determine the location of uniform masses of air. Examples of permanent air masses are the warm, moist, low-pressure air over the tropics and the cold, dry, high-pressure air masses over the polar regions. Air masses are also affected by temperature differences, as well as pressure gradients.[55–57]

Humidity

Humidity is the amount of water vapor (moisture) that has evaporated from surface water and now resides in the atmosphere as a gas.

1. *Absolute humidity* is the total mass of water vapor in a given volume of air. Absolute humidity is usually expressed as grams of water vapor per cubic meter of air (g/m^3).
2. *Relative humidity* is the amount of water vapor in the air relative to the maximum amount the air can hold at a given temperature. It is usually expressed as % Relative Humidity = amount of water vapor per unit of air ÷ maximum amount of vapor at the same temperature × 100.
3. When air reaches 100% relative humidity, it is said to be *saturated.* In unusual cases, when air cools below its **dew point**, the relative humidity can exceed 100% and become supersaturated. This condition usually occurs in the upper troposphere or, sometimes, in the humid tropical climates and can lead to extensive precipitation when cooling increases.

Humidity is related to three important conditions that take place when the temperature of air decreases (cools) to its dew point, causing condensation followed by precipitation.

1. *Radiation cooling* in rocks, soil, and water results in a loss of heat after their temperatures have increased (warmed) when they are exposed to the sun's heat. During nighttime, the absorbed heat is radiated back into the atmosphere, and condensation may take place.
2. *Contact cooling* occurs when moisture-laden air contacts a colder surface and the water vapor condenses into liquid water. Most people have experienced contact cooling many times in many different situations, for example, a cool, steaming windowpane may result when cooking, or dew or frost

may form on plants or the ground after a humid night has changed into a cold morning.

3. *Cooling by rising air* occurs as wind and cloud formations control much of our weather. This type of cooling involves the exchange of heat or differences in temperatures between air masses in the lower atmosphere. One interesting concept is the **adiabatic** temperature rate of change. The term means a change without the gain or loss of heat. For example, when you pump up a bicycle tire, the air in the tire gets hotter, making it appear that by just increasing the amount of air in the tire, the air temperature increases. However, the act of pumping the air is work, and it is this work that has increased the temperature of the air. (Work may be defined as the transferring of energy from one system to another, as in [$W = fd$], where *W*ork = the *f*orce required to move an object multiplied times the *d*istance the object moved). The same is true for the adiabatic temperature rate change for cooling air. The air is not cooled by removing heat. Instead, because the cool air is doing work, the temperature is lowered. The adiabatic rate at which dry air rises and cools is about 10°C/1000 m; while for rising moist air the rate is about 5°C/1000 m. This difference is an important factor in the formation of tall, billowing, rain-forming clouds.[58]

Clouds

As previously mentioned, very small microscopic particles of dust (less than 2.5 microns) act as nuclei on which water vapor condenses into liquid droplets. And, it is the cooler temperatures of these microparticles and water molecules, not the cooler air molecules, that result in water vapor condensing on particulate nuclei to form clouds. As the heat of the sun warms moisture-laden air, it becomes less dense. Thus, it rises, and as it rises, it becomes less dense and cooler as the altitude in the troposphere increases. Clouds are not formed by contact or radiant cooling but by rising air. Remember, higher altitudes with lower pressures result in cooling of the lower atmosphere, which is where clouds form and weather exists.

Three factors cause air to rise:

1. *Density differences* between warm and cool air are the most common reason air rises. Density of a substance is its mass divided by its volume ($D = m/v$). As air is warmed, it expands due to the heat that increases molecular motion that results in greater spaces between molecules, thus reducing its density. Cooler, more dense air below warm air replaces the warmer, less dense air, forcing the warmer air upward where it starts to cool. This assumes that the area where the air becomes hotter is bordered by cooler air masses. This is similar to the concept of a hot air balloon that contains warmer, less dense air inside as compared to the cooler, more dense air outside the balloon. The result is that the higher air pressure outside the balloon, so to speak, pushes the balloon upward.
2. *Orographic lifting* is another factor causing air to rise. This occurs as winds drive air up the sides of mountains, forcing it to rise and thus cool, often forming mountain-top clouds and windward-slope snow or rain.
3. *Wedging* is another cause of rising air masses, as one mass of air is wedged underneath another at a weather frontal area. This is where a high-pressure cold front slips under a humid, warm mass of low-pressure air, causing an imbalance in pressures and densities and thus pushing the warmer air upward (see Figure 9.2).

Although we cannot see humidity, we can see clouds—they are the most obvious objects in the sky. They partially regulate the amount of radiation (heat) the Earth receives from the sun and how much heat the Earth expels into space. Thus, they are a major factor in determining Earth's long-term climate, as well as its short-term weather. Some types of clouds reflect more of the sun's heat back into space than is transmitted to the Earth. Other types trap Earth's radiating heat, adding to global warming. Clouds are formed by tiny droplets of water or ice crystals dispersed over relatively large distances among air molecules. As warm, expanding, less dense, humid air gains altitude, it reduces the temperature of both the atmospheric gas molecules and the water vapor (H_2O gas molecules). Thus, the cooled water vapor condenses from its gaseous state to its liquid state onto microscopic particles (nuclei) as it reaches its dew point. The water vapor in clouds does not, as is often believed, condense on cold air molecules, but instead it forms on cold dust particles. At this point, the condensed water vapor reflects and refracts light, making the water droplets or ice crystals, and thus clouds, visible.

For thousands of years people observed cloud formations and established relationships between different cloud types and weather conditions. Over the years various systems developed to classify cloud types and their related weather factors. Many of these systems related to the shape of the clouds, their altitudes, their origins, their appearance, and the topography over which they form, as well as temperature, humidity, and atmospheric pressure conditions responsible for their formation. One system, which uses the altitudes for the various layers of the atmosphere, classifies clouds as follows:

1. Fog—clouds in contact with the ground
2. Low clouds—below 7,000 ft
3. Middle clouds—with a base between 7,000 and 18,000 ft
4. High clouds—with a base over 18,000 ft
5. Orographic clouds—formed by wind blowing up mountain slopes

While there are only three or four basic types of clouds, there are dozens of subtypes and subclassifications. Different types of clouds may be very diffuse or wispy or hair-like, while others may have sharply defined edges, appear sheetlike, or be thick and towering. The following sections provide examples of basic types of clouds and several of their subtypes based on their altitudes and shapes (see Figure 9.3).

Stratus

Stratus clouds can be thought of as stretched out horizontal layers and are usually the types of clouds found at lower altitudes. The Latin word *stratus* means "layer." This category usually includes ground fog, since a cloud cannot get any lower. When ground fog lifts, it may form low stratus clouds. Stratus clouds can be scattered, as well as appear as just an overcast sky. The edges of stratus clouds are not as well defined as they are for cumulus clouds. A subtype is the *nimbostratus* clouds, which are low and dark clouds often classified under multilayered clouds. ("Nimbo" is the word used to indicate that a cloud may bring rain.) They produce lightly falling rain, or if the temperature is below freezing, they produce snow or ice particles. Nimbostratus clouds usually form below 2,000 m (6,500 ft) and produce many dark, overcast, gray and rainy days with continuous precipitation. Nimbostratus clouds are not associated with violent storms.

Cumulus

Cumulus clouds are fluffy, white clouds with a relatively flat bottom and puffy top. The word *cumulus* is Latin for "heap" or "pile," which is a good definition for these billowing masses of condensed water vapor. Cumulus clouds are usually viewed on a hot summer day with tops reaching as high as 10 km (6.2 mi), but they also exist below 2,000

Figure 9.3
Clouds form when warm, moist air rises and water that evaporated from the surface of the Earth condenses as gaseous water vapor. Clouds are formed by several processes: (1) *convection* occurs when one area of the atmosphere becomes warmer than another mass, causing the warmer air to rise and form clouds; (2) *orographic lifting* forms clouds as warm moist air flows over mountaintops; and (3) *frontal wedging* forms clouds as cold air masses wedge under and force warm, moist air upward, resulting in water vapor condensing to form clouds. There are many different types of clouds that are named and classified by their shapes, densities, and altitudes.

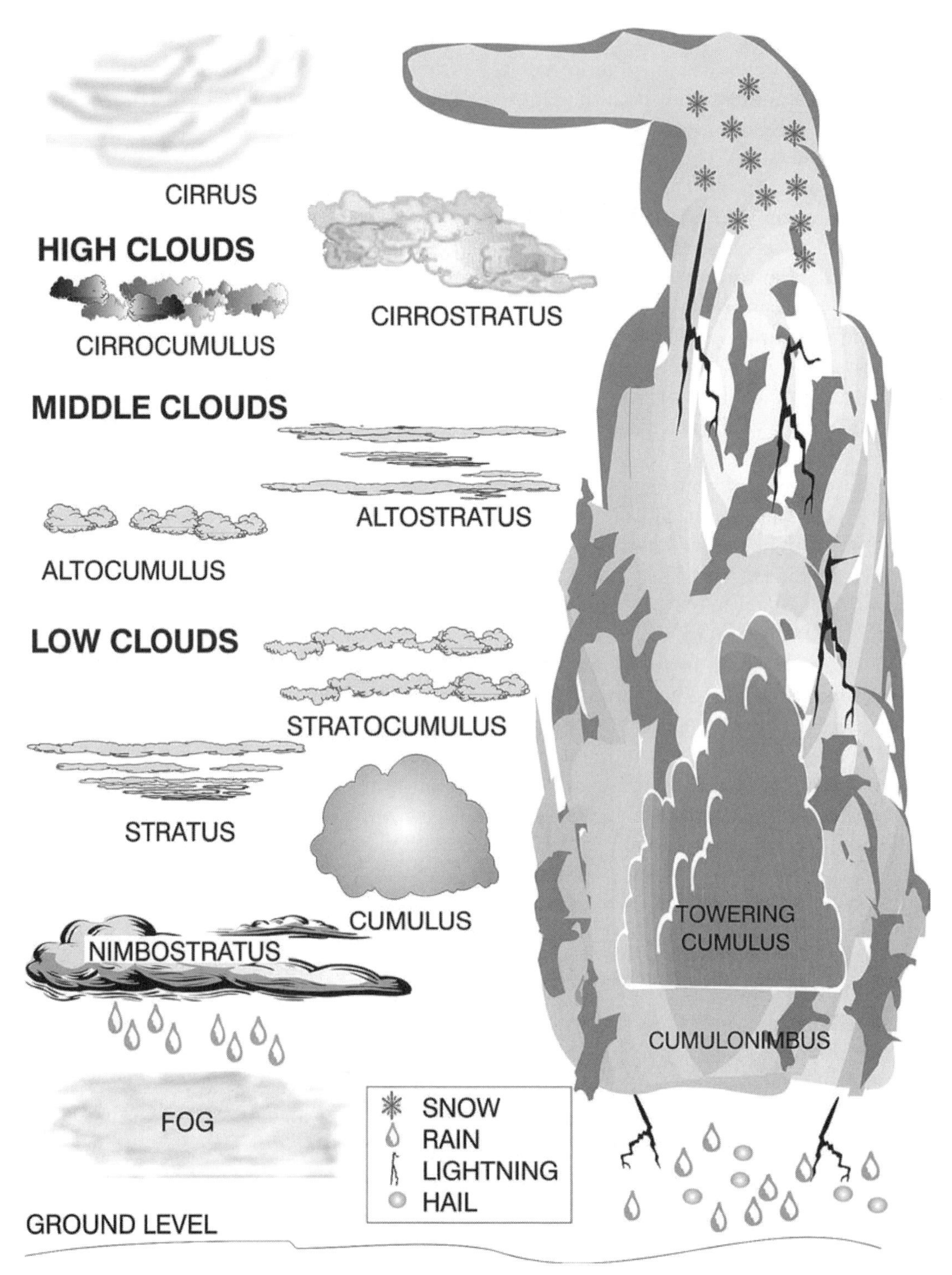

m (6,000 ft). The temperature point at which condensation of water vapor begins determines the altitude of the base of cumulus clouds. These are the types of clouds that appear to assume ever-changing forms of different types of animals or figures.

Three subtypes are (a) *stratocumulus* clouds, low, widely scattered, lumpy, layered clouds that do not attain great altitudes as compared with regular cumulus types; (b) *altocumulus* clouds, which often appear as parallel bands, are located at a lower altitude than are *cirrocumulus* clouds, are formed by convection on a warm sunny day, and often foretell of an imminent thunderstorm; and (c) *cumulonimbus* clouds, which are larger, towering higher than regular fair weather cumulus clouds. Depending on temperature changes, cumulonimbus clouds can produce lightning, thunder, heavy rain, hail, and strong winds and can even spawn tornadoes. Their tops can reach 12,000 to 19,000 m (40,000 to 60,000 ft). At this altitude, the temperature is below freezing and water vapor forms ice crystals. *Supercells* of air masses often form a squall line with strong winds that precede violent thunderstorms. Another formation of cumulonimbus are the very dark cumulus clouds that usually indicate a severe storm. Sometimes high winds in the upper layers of the troposphere deform the tops of cumulus clouds to form flat anvil shapes.

Cirrus

Cirrus clouds are thin, wispy, curly-haired formations found at very high altitudes between 6,000 and 15,000 m (20,000 and 50,000 ft). Ice crystals produced at high altitudes by supercooled water droplets are responsible for the formation of cirrus clouds. Cirrus clouds are associated with fair weather and take shape in the direction of the wind currents found at high altitudes. They often appear as long, streaming formations.

Two subtypes are (a) *cirrocumulus,* which have a scattered or wavelike shape, often look like many individual clouds in the same area, and may also appear as banded linear shapes; and (b) *cirrostratus* clouds, which form at high levels and cover large areas of the sky as great sheets. Cirrostratus clouds are formed when large, broad masses of air are lifted as high as 75 to 90 km (47 to 56 mi), or higher, by large convergences of lower air masses of higher temperatures. They are thought to consist of fine ice crystals and possibly meteoric dust. They have no distinct shape, have a low density, and do not block out the light from the sun or moon. Cirrostratus clouds produce a phenomenon known as "halos," which appear around the sun and moon on clear days and nights. As a warm, humid front approaches, water or, more likely, ice is added, causing these clouds to become thicker and denser, which may lead to some precipitation.

Other Types

Other types of clouds include the following.

1. *Orographic* clouds are produced as wind creates air flows up the side of mountains, causing water vapor to condense, thus becoming visible. Orographic clouds not only create precipitation on the windward sides of mountains but also create a decrease in precipitation for the dry areas on their leeward sides.
2. *Cap* clouds are a form of orographic clouds that forms a halo or doughnut cloud that just hangs over or around the top of high mountain peaks, capping their summit. They form when warm, moist air is uplifted on the windward side and becomes cool enough for condensation to take place, but the wind is not strong enough to push them to the leeward side of the mountain.
3. *Lenticular* clouds take the shape of a plano-convex lens. They are usually shaped by strong winds flowing over rugged land-

scapes. At times they can be layered or stacked. Some people have mistaken their round, circular shapes with sharp edges for flying saucers, or stacks of them for a fleet of flying saucers.

4. *Contrails* (short for "condensation trails") are a type of artificial clouds. They are produced by the water vapor that is generated by the combustion of petroleum fuel as its exhaust is spewed out from an airplane's engines. The warm water vapor condenses as it hits the cooler particulate (microdust or sea salt) nuclei suspended in the air. If the airplane is high enough and the temperature is low enough, the water vapor forms ice crystals. Contrails can also be formed from the wing tips of airplanes as warm air rapidly passes over the wings' surface, lowering the air pressure (Bernoulli's principle) and thus cooling the warm water vapor in the air to form wing contrails.
5. *Mammatus* clouds are named after the melonlike tropical fruit of the mammee tree,[59] or rounded, dome-shaped stones similar to small breastlike pouches that hang down.[60] Mammatus formations are sometimes found in several of the low and mid-level common types of clouds.
6. *Thunderheads* are rising cumulus storm clouds that produce thunderstorms with much lightning and rain. Often the rising water vapor forms ice in the form of hail, but usually these small hailstones soon melt, ending up as large raindrops. If small hailstones are caught in an updraft, more water vapor may condense on them to form larger ice stones. If the updrafts are strong enough, this process can be repeated many times and last over a period of time, often producing hailstones as large as baseballs before gravity brings them down to Earth. Thunderheads can generate winds up to 60 miles per hour (mph), which is just below hurricane levels.[61–68]

Winds

Wind is responsible for the visible dynamic nature of the atmosphere that we experience. As a factor of weather, wind is the direct result of the atmospheric temperature and pressure gradients. The *cause* of wind is the horizontal and vertical temperature/pressure differences (i.e., gradients). The *effect* is the movement of air (and weather) over local or regional land and water. Winds are usually named for the direction from which they blow, for example, a cold north wind is blowing from the north whereas westerlies are blowing from the west. Zonal winds are the result of uneven heating of the land (or water), which also affects the temperature of the air above the surface. As air is heated at the equator, it rapidly rises and then spreads out to both the Northern and Southern Hemispheres. This creates an imbalance (lack of air) at the equator, so cooler air from the north and south rush in to fill this partial void. This motion of hotter and colder air creates three major circulating cells of air masses: (1) the *Hadley* cells, one on each side of the equator, which move air masses from the equator to the north and south tropical regions; (2) the *Ferrel* cells, each located between 30° and 60° north and south latitude, which circulate air from the tropics to both polar regions; and (3) the *Polar cells,* each located between 60° latitude and the Earth's poles, which circulate air between the Arctic and Antarctic Circles and the North and South Poles.

How the Coriolis force affects the hydrosphere and atmosphere was discussed in chapters 7 and 8. It also affects winds that are deflected at the equator. In the Northern Hemisphere winds are deflected to the right, while in the Southern Hemisphere they are skewed to the left. This results in wind headed toward the equator in the north to arrive from the northeast, and contrawise air movement south of the equator to come from the southeast. Thus, these winds are named the "northeast trade winds" and the "southeast trade winds." At the equator, the circulation of air is more vertical than horizontal, which results in calms, referred to as the "doldrums" (see Figure 7.2). These two basic Coriolis-formed patterns of wind circulations are divided into six wide bands of winds; three of these bands are located in each hemisphere. In the Northern Hemisphere these bands are the *polar easterlies,* the mid-latitude *prevailing westerlies,* and the *trade winds* located closer to the equator.

Several other types of wind movement are related to local and regional weather.

- *Sea breeze* flows inland from the oceans during the day and reverses to flow out to sea at night. This wind is caused by the land heating up to a greater extent during the day than does the ocean. Sea breezes are caused by a small pressure gradient difference where the cooler air over water rushes inland to replace the rising warm air over the land, causing an on-shore sea breeze. The opposite pressure difference occurs at night, since water retains its heat much better than does land, resulting in an off-shore breeze.
- *Monsoons* are predictable winds that bring summer rain to India and Southeast Asia.
- *Chinook* winds are the result of adiabatic heating as air descends from the leeward side of mountains. Chinook winds often produce unusual but short warming periods during cold winters east of the Cascade and Rocky Mountains in the United States.
- *Bora, buran,* and *foehn* winds are local winds of the Balkans, Siberia, and the Alps, respectively.
- The *haboob* is a dust storm in northern Sudan. It can pile up sand for dozens of feet.
- The *sirocco* winds are the result of eastwardly moving dry winds that move northward from the Sahara Desert.
- *Sukhovey* winds usually last for short periods of time, but they are dusty, hot, and dry winds in southern Russia.
- The *Santa Ana* winds are created by a high-pressure gradient of hot and dry air that descends from the regional mountains into southern California.
- The *mistral* is a cold and strong local wind common to the Rhone valley in France. It can reach speeds of 200 km/hr (125 mph).
- The *nor'eastern* wind of the northeastern United States is created by a large, low-pressure cyclonic area off the north Atlantic coast. Northeastern storms can extend as far south as the Gulf of Mexico. They can be violent and produce a great deal of rainfall.

Another type of wind was not identified until about World War II. Air rises as a result of changes of temperature and as high-pressure air masses push up low-pressure air masses (pressure gradients). This movement of air results in friction between air molecules. To some extent, air friction slows down wind located close to the surface of the Earth due to air's density at low altitudes. This friction factor is not apparent at higher altitudes where air is not very dense, resulting in increasing wind speeds at greater heights. This phenomenon is known as the "jet stream," which is a narrow band of rapidly moving high-altitude wind. These streams of wind can reach speeds from 120 to 240 km/hr (75 to 150 mph), giving airplanes flying west to east a boost that saves fuel. However, pilots flying east to west attempt to fly south or north of the narrow jet stream. As jet streams flow eastward, they constantly change their north-to-south positions, resulting in changes in local and regional weather patterns. In North America jet streams are pushed farther south during the winter by cold Arctic air masses, but they partially recede northward during the summers.

Several types of wind are particularly hazardous to aircraft. Wind shear is a form of clear-air turbulence that can occur at both high altitudes and ground level. High-altitude wind shear is found in the regions of the jet stream when there is a sudden, abrupt, narrow change in the velocity of wind perpendicular to the normal direction of the jet stream. Clear-air turbulence of this kind can also occur near ground level. This can cause aircraft to lose altitude, but pilots can usually regain control of the airplane. Wet microbursts are intense winds resulting from local thunderstorms where high winds descend from the clouds and spread out horizontally in short intense bursts of air. Dry microbursts occur in very dry areas where rain evaporates before it reaches the ground. Since the rapidly descending air is clear, they provide no clue as to their presence. Both types can seriously affect the control of aircraft during take-off or landing. Microbursts only last a few minutes and cover a small area (about .5 to 1.5 mi); however, when they occur near airports, pilots may have difficulty regaining control of an aircraft, which can result in serious injuries and/or crashes.[69] Table 9.2 identifies the types of wind and their atmospheric effects.

The term "wind chill" was coined in 1939 by Paul A. Siple, an Antarctic explorer. The concept was based on how fast water would freeze depending on the water temperature, the outside temperature, and the speed of the wind. His idea and formula for calculating wind chill is now used by most meteorologists when reporting winter weather and storms.[70–75]

Storms

Storms are the result of natural changes in the weather factors discussed above. These changes, and the resulting storms, can be mild or very violent, causing much damage on the surface of the Earth. The various interactions of temperature (heat), pressure (air density), and humidity (moisture) produce low- and high-pressure air masses over large areas that result in a variety of cold and warm fronts.

Frontal Weather

Frontal weather, in meteorological terms, occurs when two air masses of different temperatures and with varying pressures and humidity meet and clash at their interface.

Table 9.2
Wind Types and Their Effects

Geographic Scale	Areas Affected	Atmospheric Motion Areas	Wind Life Span
Global (worldwide, continents)	Over 3,000 mi.	Large waves & jet streams	Weeks or more
Synoptic (large areas, states)	Over 1,000 mi.	High-low pressure fronts, hurricanes	Days to weeks
Mesoscale (small areas, cities)	10 to 15 miles	Breezes, T-storms, tornadoes, dust devils	Minutes to hours
Microscale (immediate area)	5 to 10 feet	Small turbulent eddies	Seconds to minutes

Source: April L. Umminger and Sam Ward, "Meteorology Today," *USA Today*, 20 September 2001.

- *Cold fronts* occur when a high-pressure, cold air mass slides under a low-pressure, warmer, humid air mass, causing the less dense air mass to rise and thus cool, creating an imbalance. Cold fronts tend to move close to the ground, somewhat eastward toward the equator. A cold front can be as broad as 1500 km (950 mi) and as high as 1.6 km (1 mi), but the main mass of cold air following the front may be 80 to 160 km (50 to 100 mi) behind the actual frontal zone. The wedge of cold air slipping under the mass of warm air often forms a squall line of rain or thunderstorms. Preceding a cold frontal zone, the warm, less dense air is marked by a lowering of barometric air pressure that can be used to predict the arrival of a storm. When a mercury barometer reads less than 30 in. of mercury (e.g., 28 to 29), a storm may be imminent. If it reads lower than 28, a violent thunderstorm or tornado is possible. If the barometer reads higher than 30, clear weather can be expected.
- *Warm fronts* can also occur when an advancing warm mass of air follows a retreating mass of cold air. If wind pushing the warm front is strong enough, it will override the cold air mass because warm, moist air is less dense than cold air. Most warm fronts move slowly and seldom result in storm conditions, but they can produce widespread low-pressure (cyclonic) rainy areas. Note that the term "cyclone" is often misused for tornadoes, hurricanes, or even severe thunderstorms. Any low-pressure area, large or small, severe or mild, is a cyclonic region. Thus for meterologists, cyclone means a large low pressure area that may or may not be stormy.
- *Occluded fronts* occur when a slow-moving warm front is followed and overtaken by a faster moving cold front. This results in the mass of warm air being lifted aloft to form an occluded front that can result in extensive mid-latitude rain, thunder, or snowstorm.
- *Stationary fronts* form when either a cold or a warm air mass is not moving, or moving very slowly. Sooner or later, as a colder air mass slowly slips under the warmer, humid air mass, widespread cloudiness and rain result for long periods, even a few days, until a stronger air mass arrives to move things along.[76–78]

Rainstorms

Rainstorms are the most common type of precipitation in the zones between the Arctic and Antarctic Circles and the equator. As mentioned previously, when warm, moist air rises and cools, the water vapor condenses on cold submicroscopic particulate matter (nuclei) to form small water droplets. As condensation continues, these droplets become too heavy to be maintained in the cloud formation, and gravity brings them to the Earth's surface. Depending on the temperature differences on the ground and in the atmosphere, as well as the type of cloud formation, humidity, and atmospheric pressure, the condensed water vapor can hit the ground as rain, freezing rain, hail, or snow.

Flooding is the main cause of damage and death from excessive precipitation. The melting of winter snow followed by spring rains produces floods in many areas of the world. According to a study by the U.S. National Oceanic and Atmospheric Administration (NOAA), flooding causes more deaths than do lightning, tornadoes, or hurricanes (see Table 9.3).

Thunderstorms

Thunderstorms are the most common type of extreme weather that occurs in most regions of the world. Obviously, a storm

Table 9.3
Deaths in the United States from Severe Weather over a 30-Year Period

Types of Weather	Annual Average No. of Deaths
Flooding	127
Lightning	73
Tornado	68
Hurricane	16

Source: April L. Umminger and Robert W. Ahrens, "Severe Weather Can be Deadly," *USA Today,* April 2001.

must actually produce thunder before it is classed as a thunderstorm, and to produce thunder that is heard during a storm, there must be lightning generated during the birth and growth of a thunderstorm. Here's what happens. When a localized mass of warm air expands, it rises and cools. This is followed by the cool air sinking back to the bottom of the warm air mass, where it is again sent aloft. This forms a *convection cell* where air reaches temperatures high enough to rapidly send the rewarmed air aloft to heights of 8 to 10 km (5 to 6.3 mi). After this air cools again, it sinks back down to the lower level of the cumulus cloud. The updraft of the convection cell supports water droplets and ice crystals as they grow in size. At some point in this cycle, the droplets of water or ice crystals become heavy and can no longer be supported by *updraft* currents. When this happens, they fall to Earth as rain or hail. This falling rain, ice, and hail set up opposing *downdrafts* that result in friction inside the cell. This internal friction builds up a static electric charge that may result in lightning. Most lightning is discharged within the thunderhead clouds. However, if the electric potential becomes great enough, the lightning passes from the cloud to the ground and vice versa. The warm updrafts move positively charged particles from the lower region of the cloud to the top of the thunderhead, thus leaving the bottom of the cloud with negative charges. This creates an electrical potential difference between the top and bottom of the cloud, as well as between the negative bottom of the cloud and the more positively charged surface of the Earth. Therefore, lightning takes place in both the cloud itself and in both directions between Earth and the bottom of the cloud (i.e., from the cloud to Earth, and from the Earth to the cloud) (see Figure 9.4).

It is not the rapid heating and expansion of air along the path of the lightning bolt that creates the thunder sound. Instead, the great

Figure 9.4
The friction of rain, ice, and hail builds up opposite electrical charges within the clouds, resulting in lightning discharges inside or between clouds. An electrical potential also exists between the charge at the bottom of clouds and the ground, which can cause lightning discharge either from the cloud to the ground or the ground to the cloud.

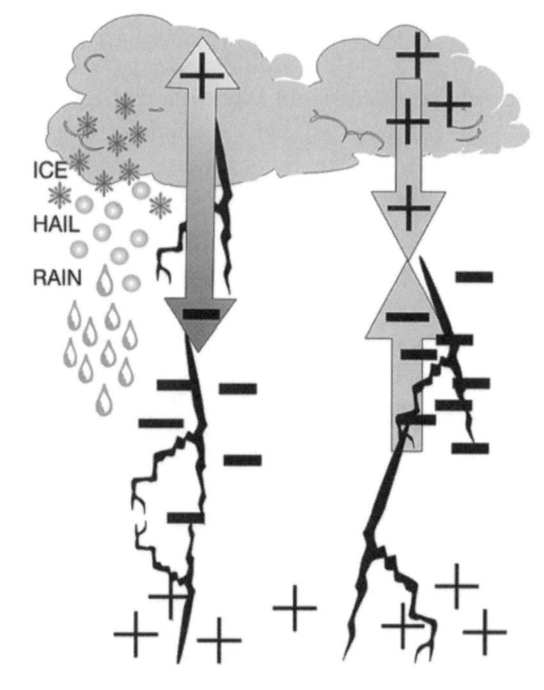

rush of air that rapidly fills this void created by expanding air creates the clap of thunder. In time, the downdrafts cool the air and violent rain and lightning taper off, resulting in rainfall that becomes less intense. So-called heat lightning is nothing more than a very distant thunderstorm, when the lightning can be seen but the sound it creates is dampened by the distance it travels, so no thunder is heard. Most thunderstorms develop when the Earth's surface receives enough heat to form localized convection cells. As indicated in Table 9.3, lightning is the second most common cause of human deaths as a result of severe weather hazards. The odds of being struck by lightning in a given year (both injuries and deaths) are 1 in 700,000 people. The odds that a person will be killed by lightning in his or her 80-year life span are 1 in 48,000, while the odds of being struck but not killed are only 1 in 3,000.[79] Although the odds for being killed by lightning are better than winning the lottery, other weather-related disasters can also occur. There is death and injury to livestock and other animals, forest and brush fires are started, and millions of dollars of damage occurs to buildings and communication/power systems as the result of lightning.[80–85]

Tornadoes

Tornadoes usually form as the temperature rises on hot summer days, often as a result of the strong convection currents associated with a thunderstorm. "Supercells" are long-lasting thunderstorms that produce the most violent tornadoes. Gusty winds precede supercells and can be used to predict the likelihood of tornadoes. Most tornadoes are relatively small, about 100 m in diameter (330 ft), and only last for a few minutes. An extreme low-pressure area in an atmospheric convection current sends air upward very rapidly, the result of which is a very accelerated movement of air outside this cell that rushes into this void. The Coriolis force deflects the rising air, which causes it to spin, decreasing the pressure in the cell even more and feeding the funnel's updraft. When the funnel touches the ground, the wind speeds inside the funnel can be very powerful, exceeding 500 km/hr (310 mph). While traveling across the landscape at about 40 to 65 km/hr (25 to 40 mph), tornadoes can leave a path of destruction before the funnel moves off the ground and disappears into the clouds. In the United States an average of 700 to 1000 tornadoes occur east of the Rocky Mountains every year. Even so, when there is a large difference in atmospheric temperatures, humidity, and pressure, tornadoes can occur almost anywhere. Texas had an average of 168 tornadoes per year (1989–98), more than any other state experienced. Florida and Kansas followed with 79 and 75 respectively. For the United States, about 42% of all tornadoes occur in the months of May or June and only about 4.3% occur in December and January.[86] (See chapter 2 for additional information and descriptions of the dangers of tornadoes and hurricanes.)

Hurricanes

Hurricanes are also called "tropical cyclones" due to their origin in tropical regions and their extensive low-pressure air masses that cover many square miles. Hurricanes are formed over oceans where solar radiation increases the water's surface temperature, promoting evaporation that adds to atmospheric humidity. (They are known as "typhoons" in the western Pacific area, and "cyclones" in the Asian/Indian Ocean region.) They start from low-pressure areas called "tropical depressions," which intensify to become "tropical storms," and after further development are classed as "hurricanes." Several storm areas can combine to form a massive cyclonic (low-pressure) convection cell of rising warm air that further reduces the pressure inside the cell. This causes winds outside the cell to blow toward the cell's low-pressure center, similar to a tornado but over a much larger area. Again, the Coriolis force deflects the wind, forming a swirling air mass that covers many square miles. In the Northern Hemisphere hurricanes rotate counterclockwise, while in the Southern Hemisphere the rotation is clockwise. This swirling mass of moist air surrounds a small, central, calm region called the "eye." The greatest rainfall occurs just outside the eye while large bands of clouds and precipitation form rain bands that cover many square miles. When wind speeds reach 119 km/hr (74 mph or 64 knots), the storm is classed as a hurricane, and winds can exceed 155 mph.

Table 9.4 uses the Saffir-Simpson scale to rate the categories, wind speeds, and potential damage related to hurricanes. Table 9.5 lists a small sample of some of the most recent major hurricanes that have hit the Western Hemisphere.

Table 9.4
Categories of Hurricanes

Category No.	Wind Speeds (mph)	Estimated Damage
1	74–95	minimal
2	96–110	moderate
3	111–130	major
4	131–155	extensive
5	156 and over	catastrophic

Source: Allison Landers, "Things to Know About Storms," *Valley Morning Star* (Harlingen, Tex.), 31 May 2001.

Once a hurricane forms, heat and moisture from the ocean feeds its growth, intensifying the storm until it reaches land, where extensive rain occurs as cooling of the hurricane takes place. Once inland, the hurricane degenerates into a less severe storm with high winds, rain, and possible flooding. Hurricanes play a major role in the general circulation of the atmosphere by transporting large amounts of warm, moist air from the low latitudes to the middle latitudes. Scientists estimate that a mature hurricane can carry more than 3.5 billion tons of air per hour that transports and releases a great amount of water and energy through the troposphere.[87] If the air mass is stalled, rain and wind can persist over a local region for many hours. Hurricanes spawned during late summer in the south Atlantic Ocean are driven northwestward by prevailing southeasterly winds. The hurricane path is similar to a curved parabola. In the northern hemisphere the curved path starts out in a northwestern direction, and as the hurricanes proceed north. it changes direction to the northeast. When the hurricane is formed at a low latitude in the south Atlantic Ocean, its traveling speed is only about 5 to 20 mph; at higher latitudes its speed can reach 50 mph or more. For an average year, the southeastern coast and the Gulf coast areas of the

Table 9.5
Devastating Hurricanes in the United States

Date	Category	Name	Location	Damage
1957	4	Audrey	Sabine Pass, Texas	unknown
1967	3	Beulah	Brownsville, Texas	unknown
1972	~5	Agnes	U.S. East Coast	134 deaths
1989	~5	Hugo	U.S. East Coast	50 deaths
1992	~5	Andrew	South Florida	~50 deaths

United States can expect about 20 to 30 or more named hurricanes (given male and female names equally). Not all of these named hurricanes will actually make landfall, and most will be small, category 1 or 2 hurricanes, or downgraded to tropical storms or disturbances, but have large amounts of rainfall with the potential to cause floods. Five to ten of these may be major hurricanes, including two or three that can be very destructive.[88–90]

Issues: Atmosphere, Climate, and Weather

Most of the issues related to the atmosphere, climate, and weather are connected to both nature and humans. However, we have a difficult time understanding cause and effect and other interrelated factors when addressing air quality, global climate, and related issues. The following issues are based on the concept that science, rather than politics, should be the criterion for dealing with the problems related to our atmosphere and subsequently developing polices to manage these challenges.

Air Pollution

Historically, air quality has always varied from local region to local region, most often from gases and particulate matter from volcanoes and dust storms, but also in times past from matter sent into the atmosphere by the impact of asteroids. It is now well established that an asteroid or comet smashed into Earth about 65 million years ago. It is theorized that this impact sent enough gas and particulates into the atmosphere to create a blanket thick enough to block out the sun for an extended period of time. The primary result was that plants died from lack of sunlight. Subsequently, this caused the demise of many species of animals, including dinosaurs that depended on plant life for food. It is assumed that some small forms of life survived, including small mammals that in time evolved into *Homo sapiens.*

Since the days when humans learned how to burn wood, and later coal and petroleum, for heating and industrial uses, local air pollution increased—often with disastrous results. During the period when England and other European cities burned coal as a source of energy for homes and industries, local air pollution was a serious problem. In 1952 a dense, polluted fog killed between 3,500 and 4,000 people in London. In 1948, the air pollution created by industries in the small steel town of Donora, located south of Pittsburgh, Pennsylvania, killed 20 people after approximately half the town became ill from sulfur fumes and particulate matter (PM). Air pollution, particularly from lead, has been traced back as far as 6,000 years, during the Greek and Roman civilizations. As early as 500 B.C.E. European countries that smelted metals experienced air pollution. In the thirteenth century, England was plagued with the stench of smoke from burning coal. In 1285 King Edward established the first air pollution commission and made it illegal to burn coal in London. Even so, people ignored the ban and used coal for heating and cooking well past World War II. Currently, the average concentration of SO_2 and smoke in London is less than it was in the year 1600. Air quality in England, Europe, and the United States has greatly improved over the past several decades. Improvements in air quality can still be made, particularly in the developing countries where pollution controls are practically nonexistent.[91]

Today, we continue to experience natural air pollution from volcanoes and other natural sources. For example, the 1991 eruption of Mount Pinatubo on the island of Luzon in the Philippines spewed more than twice the amount of sulfur dioxide into the atmosphere than did Mexico's El Chichon volcano in 1982. The cloud of gas and particulates from Mount Pinatubo completely encircled the Earth in just 3 weeks and was still evident months later.[92] These clouds of gases and particulates cooled the atmosphere for a short period of time.

The U.S. Environmental Protection Agency (EPA) used the following six types of atmosphere pollutants as criteria to establish national air-quality standards:

1. Particulates (smoke and soot as small as 2.5 microns)
2. Sulfur dioxide (SO_2)
3. Ozone (O_3)
4. Lead (Pb)
5. Nitrogen oxides (NO_x) (the "x" stands for various molecular forms)
6. Carbon monoxide (CO)

In 1963 the U.S. Congress passed the 1963 Clean Air Act. This act has been revised and updated several times. In its current form, its effectiveness and practicality have become an issue, but there is no debate that the air quality in the United States (as well as most of Europe) has improved over the past years.

In 1970 the EPA was established by the U.S. Congress to monitor the Clean Air Act and enforce many other EPA-designed environmental rules and regulations. The 1970 version of the air-quality standards includes the following pollutants: gases, such as sulfur oxides, carbon monoxide, and nitrogen oxides; hydrocarbons; photochemical oxidants; and PM. Of some interest is that these standards include hydrocarbons, whereas the earlier national air-quality standards did not. The current law addresses two basic sources of air pollution: (1) air pollution arising from individual sources, for example, automobile and industrial smokestack emissions; and (2) the ambient (surrounding) quality of air, even when human pollution is not in evidence.[93]

The smallest particles in the atmosphere (2.5 microns or less) come from combustion in motor vehicle engines, power stations, and industry, as well as from fireplaces and wood-burning stoves. Slightly larger particles come from dust and natural sources. Even so, only about 10% of all particulates in the atmosphere are man-made, and most of these are found in urban environments.[94]

By increasing the standards for air pollution (as well as other types of pollution), a point is reached where a reduction of the risks increasingly reduces the expected benefits while creating additional problems. Several techniques can be used to clean up air pollution from stationary sources. One removes PM, another scrubs gases from exhausts. Great progress has been made in developing techniques to address these types of air pollution (e.g., cyclone separators, wet scrubbers, and electrostatic precipitators) that has greatly reduced air pollution in cities and local areas.[95] Even so, meeting the standards for PM set by the EPA for ambient air (normal air that surrounds us) is not only an expensive proposition but impossible to achieve. In the early 1990s, Harvard University submitted evidence that soot (PM) was implicated in the deaths of more than 60,000 people in the United States. Without checking this data or conducting their own research, the EPA used this study to set up new regulations for the control of soot and smaller size PM. Several things about this episode are questionable. First, the Harvard study did not control for the fact that other pollutants and gases may have been in the same areas and contributed to the deaths. Second, although excessive soot is dangerous for people with respiratory problems, there was no control for those with related health problems as compared to those without respiratory problems. Third, no one replicated or followed up this study with adequate research. Other scientific groups objected to the lack of good science related to the Harvard data and the EPA's new regulations. Other unrelated studies using animals found that PM of metals, such as vanadium and nickel, may have been the sources of the air's toxicity rather than soot.[96] Regardless, PM, whether it is of human origin or natural, is a respiratory agent and should be as low as possible. Another study, conducted by Dr. Murray Mittleman, a Boston epidemiologist, related 2.5 micron PM in air to cardiovascular diseases. His five-year study did not include data for healthy people, but instead involved people already at risk for heart attacks. His population consisted of obese, inactive people with a history of heart problems. His data indicated that the risk for heart attacks for these individuals peaked at 2 hours and again at 24 hours after they were exposed to an increase of these invisible particulates. This epidemiological study statistically linked deaths of at-risk heart patients with exposure to PM-2.5, but data were not reported for any possible deaths to healthy people not at risk for heart attacks.[97] Thus Dr. Mittleman's study did not determine if PM-2.5 was also harmful to people without heart problems. The EPA proceeded to propose curbs on PM at the 2.5 micron level or less in the atmosphere. (Note that soot is 10 microns in size, whereas particulate at the 2.5 microns or smaller size is almost impossible to completely remove from very clean ambient air [one micron is one millionth of a meter, and a human hair is approximately 100 microns thick].) In February 1997, the then director of the EPA, in a report to a U.S. Senate Committee, summarized selected studies making a case for the 2.5 micron PM standard. Several other studies indicated that there was little association between 2.5 micron (or smaller) PM and illnesses/health. These studies were ignored by the EPA. For instance, most of the PM in smoke from home fireplaces ranges from less than 0.03 microns (classed as microparticulates) to about 10 microns (soot).[98] At one time the EPA proposed a ban on burning wood in fireplaces, but after public complaints and reevaluation, the regulation was not implemented.

In 1997 the EPA issued a report based on studies that attempted to separate various factors responsible for air pollution and health effects. It is extremely difficult to isolate exact causes and effects related to types of air pollution and health risks because these types of studies are very complex, involving many interrelated variables. The report concluded that PM in the air more or less accounts for the entire risk of death from air pollution. This conclusion is based on the fact that larger particles are captured in the nose and throat, while extremely small particles (2.5 microns or less) are transported to the alveoli of the lungs, thus causing the greatest health risks.[99]

Not only are gases and PM sent into the atmosphere by volcanic activity, storms, and winds, plants and animals also contribute to the pollution of ambient air. All organisms, just by living and dying, contribute to air pollution. For example, the mountain range extending from the western side of North Carolina through Virginia is covered with conifer trees (various species of pine trees), as well as some deciduous trees. Due to the natural processes of respiration and transpiration, pine trees emit volatile molecules from resins and so forth. These volatile molecules hang in the air, and as light penetrates them they give off a bluish hue, creating smog that looks similar to smoke. From these came the name "Smoky Mountains." Although these volatile pollutants are less than or about 2.5 microns in size, not many people would consider the air of the Smoky Mountains polluted. The same natural conditions led to naming the "Blue Ridge Mountains" in West Virginia and Pennsylvania. Cosmic dust from space and micrometeors ("shooting stars") that disintegrate on entering the Earth's atmosphere also contribute to upper air pollution. The most pristine air on Earth contains some PM and gases that are classified as pollutants by the EPA. At one time the EPA, following the zero tolerance policy for food additives of the U.S. Federal Drug Administration (FDA), proposed a zero tolerance for air and water pollution. Since chemicals can be detected at the parts per billion (ppb) or even parts per trillion levels, any zero tolerance policy is an impossible goal. A few molecules of almost any substance can be dispersed in large volumes of all air and water. For instance, if you mix one pint of fresh water in all the oceans of the Earth, and years later collect one pint of the mixed water, there will be more than 5,000 molecules of the original pint of water in the one collected.[100] A similar example would be that with every breath of air taken in, at least a few of these air molecules are the same as the ones that a Roman soldier exhaled. The issue is not which excessive air pollution should be curbed, as it has been over the past several decades, at least in the United States. The issues relate to determining at what level normal, as well as manmade, pollutants are, or are not, harmful to humans and what is the acceptable cost that society is willing to bear (cost vs. benefits economically, socially, and health-wise) to enforce increasingly stringent environmental regulations that may result in little or no increase in benefits. It is doubtful that many U.S. citizens would be willing to greatly reduce their standard of living based on current knowledge of risks versus benefits related to air pollution. Environmental policies and regulations should be based on the best science available. This does not mean that pollution should be ignored. Instead, pollution should be addressed rationally, using accepted processes of science, while considering all identifiable variables and using the best data available to determine related policies and regulations. The good news is that most industries in the United States realize that reducing excessive pollution is good business. This results in the reduction of both air and water pollution to the extent that the quality of our air and water is very much improved from fifty years ago. Pittsburgh, Pennsylvania, is no longer the great smoggy city of the past. It, and the air above most other industrial regions, has improved remarkably.

There may be a positive side to a small increase in PM such as smoke, soot, dust, and other microparticles in the atmosphere. The more PM in the atmosphere, the greater the increase in the reflectance (albedo) of the atmosphere, resulting in a higher percentage of solar radiation being reflected back into space. It has been suggested that atmospheric PM caused cooling of the Earth's surface that resulted in a few cooler years several decade ago. But for the present, the balance seems to be for increased regional and global warming, at least at ground level and in the lower atmosphere.[101]

Acid Rain

Natural precipitation from the atmosphere that is not polluted is normally slightly acidic, with a pH of 5.5 or 6. Note that the pH scale is a measure of the concentration of the H^+ (hydrogen ion or proton) in solution. It is a logarithm scale from 0 to 14, with 7 being neutral. An acid ranges from 0 to 7, with low numbers being the strongest acid. A basic (caustic alkali) solution ranges from 7 to 14, with 14 being the most caustic. Thus, an acid with a pH of 5 is 10 times stronger than one with a pH of 6; and a base with a pH of 9 is 10 times stronger than one with a pH of 8.

0	↔	7	↔	14
strong acid		neutral		strong base

The acidic level of natural rainwater is due to the formation of carbonic acid as water combines with carbon dioxide. Other gases, such as sulfur oxides (So_x) and nitrogen oxides (NO_x), can also combine with water to form acids. These gases, when combined with rain, form their respective acids in solution:

1. (carbon dioxide) $CO_2 + H_2O \rightarrow H_2CO_3$ (carbonic acid in solution)
2. (sulfur dioxide) $SO_2 + H_2O \rightarrow H_2SO_3$ (sulfurous acid in solution)
3. (nitrogen dioxide) $NO_2 + H_2O \rightarrow HNO_3$ (nitric acid in solution)

Both sulfurous and sulfuric acids and nitric acid have a lower pH than carbonic acid, which means they dissociate to a greater extent in water and are thus stronger than is H_2CO_3; thus they produce more acid rain when present in the atmosphere in equal concentrations. The burning of fossil fuels, particularly soft coal with high sulfur content, in electric power generating plants is associated with emissions of sulfur dioxide gas into the atmosphere. Motor vehicles that use petroleum products (gasoline and diesel) produce a number of oxides of nitrogen and some other exhaust by-products due to the incomplete internal combustion of fuels in their engines. Note also that worldwide volcanic activity spews tremendous amounts of sulfur oxides, and other noxious gases, into the atmosphere just about every day of the year somewhere in the world. These gases also produce acid rain. So, human activity is not the only source of gases that produce acid rain. Sunlight interacts with the oxides of sulfur and nitrogen and, along with water vapor, is converted into weak carbonic, sulfuric, or nitric acids. They are produced in a gaseous state that combines with water to form minute liquid particles that mix with the air as they are whiffed aloft into the atmosphere. The acid/air mixture is carried some distance along with the prevailing winds—which usually blow from west to east. When it rains, the carbon, sulfur, and nitrogen oxides combine with water vapor and are washed out of the air. After falling as acid rain, with pH ranging from about 4.5 to 6.0, they usually do not affect living organisms. But if the concentrations of these gases are between pH 3 and 4, the result is acid rain, which can adversely affect trees, fish, and insects. These weak acids are also present in moist air that can slowly erode stone buildings and exposed works of art,[102] as well as further damage unhealthy trees that have been attacked by disease or insects. Even natural rainfall of pH 6 produces acid precipitation that can react with the limestone and marble of buildings and monuments. If the pH is much lower than 6, the process is accelerated. The following is a typical reaction of sulfuric acid on limestone:

$$\text{(limestone) } CaCO_3 + H_2SO_4 \rightarrow \text{(solution) } Ca^{++}SO_4^{-} + H_2O + CO_2\uparrow \text{ (gas)}$$

A similar reaction over the past millions of years resulted in carbonic acid, and other acids, dissolving bedrock to form the geological feature known as a karst landscape, with surface features including sinkholes and caves formed by flowing water.

The emissions from power plants in the upper Midwestern region of the United States, as well as those from northern Europe, have been blamed for the acid conditions of lakes and the death of forests that are located downwind from the sources. Slow-moving bodies of water east of the sources of pollution were tested and found to be slightly acidic; some trees downwind were also found to be dying. Environmental activists have lobbied for more controls and regulations to reduce sulfur and nitrogen oxide emissions. The United States government instituted the National Acid Precipitation Assessment Program (NAPAP), which spent more than ten years and $500 million to study the alleged destructive impact of acid rain on lakes, fish, crops, and forests in southern Canada and in the northeastern United States. Their report concluded, "There is no evidence of widespread forest damage from current ambient levels of acidic rain in the United States." NAPAP also found no evidence that acid rain had harmed crops or caused any human health problems. They also stated, "We found that the average Adirondack lake is no more acidic now than prior to the Industrial Revolution, and certainly not 100 times more acidic as claimed by the U.S. Environmental Protection Agency."[103]

Some actions were taken by industries to reduce the sulfur emissions from coal power generating plants, as well nitrogen oxides from automobile exhausts. But, as scientists continued to study the problem of acid rain, they realized that the water in the ponds and lakes in question was naturally acidic. They also found that in some ponds and small lakes, fish and insects cannot survive with water that has a pH much lower than 5.0. Soils located on the floors of pine forests have a higher acid content than do most farmlands. A high acid content is required for the growth of some of the plants and trees found in southeastern Canada and the northern part of the United States. One study showed that the increase in nitrogen from acid rain actually allows trees to take advantage of extra carbon dioxide in the atmosphere, which makes trees grow faster, particularly in wetter years. The study cautions that not very much is known about the myriad interactions between plants and their environments to draw final conclusions about trees as carbon sinks or trees and acid rain.[104]

Further investigations found that many of the trees that were dying were not killed by acid rain but instead by diseases and insects—particularly the gypsy moth in the northeastern forests. Some of the diseases and insects, in addition to the gypsy moth, that cause even some healthy trees to die are White pine blister, Fusiform rust, Hemlock woolly adelgid, Dutch elm disease, Chestnut blight, and the Bronze birch borer.[105] Even with all this information, environmental advocacy groups have continued to lobby for increased environmental regulations and rules related to acid rain. Industries also are continually improving their techniques for scrubbing their emissions to remove toxic gases and PM. The quality of the air (and water) in the United States has improved several-fold in the past several decades due to both reasonable regulations and actions taken by individuals and industries, but improvements can still be made.

Global Warming

The issues related to global warming are probably the most misunderstood earth science and environmental contentions for both scientists and laypersons. There is no question that the so-called greenhouse effect exists and that global temperatures (on the average) have increased a few degrees in recent years. There are questions related to how much of the increase in global warming in the past, the present, and possible future, were, are, or will be the result of natural climate cycles or the results of human interactions (i.e., anthropic principle). Some evidence suggests that the general public is not as concerned with global warming as are environmental activists, the media, and some scientists. In 1997 the *New York Times* conducted a poll asking respondents to rank environmental issues. Only 7% listed global warming as their main concern, while 47% listed air and water pollution. A similar poll by CNN the following year listed only 24% of the respondents as being more concerned with global warming than other environmental issues. An ABC News survey the same year found that two-thirds of the respondents thought government should wait for more conclusive scientific evidence before they themselves were willing to make personal sacrifices in their lifestyles. Sixty-five percent did not believe that the 1997 summer heat wave was due to global warming.[106] However, the year 1998 was the warmest year of the past millennium in the Northern Hemisphere.[107] There is an abundance of both good and bad science espoused by proponents from both sides of the issue concerning the causes and effects of global warming. Let's look at some of the factors that are involved.

First, climatologists, paleoclimatologists, meteorologists, and most other scientists are well aware of past periods of global cooling and warming. Over the past 1 million years there were a series of eight glacial/interglacial cycles that were driven by the Earth's orbital changes around the sun. It is well established that the last major interglacial period (the Holocene, which is the present period) began ~10,000 years ago, and even before this last ice age, there were several periods of very warm and very cold temperatures. To date, the Holocene is the longest stable warm period and may well be a factor in the development of human civilization. Also, since the end of this last major ice age there have been cycles of mini warming and cooling periods. Some indicators showed temperature changes of from 5 to 8°C over the past 1,500 years, and one study indicated a slight cooling trend between the years 1000 and 1900, followed by a warming trend.[108] It is true that the average annual global temperature has increased slightly over the *past century.* Also during the past one-hundred-year period there were several mini cooler periods, which means that there is a "but" to this *average* figure, since obviously some years were cooler and some warmer, as has been the case for any extended period of time (see Table 9.6).

The coldest summertime temperature in the northeastern United States was in 1903, when the annual average temperature was 64.4°F (the average over longer periods is 67.4°F), and for 1992 the average was 65.2°F. Both of these lows indicate significantly lower summertime temperatures than the average for this region of the country.[109] Even though these figures are averages for temperatures for specific years, it is more important to look at trends in temperature changes over longer periods of time.

Over shorter time periods, as temperatures are measured near ground levels by instruments carried aloft by balloons and at much higher altitudes by satellites, a more accurate picture evolves. At *ground level* there has been a slight warming trend of 0.1°C to 0.15°C (0.18°F to 0.27°F) for the *decade* of 1989 to 1998, whereas NASA's weather satellites have measured a slight cooling trend in the upper atmosphere of about 0.04°C (0.72°F) since 1997. It is not known whether these slight variations in temperatures over a ten-year period are significant, or whether these are part of a natu-

Table 9.6
The Five Coldest and Hottest Summers in the Midwestern United States in the Twentieth Century

Coldest Summer Temperatures		Hottest Summer Temperatures	
Year	Temp. (°F)	Year	Temp (°F)
1903	68.0	1901	73.8
1915	66.4	1921	73.7
1927	67.6	1934	73.8
1950	68.0	1936	74.3
1992	66.8	1988	73.5

Source: Mike Palecki, Midwest Regional Climate Center, "USA Today Weather Focus," *USA Today,* 17–18 July 2001.
Note: The average summertime temperature in the Midwest is 70.8°F; the states included are Illinois, Indiana, Iowa, Kentucky, Michigan, Minnesota, Missouri, Montana, Ohio, and Wisconsin.

ral cycle or in some way related to human activity.[110]

Myth: An increase in 1.5° F in global temperatures since 1850 is proof that all types of carbon dioxide emissions are increasing and thus dangerously heating up the planet. **Fact:** This myth ignores the fact that the Earth's temperature naturally rises and falls over the course of several centuries. Since the last ice age, ~10,000 years ago, there have been seven major warming and cooling periods. Of the six periods preceding the current period of warming, three produced temperatures warmer than today while three produced temperatures colder than today. The temperature increased 1° F between 1850 and 1940 and only 0.5° F between 1940 and 1979 when the rapidly rising amounts of carbon dioxide emissions should have been causing warming to accelerate. Also, NASA's satellites indicate that there has even been a slight cooling trend of 0.02° F since 1979. These trends have been corroborated by other methods that measure atmospheric temperatures.[111]

Another problem deals with the use of computer models for predicting global climates—past, present, and future. In the early 1960s, a scientist at the Massachusetts Institute of Technology who developed computer programs associated with atmospheric conditions made an important discovery. Climate is a complex of systems involving the oceans, land areas, weather, and the sun, as well as the atmosphere. He discovered that such systems do not follow a straight-line linear equation. Instead, these equations are nonlinear. The main difference is that a linear equation can easily be solved and extrapolated on a graph to predict a long-term outcome, whereas a nonlinear equation is not easily solved or extrapolated on a graph, or depending on its complexity, it may be unsolvable even by the best of programs and supercomputers. One reason climate systems are nonlinear is that they are sensitive to initial conditions in all parts of their complex nature and the interrelatedness of causes and effects (see the previous discussion of chaos theory). Thus, even very small differences and connections between the factors that make up climate can greatly affect the final predictions as to the amount of global warming, and the longer the time frame, the larger the error in predicting global temperature changes. Also, the greater number of inaccurately known variables associated with a complex system, the greater are the odds of an incorrect outcome or prediction. Over the past few years, techniques for computer modeling of the climate have improved with the use of more sophisticated software and supercomputers, but no matter how capable the system, if the input data are inaccurate, errors will just be compounded. Computer scientists have a saying for this, GIGO, meaning "garbage in—garbage out." In other words, if you enter incomplete or imperfect data, regardless of the elegance of the program or computer, the output will not be reliable for making long-range predictions.[112]

Greenhouse Effect

What is the greenhouse effect that prevents our globe from becoming a frozen ice ball? A plant nursery greenhouse allows the light and infrared heat from the sun to enter the building through the glass roof, while at night the glass keeps the heat entrapped like an insulated container. The atmospheric greenhouse effect is somewhat different. More than eighty years ago, Svante August Arrhenius (1859–1927) related carbon dioxide to global climate changes. He was unable to establish a direct cause-and-effect between CO_2 and atmospheric temperatures, but he did consider both the cooling and warming effects of CO_2 on the atmosphere as evidence for the ice ages.[113] The so-called greenhouse gasses allow the sun's light and infrared heat that is not reflected back into space by clouds to bathe and warm the Earth's rocks, soil, sand, and oceans. At night this heat is radiated back to space, where some of it is reradiated back to the Earth by the greenhouse gases. Thus clear nights are cooler, because more of Earth's radiated heat goes into space rather than being reradiated back to the surface. The amount of heat that is trapped near the ground by clouds and the lower atmosphere, as well as some residual heat from the Earth, determines over time to some extent regional and global temperatures. This brings up two questions: What are these greenhouse gases? and What are some of the many other variable conditions that affect global warming?

Several gasses contribute to the Earth's greenhouse effect. The following are listed in order of their relative man-made influence on greenhouse temperature changes: carbon dioxide = 60%; methane = 20%; halocarbons (CFCs and ozone) = 14%; and N_2O = 6%.[114] In addition to these man-made greenhouse gases, natural water vapor in the atmosphere is by far the most important greenhouse factor.

Most scientists consider that carbon dioxide was a major gas in the early atmosphere when life started ~3.5 billion years ago. Much later, bacteria developed pigments of photosynthetic molecules capable of absorbing light energy. In time, this resulted in plant photosynthesis that consumed carbon dioxide and created free oxygen. Over eons of time, simple plants generated oxygen that began to replace carbon dioxide in the primeval atmosphere. Today, carbon dioxide comprises only ~0.034% of the Earth's atmosphere. Even with the large amount of CO_2 pumped into the atmosphere from human activities since the industrial revolution, the percentage has not increased dramatically (to ~0.035 or 0.036%). Percentage increases are not the same as actual total increases in the amounts of CO_2 emissions. The Climate Monitoring and Diagnostic Laboratory of NOAA compiled several studies to calculate the increase in annual global emission of carbon dioxide between the years 1800 and 2000. The following data and years are for the global emissions of man-made CO_2, in billions of tons per year: 1850 = ~0.01; 1875 = ~0.2; 1900 = ~0.5; 1925 = ~1.0; 1950 = ~1.5; 1975 = ~5.0; and 2000 = ~6.5.[115] (Note that these figures do not include natural CO_2 emissions.)

The reason there is not a higher percentage is the same as the explanation for what happened with much of the prehistoric CO_2—it was absorbed by soils, oceans, marine life, and plants as it became chemically combined with other elements. The huge limestone and dolomite deposits located from New England to Pennsylvania to Florida are evidence that these areas were once covered by oceans.

Ancient CO_2 was absorbed by seawater and used by trillions on trillions of organisms (from microscopic to larger) that incorporated the carbon to form compounds that make up skeletons and shells. Over time these were deposited and compacted on the ocean beds, ending up as limestone, as well as other carbon-containing minerals and rocks. An article in *Science* magazine suggested that the carbon from atmospheric CO_2 that is absorbed in seawater is not all tied up in living organisms. Instead, it is present in marine-originated dissolved organic matter (DOM), which is nonliving organic matter containing carbon. The study indicated that microbial processes affect the molecular structure of the DOM, thus preventing it from breaking down and thereby preserving the fixed carbon that was absorbed from the atmosphere as CO_2.[116] When these DOM carbon sinks, plus the huge amounts of CO_2 used by plants, are buried as sediments and are incorporated in the rock cycle, the carbon is removed from the atmosphere. At one point in geological history even England and other northern regions were tropical, as evidenced by large coal deposits formed by ancient swampy vegetation. These regions also acted as sinks for CO_2 as the carbon was buried with the coal-forming vegetation. Even so, the Earth can only act as a sink to absorb so much CO_2, and we are producing more than it can handle by natural processes.

Recent research indicates that carbon dioxide is not the only, or even the main, greenhouse gas that affects regional and global temperatures. Water vapor is far more prevalent in the atmosphere than is carbon dioxide, and due to its high latent heat characteristic, it also affects global temperatures. Cloud formations are the result of condensation of evaporated water (water vapor) that forms tiny droplets in the clouds. (Note that when water changes its state from liquid to gas to form water vapor, heat is absorbed. This is why you cool off after sweating. In contrast, as water vapor condenses on nuclei of microparticles to form liquid droplets, heat is given off.) Another factor to consider is that the more clouds in the sky, the more solar radiation is reflected back into space. It might seem that a catch-22 situation exists with water vapor's role in global warming. The warmer the temperatures, the more surface water (liquid) evaporates, thus the more water vapor (gas) is formed. The more evaporation, the greater is the amount of water vapor rising into the atmosphere. The more water vapor stored in the atmosphere, the greater the extent of cloud formation, resulting in a net increase in the greenhouse effect. The greater the greenhouse effect, the greater global temperatures and thus more evaporation of surface water, and so on, and so on. In addition, greater cloud cover over parts of the Earth may alter the patterns of rainfall and thus affect agriculture.

There are other greenhouse gases, including methane and oxides of sulfur and nitrogen. Methane (CH_4) is natural gas that has seeped out of the ground and ocean floors for eons of time. Scientists consider methane 60 times more effective as a greenhouse gas than CO_2, and methane may have been underestimated as to its importance in the cyclic nature of global temperatures. Recently discovered methane-hydrate ice crystals on the ocean's floors worldwide may be tapped as a future source of energy. These methane-hydrate ice deposits constantly emit methane gas into the oceans where the solid form of methane becomes unstable when the surrounding water temperatures rise or when pressure decreases. When this happens, the methane ice crystals change to gas (sublimate) and escape from the ocean's surface into the atmosphere.[117] There is also evidence that massive, worldwide methane ice deposits on the seafloor may have been responsible for the thawing out of the ancient ice-bound Earth, as well as partially responsible for the decrease in carbon dioxide and increase in oxygen in the primitive atmosphere (see chapter 8).

Recent research indicates that during the Holocene period (the past 8,000 to 10,000 years) the levels of carbon dioxide gas did not remain constant until the beginning of the industrial age, and during this same period, temperatures remained relatively stable. In other words, the level of CO_2 in the atmosphere has only been relatively stable for the past hundred years or so. Before that period, temperature levels were not greatly affected by swings in the levels of carbon dioxide. Studies also show that there is an ice age/warming cycle that occurs about every 100,000 years, which leads some scientists to speculate that warming comes first and a rise in carbon dioxide takes place about 400 to 1,000 years later. This suggests a chicken-and-egg situation in which cause and effect need to be reexamined.[118] One climate model suggested that 145 million years ago, duck-billed dinosaurs roamed the northern areas of Alaska, and even the deep seawater of the area reached 15ºC. At this time Alaska was near the equator, and only later through plate tectonics did it reach its present latitude. The model explained how geoproxies are involved with the carbon cycle. It also explained the ancient age downward trend in CO_2 as reflected in the appearance of vascular land plants ~380 to ~350 million years ago. This resulted in the accelerated silicate weathering and created a new sink of more bacterially resistant organic matter in marine and nonmarine sediments. This paleoclimate model for CO_2 also suggested that periodically the ice caps and glaciers melted as the northern latitudes became warmer, but at the same time equatorial climates stayed at about the same temperatures.[119] Another study used deep-sea sediment cores to examine the complex evolution of global climate since 65 million years ago. The scientists determined that gradual warming and cooling of the climate follows trends and rhythms and exhibits aberrations caused by Earth's orbital geometry, which results in shifts in incoming solar energy, and that boundary conditions caused by plate tectonics (see chapter 2) gradually alter continental geography, topography, and the oceans, as well as the concentration of atmospheric greenhouse gases.

A research report in *Science* magazine stated that although the sun is the controller of climate changes there is some question as to its nonlinear feedback role that can amplify or diminish the forcing of climate changes. Other factors that can be considered as part of solar variability are related to orbital variations that force climate by altering the solar input. They are *precession, obliquity,* and *eccentricity* in the Earth's orbit around it's axis and about the sun. These cyclic perturbations were major factors in determining paleoclimates and ice ages. The study was interested in mechanisms operating on time scales of billions of years to just a few decades that may assist in understanding feedbacks and climate forcing for both past and future greenhouse gases.[120]

An additional study of conditions over the past 65 million years led to the conclusion that these physical changes (Earth's orbital oscillations and plate tectonics) triggered large-scale biogeochemical feedbacks in the carbon cycle that initially amplified the climatic changes, although over time equilibrium would be established. These scientists also concluded that "[o]rbital forcing of solar heat on Earth may have had a hand in triggering these events as well, possibly as a means of providing the climate system with a final and relatively rapid push across a climatic threshold resulting in global warming."[121]

Another study in *Science* magazine, "Solar Forcing of Drought Frequency in the Maya Lowlands," concerned a more recent period of history and reported that over the past 2,600 years or so the climate in the Yucatan region of Mexico changed drastically. The authors' theory was that climate changes led to droughts and the demise of the Mayan culture. The scientists used sediment proxies sensitive to oxygen isotopes and gypsum precipitation, as well as carbon-14 and beryllium-10 radioactive isotopes, to establish 208-year variations in solar activity over the past 2,600 years. These cycles reflect variations in solar activity and are not related to human activity; these variations in solar activity are not only natural but have affected global warming for thousands of years before the industrial revolution.[122] Another investigation integrated archaeological and paleoclimatic records to study four cases of Old World and New World civilizations and their responses to periodic drought due to climate changes. Over the past 8 to 10 thousand years (Holocene epoch), persistent shifts in climates resulted in whole empires collapsing, as well as resulting in much lower standards of living for the residents of affected regions. If people migrated out of the regions, there were stressful interactions between cultures, for example, socioeconomic, political, and religious stresses. The four empires studied were the Akkadian, c.4200 yr. B.P.; Classic Maya, c.1200 yr. B.P.; Mochica, c.1500 yr. B.P.; and Tiwanku, c.1000 yr. B.P.. By studying how ancient cultures handled dramatic climate changes, scientists hope to gain a perspective on the possible responses of modern societies to future climate changes.[123] The recent book *Nature's Capitalism* described ways that the environment can be saved and how to get rich doing so. The premise is that pollution and waste are inefficient and expensive and that entrepreneurs and capitalists are clever enough to tap into the massive potential of the free markets related to clean water and air. The authors give examples of how large amounts of resources were saved by planning new enterprises to be energy efficient and reusing or selling waste products.[124] One proposal was a market approach of trading CO_2 and other pollution credits like stock and putting caps on how much pollution an industry can produce. If they go over their cap, they can then buy credits from industries and countries that have not met their caps. Several countries (particularly China and India) have objected to this trading scheme solution because they believe that the more developed countries would buy up their CO_2 and pollution stocks.[125] At the present, the United States has not demonstrated much interest in this concept, because many U.S. industries continue to make progress in reducing both air and water pollution, although in general less progress is being made in reducing excessive CO_2 in the atmosphere.

What can be done to decrease greenhouse gases and thus reduce global warming? First, even though some (but not all) recent and current increases in the levels of atmospheric CO_2 are **anthropic**, the extent of human versus natural causes is still being investigated. In the long run, nature has its own agenda and often over time does not pay much attention to our activities.

Second, there are ways to reduce levels of greenhouse gases without seriously disrupting civilization as we know it. Although a significant reduction of CO_2 might not make much difference in future global climate changes, one possible solution is to develop better economic models for reducing environmental pollution. The development and implementation of new cost-effective technologies for controlling or limiting the production of man-made greenhouse gases is always a possibility. (Not much can be done to control nature's production of greenhouse gases.) Another possible solution is to derive more effective computer climate models to provide more reliable information on which to base policies and regulatory programs. As evidenced by cleaner streams and rivers and urban atmospheres in western countries over the past several decades, many industries in the developed world already have made progress in reducing many of their pollutants and are making progress on reducing levels of CO_2. An economic pollution credit system that would implement market forces may provide more incentives. Almost all man-made environmental and ecological problems are also economic problems with known solutions.[126] An interesting solution for the storing of excess CO_2 by burying it is used in Norway. About 240 km (150 mi) off the coast of Norway an oil rig pumps more than 20,000 tons of carbon dioxide gas each week into deep sandstone layers located 1,000 m (3,280 ft) beneath the ocean floor. This project, begun in 1996, is the first example of injecting CO_2 underground for permanent storage. There are other ways of doing this as well. The CO_2 can be pumped into old, depleted gas or oil deposits; abandoned, mined-out salt domes or coal beds; underground rock caverns, saline aquifers, or deep oceans where it would be contained by the deep-water pressure. These underground structures are found all over the world and, if used, could dispose of unwanted gigatons of carbon. Pumping waste CO_2 to the bottom of deep oceans, where the temperature is low and pressure is high, will form a deep carbon dioxide-lake that will stay in place. In addition, some oil companies in the United States are pumping CO_2 into oil wells to enhance productivity in the recovery of crude oil. As the pressure of the gas helps force oil to surface, the CO_2 also becomes buried in the strata that contained the oil.

A natural solution for use of carbon dioxide is the uptake of the gas during photosynthesis in green plants. It has been demonstrated over the past few years that most plants thrive in an atmosphere rich in CO_2, particularly when they are young and growing fast. Some grasses and, in particular, trees are excellent sinks for CO_2. According to recent estimates, about one trillion tons of carbon are stored in the forests that now exist on the Earth. Europe and the United States have not only maintained their forests but have increased the planting of trees over the past 50–75 years. Today, about 35% more trees are growing in the United States than there were 100 years ago. In the United States, industry and government have combined to plant four trees for every three harvested. Tree farmers, who depend on the market for their existence, are well aware of the economics of planting more trees than they harvest. Unfortunately, some developing nations are using more of their trees than they are replacing. Norway, Brazil, Malaysia, Russia, Australia, and others are involved in projects to plant millions of hectares of forests, while India, cognizant of its problem with denuded forestland, is attempting to replace some of its trees. In many developing nations, however, people use wood as their main fuel for cooking and heating (80% or more of trees cut in the Third World are used for firewood), which makes significant third-world reforestation problematic. A possible solution for using some of the excess CO_2 would be to replant marginal agricultural lands on a global scale with grasses or trees.[127–130] A report in *Science* magazine stated that over the past half century in China, reforestation may be contributing to the global terrestrial carbon sink.[131]

There is some concern that trees may not be as an efficient sink for as much carbon dioxide as once believed. Young trees use a great deal of CO_2, as they are growing, but as they mature and become older, growth slows and they use much less CO_2. When trees are burned, or die and decay, they release their elements and compounds, including CO_2 and CH_4, back to the soil and atmosphere. Therefore, the net gain of sequestered carbon is less than originally expected. The net gain in stored carbon could be reversed if mature trees were buried and became part of the long-term rock cycle.

El Niño and Global Warming

Philip Ball, in the article "Cycle Lock Causes Quick Change" in *Nature* magazine, stated that the process that produces El Niño events may also cause abrupt shifts in global climate and that this could explain why the thaw after the last ice age was interrupted by a frigid spell lasting hundreds of years. Ball reported on a research study that challenged the conventional wisdom about the reason for this event in which Earth returned to near-ice-age conditions about 11,500 years ago. Sudden shifts in temperature can occur across the globe due to melting of ice sheets caused by warming, which discharges fresh water into the North Atlantic Ocean. Since fresh water is less dense than salt water, it stops the ocean water from sinking as it flows northward. Sinking water in the North Atlantic drives large circulation patterns in ocean water that bring heat from the tropical regions toward the poles, resulting in more ice melting. When this circulation ceases, however, the North Atlantic region cools, thus affecting global climate. It seems that El Niño events bring warm, wet weather from the southern hemisphere to the northern regions. In addition, over billions of years there has been a periodic slight change in the shape of the Earth's orbit around the sun, slight changes in the Earth's tilt on its axis, and slight changes in the sun's output of energy. These and other natural changes are all thought to have contributed to global temperature changes and climate shifts between ice ages and excessive global warming over the geological history of the Earth. It also seems that these changes from normal temperatures can happen in a matter of decades, as the Earth is currently experiencing.[132]

International Global Warming Assessments

The Intergovernmental Panel on Climate Change (IPCC) was established in 1988 by the World Meteorological Organization (WMO) of the United Nations Environment Programme (UNEP) to assess the scientific, technical, and economic information relevant for understanding the risk of human-induced climate change. The IPCC is charged by a number of governments to report the state of knowledge and the issue of uncertainty in scientific reports. The IPCC is staffed by scientists with a variety of specializations. The organization itself is not a research organization. Instead, it monitors climate-related data and publishes assessments of climate literature.[133] It is one of the less political organizations of the United Nations. The IPCC charges special assessment panels of scientists to review published research conducted by other scientists. The Third Assessment Report Group (TAR) examined the uncertainty in the IPCC's estimations that the Earth's climate would rise 2 to 6°C over the twenty-first century. This estimate was raised to double the range, to possibilities of anywhere from 3 to 11°.[134–135] Both scientists and governments were dissatisfied with such a high degree of uncertainty for climate predictions for the next century. Therefore, they established TAR as an intergovernmental panel to examine the lack of a reliable estimate of the probability that human-induced global warming over the 1990–2100 period will lie either above or below the projected range of 1.4° to 5.8°C. The members of TAR asserted that any method of uncertainty analysis should be both documented and reproducible. It was their unanimous view that no method of assigning probabilities to a 100-year climate forecast is currently available to pass the extensive IPCC review process. The authors of the TAR document gave the following reasons for this lack of certainty in predicting climate temperatures for the next century:

1. The difficulty of assigning reliable probabilities to socioeconomic trends (and hence emissions) in the latter half of the twenty-first century
2. The difficulty of obtaining consensus ranges for quantities like climate sensitivity
3. The possibility of a *nonlinear* response in the carbon cycle or ocean circulation to very high late-twenty-first-century greenhouse gas concentrations.

The panel felt that that the nonlinear relationship of carbon to the environment cannot be predicted with any degree of accuracy for the latter 50 years of the twenty-first century and that current climate models, being inadequate to establish these nonlinear relationships, may underestimate the range of uncertainty of human activity on the future climate. In other words, based on the myriad factors involved and the high degree of uncertainty, no one has reliable data or sufficiently accurate models to predict the future of the Earth's climate.[136]

Dixy Lee Ray, the late Governor of Washington state and former chair of the Atomic Energy Commission, emphasized the complex science aspects of climate change:

> We do not know what caused severe climate changes in the geological past, but we can be sure they were not due to human industrial activity. Most likely, the causes were and still are colossal cosmic forces, quite outside human ability to control. Now that we live in an industrial, high technology society, there is no reason to believe that such cosmic forces have ceased to exist.[137]

In other words, the Earth has experienced significant changes in its climate and will continue to do so in the future regardless of the existence of humans, past, present, or future.

Early in 2001, the White House requested that the *National Academy of Sciences* develop a report on climate change. The eleven scientists who produced the report did not agree on all aspects of the global warming issue. Nevertheless, the press and media implied otherwise. To correct the media's coverage of the report, Richard S. Lindzen, one of the eleven scientists involved in the study, summarized the debate on global warming in an opinion piece in the *Wall Street Journal.* He stated that, contrary to press reports, there was no unanimous consensus among the eleven scientists and that using records for a 20-year period was too short a time for estimating long-term trends. Lindzen continued,

> Our primary conclusion was that despite some knowledge and agreement, the science is by no means settled. We are quite confident (1) that the global mean temperature is about 0.5 degrees Celsius higher than it was a century ago; (2) that atmospheric levels of carbon dioxide have risen over the past two centuries; and (3) that carbon dioxide is a greenhouse gas whose increase is likely to harm the earth (one of many, the most important being water vapor and clouds). . . . I cannot stress this enough—we are not in a position to confidently attribute past climate change to carbon dioxide or to forecast what the climate will be in the future . . . and agreement on these three basic statements tells us almost nothing relevant to policy discussions. . . . One reason for this uncertainty is that, as the report states, the climate is always changing; change is the norm. Two centuries ago, much of the Northern Hemisphere was emerging from a "little ice age." [A] Millennium ago, during the Middle Ages, the same region was in a warm period. Thirty years ago, we were concerned with global cooling. Distinguishing the small recent changes in global mean temperature from natural variability, which is unknown, is not a trivial task. . . . We simply do not know what relation, if any, exists between global climate changes and water vapor, clouds, storms, hurricanes, and other factors, including regional climate changes, which are generally much larger than global changes and not correlated with them. Nor do we know how to predict changes in greenhouse gases. . . . What we do know is that a doubling of carbon dioxide by itself would produce only a modest temperature increase of one degree Celsius.[138]

Coral Bleaching

Coral reefs consist of calcium skeletons of millions of polyps. Coral can be red, pink, green, or blue according to the type of algae that lives on the reefs and that provides food and nutrients for the polyps that form the coral skeletons. Coral bleaching happens when the algae coating on the surface of the coral dies because of temperature changes in the water. Coral can live for some time without their usual algae but not over long periods of time. The bleaching of coral reefs, including those off Florida's southern coast, has increased over the past several decades. Many environmentalists blame global warming and predict that coral reefs are headed for extinction.

Andrew Baker of the New York Aquarium stated that bleaching may actually help reefs adapt to changing conditions. He set up an experiment that changed the positions of different corals to shallow and also deeper waters and found that coral bleaching and death did not go hand-in-hand. The coral that was transferred to deeper water bleached less and died more, whereas the coral moved from deeper areas to more shallow waters bleached more but died less. Baker's conclusions were that bleaching is an ecological gamble that may sacrifice short-term benefits for long-term advantages as the loss of some reef's algae coat may be an opportunity for a more beneficial type of algae to take hold. In other words, coral may be more adaptive than previously thought. This conclusion is not accepted by all scientists, nor does it deny that coral reefs are currently under stress, but it indicates that whether global warming is entirely responsible is not certain.[139] There may be other factors involved in the death of coral's algae besides global warming. Such factors as degree of salinity of the ocean water, pollution, diseases, and so on, may also be partially responsible for the bleaching. Additional research may provide some answers. In the article "The Global Transport of Dust," Dale W. Griffin reported on the massive dust storms originating in the North African Sahara Desert and other regions that transport a variety of microorganisms across the world. During some seasons, bacteria, fungi, and viruses found in these regions are adsorbed (stick to) very fine soil particles as they become airborne. A significant amount of this pollution is dumped into the shallow waters off the coasts of Florida and the Caribbean Islands where it settles onto the coral. Some of these microorganisms, as well as the settling dust, kill the coral.[140]

The issue of global warming might be summed up as follows: Yes, there is some global warming occurring, and, no, we are not sure how much of the slight increase in warming at ground level over the past 100 years is due to natural causes and climate cycles or to human influences. It is generally agreed that our current technology and knowledge (including computer models) are inadequate to make accurate future predictions on which to base public policies dealing with global warming. Even so, many environmentalists insist that corrective actions be taken as soon as possible because the forcing factor may soon reach a threshold that is irreversible. Both national and international policies have been proposed to limit the production of CO_2 and other aerosols. Unfortunately, not all of these proposals consider the cost-benefits aspects of the policies designed to regulate the production of atmospheric gases. Also, how can the worldwide aerosol pollution from dust storms be controlled?

Ozone

Issues related to ozone are as complicated as those for global warming. First, it's important to understand exactly what ozone is and the reasons for its importance. Oxygen comes in three flavors. Atomic oxygen (O) has a single atom of oxygen and is sometimes referred to as nascent, or newborn, oxygen that does not exist in nature.

Diatomic oxygen (O_2) is the molecular form of the gas that comprises about 20% of the atmosphere that we breathe. It both supports life and combines with many other elements. It is not combustible, as many believe, but oxygen is a reactive gas and thus supports combustion. Ozone is the triatomic oxygen molecule (O_3), which is relatively unstable and in high concentrations is also poisonous. Ozone is produced at ground level by electrical sparking (lightning) and burning of petroleum products (e.g., auto exhaust exposed to sunlight radiation). It can be used to sterilize air and as a bleaching agent. Since ozone produced near the surface of the Earth is heavier than regular oxygen, it tends to stay near ground level and is one of the main ingredients of smog. Ground level ozone as a major component of smog might aggravate childhood asthma, but it does not cause asthma, as is generally believed. The relationship of ozone to asthma, particularly in children, is controversial and is still being researched.[141]

On the other hand, ozone is also produced in the stratosphere, the layer of the atmosphere that extends from approximately 10 to 50 km (6.2 to 31 mi) in altitude. At the colder polar regions, ozone can form at lower altitudes, while most of the atmospheric ozone is formed above the warmer equator but at higher altitudes. If you recall, the stratosphere warms with the increase in altitude. This phenomenon is the opposite of the lower layer of air, the troposphere, which cools with altitude. This is important because the warmer stratosphere forms a region of **temperature inversion**, which traps the ozone in a thin layer, holding it for several years so that it neither sinks to the ground or escapes into space. The band of ozone in the stratosphere extends from about 15 to 35 km (9 to 22 mi) above the Earth. Its thickness and distribution depend on the dynamics of the stratospheric winds.[142] The ozone layer is not very thick and is far less dense (low pressure) than the atmosphere near the surface of the Earth. For instance, if stratospheric ozone had the same density (pressure) as surface air, it would be compressed down to only about 2 to 3 millimeters (about 0.1 in.) thick.

Even this thin layer of triatomic oxygen molecules is important to us. Ozone serves as a protective radiation shield that blocks out certain ultraviolet (UV) solar radiation that can be harmful to living organisms. Excessive exposure (sunbathing) can cause skin cancer and eye cataracts in humans and is destructive to some plants. Oddly, radiation from this portion of the electromagnetic spectrum does not affect all types of plants. Some actually thrive on excessive UV. The ozone problem is one related to chemistry and physics.

The chemistry goes something like this: First, for over hundreds of millions of years UV broke down regular oxygen in the ozone layer into two separate, single atoms of oxygen, which then combined with other regular diatomic oxygen to form ozone.

$$UV \rightarrow O_3 \rightarrow O + O_2\text{: then } O + O_2 \rightarrow O_3\text{: then } UV \rightarrow O_3 \rightarrow O + O_2\text{: then } O + O_2 \rightarrow O_3\text{, etc.}$$

Ozone, which is normally unstable, is broken up by UV followed by the recombination of atomic oxygen with diatomic oxygen to form new triatomic oxygen (ozone). Over centuries this process has achieved dynamic equilibrium (balance), forming the thin ozone layer in the stratosphere. The problem is that this equilibrium is in delicate balance and may be disturbed by other types of molecules entering this level of the stratosphere.

During observations in 1957 and 1958, scientists first discovered a hole in the ozone layer over Halley Bay in the polar region. These preliminary findings were confirmed by ground-based stations and satellite data in the 1960s and 1970s. At about this time, scientists learned that the halogens (chlorine, fluorine, and bromine) could break down the ozone molecule, at least in their laboratories. Since then, the hole has been observed to have both increased and decreased in size in response to seasonal temperatures. The ozonosphere above the Antarctic has a hole that is really a thin area that naturally expands and contracts each year, with the maximum hole occurring between September and November.[143] The hole closes as the ozone layer expands during late winter, spring, and early summer. In the late 1970s, it was discovered that chemicals called "chlorofluorocarbons" (CFCs, composed of chlorine, fluorine, and carbon), used as refrigerants in homes and industries, could evaporate and rise into the upper atmosphere, even though CFCs are heavier than air. By 1985 scientists also discovered that the ozone layer was becoming thinner over the South and North Poles. By this time, the rest of the chemical reaction had been worked out.

The sun's rays break up the CFCs, releasing chlorine atoms (Cl), which then combine with single atoms of oxygen (O) from the ozone triatomic (O_3) to form chlorine monoxide (ClO) and leave diatomic O_2. Next, when the ClO meets another oxygen atom, the ClO releases its atom to form O_2, which leaves the Cl again free to attack more ozone. Chlorine deprives the diatomic oxygen molecules access to the single oxygen atom that would have combined to form ozone:

1. UV $\rightarrow O_3 \rightarrow O_2 + O$ (Normal: UV radiation breaks up O_3) $O_2 + O \rightarrow O_3$ (Normal: O_2 combines with atomic O to form O_3) UV $\rightarrow O_3 \rightarrow O_2 + O$: etc., etc. (equilibrium)
2. UV $\rightarrow$ CFCs $\rightarrow$ Cl (free chlorine atoms liberated in ozone layer)
3. $Cl + O_3 \rightarrow ClO + O_2$ (chlorine takes O from ozone, forming ClO)
4. O_2 now has less atomic O available with which to form O_3
5. $ClO + O \rightarrow O_2 + Cl$ (Cl now free to break up more O_3)
6. O_2 + reduced amount of O yields less O_3 (to start reaction over)

The actual reaction is somewhat more complicated, but this representation gives an approximation of the process for the partial depletion of the ozone layer. The reaction is repeated over and over as Cl acts as a catalyst as it steals atomic oxygen (O) from ozone (O_3) to form ClO and diatomic oxygen (O_2). Thus, less atomic oxygen is available to recombine with diatomic oxygen to form triatomic ozone. (Note that the O in the ClO is the product of the breakup of O_3, as in the previous equation, but now the free atomic oxygen combines more readily with chlorine [or fluorine or bromine] atoms than with diatomic oxygen to reform ozone. The chlorine catalyst is not used up, so it can recombine with free oxygen atoms over and over again, interfering with the normal production of ozone molecules in the process. Due to the catalytic nature of chlorine, one chlorine atom can affect about 10,000 ozone molecules. Note also that there are sources other than CFCs, such as the oceans and volcanic emissions, that contribute chlorine and other gases for this ozone catalytic chemical reaction.)

At issue is how much of the ozone hole is part of a natural cycle and how much is the result of human activity. It is known that ozone forms over the equator, and then spreads out toward the poles where it is normally found in higher concentrations. The so-called holes are not areas where there is no ozone, but instead they are thinnings of the ozone layer above the colder regions on the Earth. These holes seem to get larger and then shrink with the seasons and may possibly exhibit longer cycles over periods of many years. It is obvious that humans living near the equator are exposed to more direct sunlight (and UV). This was also true before we knew about the ozone layer and its holes, which implies that precautions to excessive UV exposure should be observed.

In 2000, *Scientific American* magazine published some interesting statistics relating chlorofluorocarbon levels to the ozone, the exposure to the sun's UV, and melanoma (skin cancer) in the United States:

1. Melanoma rate per 100,000 persons in the United States in 1972 = 5.7, while in 1996 the rate was 13.8
2. Estimated number of persons who will be diagnosed with melanoma in the year 2000: 47,700
3. Number of deaths from melanoma expected in the year 2000: 7,700
4. Chance that an American will develop skin cancer in his or her lifetime: 1 in 5
5. Attendance at beaches in 1998: 256,721,218
6. Average percentage of lifetime sun exposure received by age 18: 80
7. Year when atmospheric chlorofluorocarbon levels are expected to peak: 2000
8. Year expected ozone recovery may first be conclusively detected: 2010
9. Soonest year by which Antarctic ozone layer will recover: 2050
10. Number of ozone molecules that can be destroyed by one chlorine atom: 100,000[144]

Most developed nations have banned the production and use of CFCs and similar compounds. This is not the case with some developing countries. For instance, just across the southern border of the United States it is still possible to buy CFCs from street vendors for automobile air-conditioners. It is estimated that the peak amount of CFCs in the stratosphere was reached in the year 1999. However, it will take another 40 or 50 years for most of the chlorine (and other halogens) of the CFCs to dissipate. Even then, not all ozone-destroying gases will be eliminated, as a number of these gases are sent into the stratosphere from volcanic activity and chlorine evaporates from the oceans. Because most of these natural gases are soluble, however, a large portion of them are brought back to Earth, in time, by precipitation. A fact not often considered when the ozone problem is addressed is that certain natural gases have always helped keep stable the ozone levels in the stratosphere. It is nature's way of maintaining the dynamic equilibrium between ozone creation and destruction, which provides adequate protection from UV for living organisms on the Earth. A study published in *Science* magazine explained how additional ozone may be produced. Scientists compared the concentrations of the hydrogen **radicals** OH and HO_2 found in the middle and upper troposphere, while at the same time measuring the radicals of NO, O_3, CO, HO_2 and CH_4 in the UV and visible radiation field. From these measurements they were able to determine the effects of the OH and HO_2 radicals on the production of ozone. The production of these radicals in the lower atmosphere and their transport to the troposphere (ozone layer) was found to be more rapid and greater than expected.[145] Along with the addition of NO resulting from the burning of biomass and aviation emissions, the increase in OH, HO_2, and NO radicals (1 part per billion by volume each day) will lead to a greater production of ozone (O_3) than expected.[146–151]

A research report, also in *Science* magazine, builds a case for a climate master switch in the atmosphere over the high southern latitudes that is driving the variable climate shifts of Antarctica. This switch is driven by man-made chemicals that form the annual Antarctic ozone hole. Scientists call this switch the Antarctic oscillation (AAO), which accounts for the different climate trends on the surface of Antarctica. The fact that the temperature of the Larsen ice shelf increased 2.5°C during the past 50 years and a large chunk of the ice shelf drifted out to sea is given as evidence that greenhouse warming causes thinning of the glaciers, but actually, "other glaciers are thickening. In places, sea ice is actually advancing, and most of Antarctica is not warming at all or it is even cooling." The thinning layer of ozone cools the lower atmosphere by 6°C each spring, which results in a shift of the winds in the lower atmosphere and cools, rather than warms, the surface landmass of Antarctica.[152]

In summary, environmentalists claim that the Earth is in serious trouble. Even the media, which like a good story, claim that everyone knows the planet is in bad shape. There is evidence that environmental conditions on the Earth are improving, even though more can be done. The following list provides some examples of how things are getting better.

- The worldwide discharge level of CFCs into the atmosphere has been greatly reduced in just the past few decades. An adequate ozone layer still exists, and normal seasonal cycling of the thin areas will continue. The so-called ozone hole is cyclic and will decrease in the coming years due to the restriction of CFCs.
- Although the level of CO_2 discharged into the atmosphere has not decreased, the developed nations are cognizant of the possible connection between global warming and carbon dioxide, methane, and other gases. Thus, several developed nations and their industries are beginning to take corrective actions, while the developing countries continue, unabated, to increase their discharges of greenhouse gases.
- The forest cover on the Earth has increased over the past 50 years, and more countries are effectively managing their forests and planting more trees. The United States plants more trees than it harvests each year.
- Average annual oil spills have been reduced from 65 million liters in the 1970s to about 11.8 million liters in the 1990s. With the introduction of new ocean-going double-hulled tankers, the number of spills continues to decrease. New chemicals and biological technologies used to clean up oil spills have greatly reduced environmental damage. In addition, it is now recognized that most damage by oil spills near coastal areas is short-term, that natural bacteria, in a few years, digest the oil, and that life returns to normal. Even the environment affected by the disastrous *Valdez* oil spill in Alaska has, for the most part, recovered.
- Toxic chemicals found in North Sea fish declined by 76% or more between 1982 and 1996. Sewage and garbage discarded at sea by fishing and tourist ships has been banned, and ocean pollution is being reduced.
- The amount of ammonia in European rivers has fallen by about half in the past 10 years. Rivers in all of the western countries have become significantly less polluted in the past several decades, but rivers and streams in some areas still need to be cleaned up.
- Many species formerly on the endangered list have been removed as they are recovering. Many endangered species are thriving (e.g., the bald eagle, wolf, bears, deer, buffalo, native grasses, etc.), while others are still threatened. Worldwide, more and more land is being set aside for parks,

reserves, preserves, and other nonindustrial use.

- The discharge of methane, one of the major greenhouse gases, by humans has decreased worldwide by a significant percentage. Sulfur and nitrogen emissions from coal burning electric generation plants have been reduced in the past several decades. The installation of catalytic converters in automobiles and the burning of cleaner gasoline have reduced the amount of smog in many of the developed countries, but this is offset somewhat by the increase in the number of automobiles and trucks.
- Due to the great increase in organic farming in the developed countries, there has been a significant reduction in the use of both pesticides and herbicides, as well as a reduction in the application of chemical fertilizers.
- Rivers, streams, lakes, and oceans in Europe and the United States are far less polluted than they were just a few decades ago. Fish are making a comeback in many of these freshwater regions, including New York Harbor.
- The substitution of new products for scarce minerals and metals has extended the supply and use of many natural resources. For instance, the use of fiber-optic cables and wireless technologies has significantly reduced the amount of copper required for the telecommunication industry. As new technologies are developed and old ones improve, so will the environment.

Even though many people assume that the environment and related issues are worsening, in many respects the opposite is true.[153] Those involved in the political processes that form environmental policies are beginning to recognize the importance of applying the best science available to the decision-making process. Many of the improvements in the environment are due to increased environmental legislation and regulations, as well as the recognition by many industries, as well as the general public, of economic and social benefits for protecting the environment. Most environmental scientists are well aware of the complexity of environmental problems and the limitations to their knowledge and continue to promote sound policies that will safeguard the Earth's environment. Unfortunately, not all politicians, regulators, media personnel, and organizations are rational when considering environmental issues. Even so, the goal of all enlightened governments is that modern civilization not leave the Earth as a rubbish heap. Instead, we must all learn how to manage our natural resources and environment more economically and effectively.[154]

Notes

1. Angus M. Gunn, *The Impact of Geology on the United States: A Reference Guide to Benefits and Hazards* (Westport, Conn.: Greenwood Press, 2001), xxii–xiii.
2. National Oceanic and Atmospheric Administration (NOAA), Paleoclimatology Program: *Study Paleoclimatology,* accessed 2001, http://www.ngdc.noaa.gov.
3. National Oceanic and Atmospheric Administration (NOAA), Paleoclimatology Program: *What Is Paleoclimatology?,* accessed 2001, http://www.ngdc.noaa.gov.
4. National Oceanic and Atmospheric Administration (NOAA), A Paleo Perspective on Global Warming: *Climate Change,* accessed 2001, http://www.ngdc.noaa.gov.
5. Richard B. Alley, *The Two-Mile Time Machine: Ice Cores, Abrupt Climate Change, and Our Future* (New York: Princeton University Press, 2000).
6. Graham R. Thompson and Jonathan Turk, *Earth Science and the Environment* (New York: Harcourt Brace, 1999), 423.
7. History of Climate Change: *Past Climate,* accessed 2001, http://www.athena.ivv.nasa.gov.

8. Larry Massett, The Weather Notebook: *Ancient Weather Theory,* accessed 2001, http://www.Mountwashington.org.

9. The Daily Revolution: *Ancient Weather,* accessed 2001, http://www.dailyrevolution.org.

10. Thompson and Turk, *Earth Science,* 427.

11. Thompson and Turk, *Earth Science,* 427.

12. Thompson and Turk, *Earth Science,* 427–28.

13. History of Climate Change: *How Do We Know About Past Climate,* accessed 2001, http://www.athena.ivv.nasa.gov.

14. Richard B. Alley and Michael L. Bender, "Earth From the Inside Out: Greenland Ice Cores: Frozen in Time," *Scientific American,* special publication, 2000, 4–9.

15. Alley, *The Two-Mile Time Machine.*

16. National Oceanic and Atmospheric Administration (NOAA), *Proxy Climatic Data,* accessed 2001, http://www.ngdc.noaa.gov.

17. *Caves Reflect Ancient Weather Patterns,* accessed 2001, http://www.uct.ac.za/depts/dpa.

18. Thompson and Turk, *Earth Science,* 428.

19. Sybil P. Parker, ed., *McGraw-Hill Concise Encyclopedia of Science and Technology,* 3d ed. (New York: McGraw-Hill, 1992), 405.

20. Roberto Sabadini, "Ice Sheet Collapse and Sea Level Change: Feedbacks and Connections," *Science,* 29 March 2002), 2376–77.

21. Edward J. Denecke Jr., *Let's Review: Earth Science* (Hauppauge, N.Y.: Barron's Educational Series, 1995), 389.

22. Robert E. Krebs, *Scientific Development and Misconceptions through the Ages* (Westport, Conn.: Greenwood Press, 1999), 194.

23. Denecke, *Let's Review,* 390.

24. John Tomikel, *Basic Earth Science: Earth Processes and Environments* (New York: Allegheny Press, 1981), 63.

25. Tomikel, *Basic Earth Science,* 63.

26. Hans Junginger, *The Human Skin: Your Largest Organ,* accessed 2001, http://www.arubaaloe.com.humanskin.

27. Denecke, *Let's Review,* 391.

28. Thompson and Turk, *Earth Science,* 399.

29. Denecke, *Let's Review,* 393.

30. Tomikel, *Basic Earth Science,* 64.

31. John Daintith, Sarah Mitchell, Elizabeth Tootill, and Derek Gjertsen, *Biographical Encyclopedia of Scientists,* 2d ed., vol. 2 (Bristol, England, and Philadelphia: Institute of Physics Publishing, 1994), 180, 377.

32. Thompson and Turk, *Earth Science,* 401.

33. Tomikel, *Basic Earth Science,* 65.

34. Daintith et al., *Biographical Encyclopedia of Scientists,* vol. 2, 495–96.

35. The following are some Web sites related to Köppen's climate zone classification system (see also Internet search engines): http://www.huizen.dds.nl/~gvg.ctkoppe2 (accessed 2001); http://www.Fao.org/WAICENT/faoinfo/sustdev/Eldirect/climate/Elsptext (accessed 2001); http://www.snow.ag.uidaho.edu/clim_map/koeppen_criteria (accessed 2001).

36. Betty Debnam, "The Amazon River Basin," The Mini Page: Syndicated News Service, *The Valley Morning Star,* Harrington, Tex., 4 April 2001, D4.

37. *1995 World Almanac and Book of Facts* (Mahwah, N.J.: Funk and Wagnalls, 1994), 181, 184.

38. Robert L. Bates and Julia A. Jackson, *Dictionary of Geological Terms* (New York: Anchor Books/Doubleday, 1984), 377.

39. *Köppen Climate Classification,* accessed 2001, http://www.huizen.dds.nl/~gvg/ctkoppe2.

40. *Environment: Specials: Global Climate Maps—Tour Guide,* accessed 2001, http://www.fao.org/WAICENT/faoinfo/sustdev/Eldirect/climate/Elsptext.

41. Thompson and Turk, *Earth Science,* 404–15.

42. Tomikel, *Basic Earth Science,* 65–73.

43. Kenneth W. Hamblin and Eric H. Christiansen, *Earth's Dynamic Systems* (Upper Saddle River, N.J.: Prentice Hall, 1998), 218–20.

44. *Climate Classification Criteria,* accessed 2001, http://www.snow.ag.uidaho.edu/Clim_Map/koppen_criteria.

45. Thompson and Turk, *Earth Science,* 384–85.

46. University of Illinois, *El Niño: Online Meteorology Guide,* accessed April 24, 2001, http://www.2010.atmos.uiuc.edu/(Gh)/guides/mtr/eln/atms.rxml.

47. *Virtual Climate Alert* #5, January 21, 2000, vol. 1, no. 5, accessed 2001, http://www.greeningearthsociety.org/Articles/2000/vca5.html.

48. U.S. Department of Commerce and the National Oceanic and Atmospheric Administration (NOAA)*What is La Nina?* Information provided by the TAO Project, http://www.taogroup@pmel.noaa.gov.

49. *Ancient Weather,* 27 April 1998, accessed 2001, http://www.dailyrevolution.org/monday/oldweather.html.

50. *How to Forecast Weather by Nature,* accessed, 2001 http://www.stalkingthewild.com/weather/html.

51. Weather Folklore: *Welcome to Weather World's Huge Collection of Weather Folklore and Weather Wits!,* accessed 2001, http://www.members.aol.com/Accustiver/wxworld_folk.html.

52. Alan Isaacs, ed., *A Dictionary of Physics* (New York: Oxford University Press, 1996), 422.
53. "Sun Signals Cycle's Peak with a Blast," *USA Today,* 17 April 2001.
54. "Sun's Magnetic Flip Affects Cosmic Ray Penetration," Science and Education, *USA Today,* 10 April 2001.
55. Patricia Barnes-Svarney, ed., *The New York Public Library Science Desk Reference* (New York: Macmillan, 1995), 434.
56. Thompson and Turk, *Earth Science,* 375–78.
57. Tomikel, *Basic Earth Science,* 47–49.
58. Thompson and Turk, *Earth Science,* 369–70.
59. *The Oxford English Dictionary,* 2d ed., CD-ROM, s.v. "mammatus."
60. Weather Basics: *Understanding Clouds,* accessed 2001, http://www.usatoday.com/weather/wcloud0.htm.
61. *Cloud Types: Common Cloud Classifications,* accessed 2001, http://www.2010.atmos.uiuc.edu.
62. *Clouds: The Earth's Energy Balance,* accessed 2001, http://www.airs.jpl.nasa.govhtml/edu/clouds/Earth.
63. Jim Koermer, *PSC Meteorology Program Cloud Boutique,* accessed 2001, http://www.vortex.Plymouth.edu/clouds.
64. *Types of Cloud Formations,* accessed 2000, http://www.met.hu/cloudalbum/alfelhok.htm.
65. Tomikel, *Basic Earth Science,* 50–53.
66. Thompson and Turk, *Earth Science,* 369–71.
67. Environmental Technology Laboratory, *Clouds and Climate,* accessed 2001, http://www.etl.noaa.gov/eo/notes/clouds_and_climate.html.
68. NASA Goddard Institute for Space Studies Research, *Clouds and Climate Change: The Thick and Thin of It,* accessed 2001, http://www.giss.nasa.gov/research/intro/delgenio_03.
69. *Encyclopedia Britannica,* CD-ROM 2001, s.v. "Wind Shear and Microburst."
70. *Encyclopedia Britannica,* CD-ROM 2001, s.v. "Climate: Atmospheric Pressure and Wind."
71. Tomikel, *Basic Earth Science,* 54–56.
72. Barnes-Svarney, ed., *The New York Public Library Science Desk Reference,* 424–26.
73. University of Illinois, Air Masses: *Uniform Bodies of Air,* accessed 2001, http://www.2010.atmos.uiuc.edu.
74. The Franklin Institute Online, *Investigation Wind Energy,* accessed 2001, http://www.sln.fi.edu/tfi/units/energy.
75. Weather: *The Wind Chill Formula, Its Development,* accessed 2001, http://www.usatoday.com/weather/wchilfor.
76. *Encyclopedia Britannica,* CD-ROM 2001, s.v. "Front."
77. Thompson and Turk, *Earth Science,* 378.
78. Denecke, *Let's Review,* 322–25.
79. April I. Umminger and Adrienne Lewis, "Summer Is Peak Lightning Season," Weather Focus, *USA Today,* 25 July 2001.
80. *Encyclopedia Britannica,* CD-ROM 2001, s.v. "Thunderstorms: Types and Physical Characteristics."
81. Thompson and Turk, *Earth Science,* 388–91.
82. Denecke, *Let's Review,* 328–29.
83. Weather Basics: *More About Thunderstorms,* accessed 2001, http://www.usatoday.com/weather.
84. *A Lightning Primer,* accessed 2001, http://www.thunder.msfc.nasa.gov/primer.
85. Ron Hipschman, *Lightning,* accessed 2001, http://www.exploratiorium.edu/ronh/weather.
86. Harold Brooks, "National Severe Storms Laboratory," Weather Focus, *USA Today,* 17 May 2001.
87. *Encyclopedia Britannica,* CD-ROM 2001, s.v. "Tropical Cyclone: Hurricanes."
88. *Hurricanes: Online Meteorologist,* accessed 2001, http://www.2010.atmos.uiuc.edu.
89. Denecke, *Let's Review,* 329–30.
90. Thompson and Turk, *Earth Science,* 391–96.
91. Bjørn Lomborg, *The Skeptical Environmentalist: Measuring the Real State of the World* (New York: Cambridge University Press, 2001), 163–64.
92. Barnes-Svarney, ed., *The New York Public Library Science Desk Reference,* 394.
93. Thompson and Turk, *Earth Science,* 444.
94. Lomborg, *The Skeptical Environmentalist,* 167.
95. Pollution Control: *Regulations* CD-ROM 1998, Version 1 (Grolier Electronic Publications).
96. Jocelyn Kaiser, "Soot's Health Effects Strengthened," *Science,* April 2000, 424–25.
97. Lisa Falkenberg, "Air Pollution Could Trigger Heart Attacks," *Houston Chronicle,* Houston, Tex., 12 June 2001.
98. *Particulate Sources in San Francisco Bay Area,* accessed 2001, http://www.burningissues.org/particle-sources.
99. Lomborg, *The Skeptical Environmentalist,* 167.
100. Michael Fumento, *Science under Siege* (New York: Quill, 1993), 55.
101. *Encyclopedia Britannica,* CD-ROM 2001, s.v. "The Pollution of Natural Resources: Climatic Effects of Polluted Air."

102. Cathy Cobb and Harold Goldwhite, *Creations of Fire* (New York: Plenum Press, 1995), 417.

103. Ronald Bailey, *Eco-Scam: The False Prophets of Ecological Apocalypse* (New York: St. Martin's Press, 1993), 160.

104. John Whitfield, "Sink Hopes Sink," Science Update, *Nature,* accessed 2001, http://www.nature.com/msu/010524/010524–14.html.

105. Wayne A. Sinclair, et al., *Diseases of Trees and Shrubs* (Ithaca, N.Y.: Cornell University Press, 1987).

106. Fred S. Singer, *Environmental Myths of 1999*, accessed 2001, http://www.sepp.org/New SEPP /environmyths.

107. Lomborg, *The Skeptical Environmentalist,* 261.

108. M. E. Mann, R. S. Bradley, and M. K. Hughes, "Northern Hemisphere Temperatures During the Past Millennium: Inferences, Uncertainties, and Limitations," *Geophysical Research Letter* 26, no. 6 (2000): 759–62.

109. Keith Eggleston, "USA Today Weather Focus," *USA Today,* 2 August 2000.

110. Earth Day 1998 Fact Sheet, *Myths and Facts About the Environment*, accessed 2001, http://www.nationalcenter.org/EarthDay98Myths.

111. Earth Day 1999 Fact Sheet, *Myths and Facts About the Environment,* accessed 2001, http://www.nationalcenter.org/EarthDay99Myths.

112. Michael L. Parsons, *Global Warming: The Truth behind the Myth* (New York: Insight Books, 1995), 104–5, 197, 204, 236–37.

113. Robert E. Krebs, *Scientific Laws, Principles, and Theories: A Reference Guide* (Westport, Conn.: Greenwood Press, 2001), 27.

114. International Panel of Climate Control (IPCC), *IPCC Third Assessment Report—Climate Change 2001,* Table 6.1.

115. Lomborg, *The Skeptical Environmentalist,* 260.

116. Hiroshi Ogawa et al., "Production of Refractory Dissolved Organic Matter by Bacteria," *Science,* 4 May 2001, 917–20.

117. Erwin Suess, Gerhard Bohrmann, and Jens Greinert, "Flammable Ice," *Scientific American,* November 1999, 76–83.

118. Curt Suplee, "Studies May Alter Insights Into Warming," *Washington Post,* 15 March 1997, A7, accessed 2001 at http://ww.globalwarming.org/science/alter.

119. Thomas J. Crowley and Robert A. Berner, "CO_2 and Climate Change," *Science,* 4 May 2001, 870–72.

120. D. Rind, "The Sun's Role in Climate Variations," *Science,* 26 April 2002, 673–77.

121. James Zachos et al., "Trends, Rhythms, and Aberrations in Global Climate 65 Million Years Ago to Present," *Science,* 27 April 2001, 686–93.

122. David A. Hodell et al., "Solar Forcing of Drought Frequency in the Maya Lowlands," *Science,* 18 May 2001, 1367–70.

123. Peter B. de Menocal, "Culture Responses to Climate Change during the Late Holocene," *Science,* 27 April 2001, 667–73.

124. Paul Hawken, Amory Lovins, and L. Hunter Lovins, *Natural Capitalism* (Boston: Little, Brown and Co., 2000).

125. Traci Watson and Jonathan Weisman, "6 Ways to Combat Global Warming," *USA Today,* 16 July 2001.

126. William Ashworth, *The Economy of Nature: Rethinking the Connection between Ecology and Economics* (New York: Houghton Mifflin, 1995), 14.

127. Dixy Lee Ray and Lou Guzzo, *Environmental Overkill* (New York: HarperPerennial, 1993), 108–116.

128. Howard Herzog, Baldur Eliasson, and Olav Kaarstad, "Capturing Greenhouse Gases," *Scientific American,* February 2000, 72–79.

129. Krebs, *Scientific Development and Misconceptions,* 221, 244–47.

130. Hamblin and Christiansen, *Earth's Dynamic Systems,* 222–24.

131. "Wither CO_2?," This Week in Science, *Science,* 22 June 2001, 2213.

132. Philip Ball, "Cycle Lock Causes Quick Change," Science Update, *Nature,* 8 June 2001, accessed, http//www.nature.com/nsu.

133. *About IPCC,* accessed 2001, http://www.ipcc.ch/about.htm.

134. John Reilly et al., "Uncertainty and Climate Change Assessments," Public Forum: Climate Change, *Science,* 20 July 2001, 430–33.

135. Elizabeth Kolbert, "Comment: Hot and Cold," *The New Yorker,* 13 August 2001, 25–26.

136. Myles Allen et al., "Uncertainty in the IPCC's Third Assessment Report," Policy Forum: Climate Change, *Science,* 20 July 2001, 430–33.

137. Dixy Lee Ray and Lou Guzzo, *Trashing the Planet: How Science Can Help Us Deal with Acid Rain, Depletion of the Ozone, and Nuclear Waste (Among Other Things)* (New York: Regnery Publishing, 1990), 5.

138. Richard S. Lindzen, "Global Warming: The Press Gets It Wrong," *Wall Street Journal,* 11 June 2001, accessed 2001, http://www.opinionjournal.com.

139. "Scientist Says Bleaching May Help Reefs," *The Associated Press,* 14 June 2001, accessed 2001, http://www.enn.com/extras/printer-friendly.asp?storyid=44002.

140. Dale W. Griffin, "The Global Transport of Dust," *American Scientist,,* May–June 2002, 228–35.

141. "Toxicologists Hit the West Coast," *Science,* 4 May 2001, 837–38.

142. *The Science of Ozone Depletion,* accessed 2001, http://www.ucsusa.org/resources/ozone.science.

143. Barnes-Svarney, ed., *The New York Public Library Science Desk Reference,* 483.

144. "No Fun in the Sun," Data Points, *Scientific American,* September 2000, 34.

145. P. O. Wennberg et al., "Hydrogen Radicals, Nitrogen Radicals, and the Production of O_3 in the Upper Troposphere," *Science,* January 1998, 49–53.

146. *The Science of Stratospheric Ozone Depletion,* accessed 2001, http://www.ucsusa.org/resources/ozone/science.

147. *Ozone Reality Check,* accessed 2001, http://www.foe.org/ozone/into.

148. Robert Parsons, "Re: Critical Analysis of *Environmental Overkill,* Chapters 3 and 4," accessed 2001, http://www.tp.alternatives.com.

149. Ray and Guzzo, *Environmental Overkill,* chapters 3 and 4.

150. *Common Ozone Depletion Myths,* accessed 2001, http://www.members.aol.com/jimn469897/common.html.

151. Thompson and Turk, *Earth Science,* 453–56.

152. David W. Thompson and Susan Solomon, "Interpretation of Recent Southern Hemisphere Climate Change," *Science,* 3 May 2002, 895–99.

153. Lomborg, *The Skeptical Environmentalist,* 4–5.

154. "The Earth is Doing OK, Actually," Features and Arts, *Smh.com.au.,* 21 June 2001, accessed 2001, http://www.smh.com.au/news/0106/21/features/features1.html.

Appendix: Earth Science Projects

The following projects are designed as hands-on activities for readers interested in investigating some of the concepts addressed in *The Basics of Earth Science.* The projects may be conducted at home or in school, without the need for elaborate laboratory equipment. Most of the materials are found in the home, or they may be purchased from pharmacy, craft, or hardware stores.

I. Earth's Spheres

The spheres or layers of the Earth are defined by their chemical composition and/or other physical characteristics. The actual separations between layers are not sharp divisions but instead are transition zones of different chemicals consisting of different densities under differing pressures and temperatures. The following projects are designed to provide a model cross section depicting several of the major spheres of the Earth.

Project 1: Inside the Earth

• Materials

1. One or two hardboiled eggs, with shells intact
2. Green or brown soft crayon
3. Small marble or ball bearing
4. Sharp knife or fine-tooth hacksaw

• Procedures

1. Gently heat crayon and rub on entire outside surface of warm hardboiled egg. Be careful not to break shell, and be sure to leave a thick but rough layer of crayon on the surface. Surface wax should be rough.
2. When egg is cooled, slice crosswise in the middle (see Figure A.1), being careful not to destroy the shell.
3. Scoop out a very small center portion of the yoke and insert small marble or ball bearing.

• Conclusions

Compare spheres of cross section of your egg with Figure A.1. Can you identify the (1) crust (crayon layer), (2) lithosphere (shell), (3) asthenosphere (membrane inside shell), (4) mantle (egg white), (5) outer core (egg yolk), and (6) inner core (marble)?

Figure A.1
If the egg is cut smoothly, the cross section of its outer and inner layers somewhat resembling the Earth's spheres are visible.

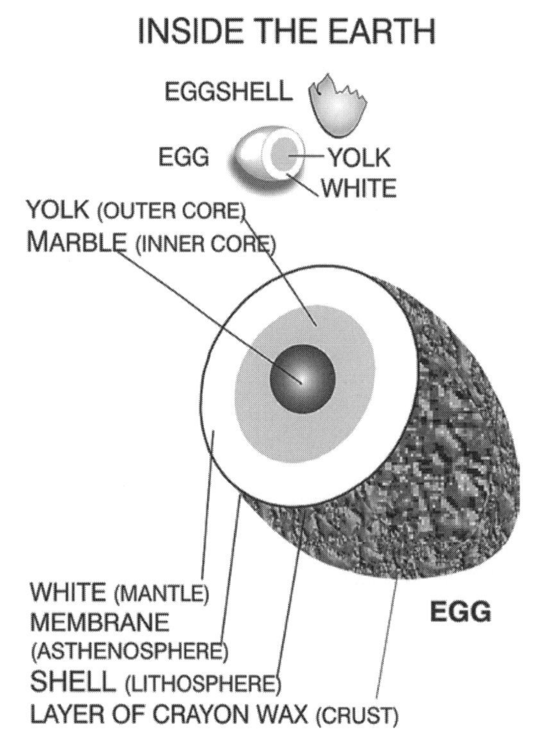

Project 2: Earth's Sphere

• Materials

1. One piece each of blue, green, brown, red, and yellow soft modeling clay
2. Metric ruler
3. Sharp knife or clay wire cutter
4. Pencil and paper (or optional hand-held calculator)

• Procedures

1. Using the following list of approximate thickness for each sphere of the Earth, calculate the proportional differences.

Crust	~25 km
Lithosphere	~75 km
Mantle	~2,900 km
Outer Core	~2,200 km
Inner Core	~1,200 km

   Earth's radius: ~6,400 km × 2 = ~12,800 km (Earth's diameter)
2. Hint: Divide each of the larger figures by the smallest figure to determine the relative thickness of each sphere to the others. Two examples: (A) Compare the relative thickness of the lithosphere to the crust. (Divide the thickness of the lithosphere by the thickness of the crust, i.e., 75 km divided by 25 km.) Thus, the lithosphere is about 3 times as thick as the crust. (B) If you divide the thickness of the mantle (2,900 km) by the thickness of the crust (25 km) you can see that the mantle is about 116 times thicker than the crust. Complete the calculations for each sphere. Hint: Reduce your final figures to manageable sizes for forming your concentric spheres. You do not want to make your final clay Earth too large.
3. Construct your spheres out of clay according to the proportions you calculated for each layer.
4. Work each colored clay stick separately with your hands until the clay warms up.
5. After the clay is soft, use the proportions calculated in step 2 and use the yellow clay for the inner core, the red clay for the outer core, the brown for the mantle, the green for the lithosphere, and the blue for the crust.
6. After forming one sphere around the other according to your proportional figures, allow the clay to cool (10–15 minutes in the refrigerator will cool it adequately) and then cut your Earth in half with a clay wire cutter or knife.

• Conclusions

Does the cross section of your clay structure clearly represent the inner and outer spheres of the Earth? Compare your cut cross section of the clay spheres with that of Figure A.2. What conclusions can you make about the relative sizes of the Earth's spheres?

Figure A.2
If the general proportions for the layers of the clay Earth are approximate, the hemisphere somewhat resembles the layers of the Earth.

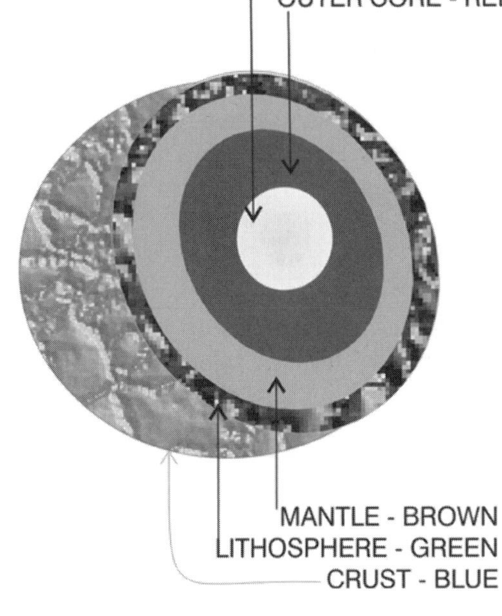

Project 3: Core Samples

• Materials

1. A layer cake made with several layers, each layer with different colored cake mixes: yellow or lemon cake mix, dark brown cake mix (dark chocolate or fudge), light brown (ginger or spice), white cake (any white), plus already prepared white icing
2. An 8–12″ long, 1/4″ or 3/8″ diameter clear plastic tube or a stiff plastic straw (a clear plastic straw is preferred, if available)

• Procedures

1. Mix different colored cake mixes in separate bowls as per instruction on the box. It may not be necessary to use the entire contents of each box.
2. Construct your layer cake in a small baking cake pan.
   (a) Start with a thin layer of yellow cake mix about 1/2″ thick.
   (b) Add a layer of light brown cake mix about 3/4″ to 1″ thick.
   (c) Follow this with a thin layer, about 1/2″, of dark brown mix.
   (d) Last, add a very thin layer, 1/8″, of white cake mix.

   The cake's layers before baking should approximate the following:

   Bottom Layer: 1/2″ yellow cake.
   Next to bottom: 3/4 to 1″ light brown cake.
   Next layer: 1/2″ dark brown cake.
3. Bake the cake according to instructions on the box. Note: When cake is baked, the batter will rise so the layers will be thicker than the unbaked batter you placed in the pan.
4. When cake is baked and cooled, top off with 1/8″ white icing. After icing has been added, press tube or straw down through top of cake (see Figure A.3).
5. Carefully withdraw tube or straw after penetrating to the bottom of the cake.
6. When using an ordinary straw, use scissors or sharp knife or razor blade and gently slice open the end containing the core sample. If you use a clear plastic tube or straw, you do not need to cut it open to see the core sample.

• Conclusions

Examine the core sample that indicates the different layers of material within the pastry. What you did was similar to the process geologists use to take core samples of soils, rocks, and sediments on the ocean floor. The same general procedure is employed when taking core samples of glacier ice, as well as boring cores through tree trunks and ocean bottom sediments. (You may now eat the cake.)

Figure A.3
The different colored strata of cake inside the straw are representative of the sedimentary strata of the Earth's crust. This is somewhat similar to what cores drilled into the Earth resemble. So far, drills have not reached the deeper spheres below the crust, and what evidence we have of these inner regions is determined by indirect means, including seismology.

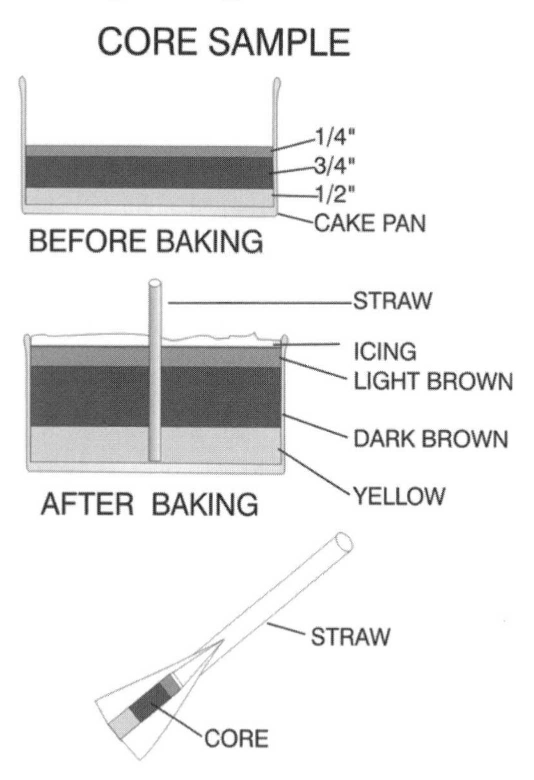

II. Rocks and Minerals

The upper mantle, lithosphere, and crust are made of many minerals with varying chemical and physical compositions. Rocks are aggregates of one or more minerals or other geological materials. As with all aspects of the Earth, rocks go through a cyclic sequence of events, such as erosion, transportation, deposition, burial lithification, metamorphism by heat and pressure, uplift, weathering, and again erosion, transportation, deposition, and so forth, completing the cycle. The density and specific gravity of minerals and rocks depends on their chemical and physical composition. Geologists use specific gravity as one way to arrive at the densities of minerals because specific gravity relies on a simple water displacement method that is easy to use. Specific gravity is a ratio. It compares weight of an object with the weight of an equal volume of water. To calculate density (mass/volume), it is necessary to determine the volume of a solid object. This is usually more difficult to determine than the volume of water the object displaces.

Note: Before starting the next several projects, it is a good idea to begin your collection of different types of minerals and rocks of different sizes. If you cannot find a variety of minerals or rocks, they are available from hobby shops or rock stores. Egg cartons are useful for storing your collection. Label or number each specimen and prepare a list of your collection. You will get a chance to use your collection more than once.

Project 4: Specific Gravity of Minerals

- **Materials**

1. A small, ~6 cm to ~10 cm (~2.5″ to ~4″) chunk of several of the following minerals or crystals: quartz, calcite (limestone or dolomite), galena (lead ore), biotite (copper ore), amphibole, pyrite (fool's gold), hematite or magnetite (iron ores), feldspar, and mica
2. A 2–3 liter (L) (2–3 quart [qt.]) wide-mouth bowl
3. A 65–75 cm (25–30″) piece of string
4. A hand-held metric spring scale (see Figure A.4 for type used, or you can use a spring scale with English units, such as ounces,

Figure A.4
This method of determining the specific gravity of an object with unknown size, shape, and density also demonstrates the buoyancy of water.

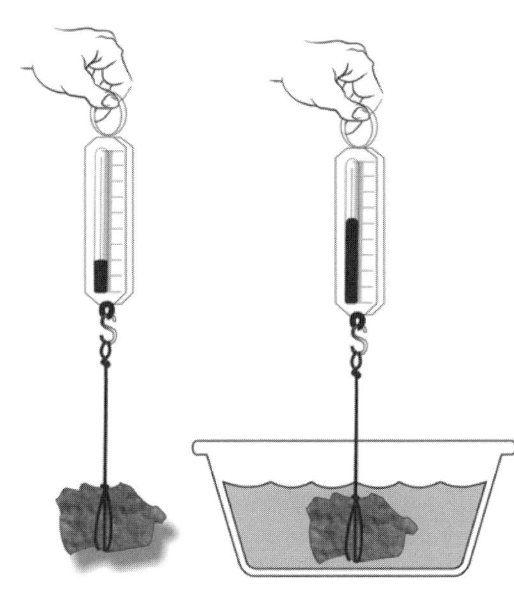

since the values calculated for specific gravity are the same no matter what system is used to measure weight)

5. Enough water to fill the bowl 3/4 full
6. Pencil, paper, and pocket calculator (if desired)

• Procedures

1. Tie one end of the piece of string securely around a chunk of one of the minerals. Tie the other end of the string to the bottom hook of the spring scales.
2. Take a reading of the weight of the chunk of mineral as it hangs freely from spring scales. Record your reading.
3. Place bowl on paper or cloth to protect table. Fill bowl 3/4 full of water.
4. While holding spring scales above bowl of water, slowly dip mineral until is it entirely submerged. Be careful not to let it touch sides or bottom of bowl.
5. Take a reading of the weight of rock as suspended under water. Record this figure.

• Results

The next step involves some calculations to determine how much water (mass) was displaced by the mass of the chunk of mineral. The equation for determining specific gravity is as follows:

$$\text{sp. gr.} = \text{Mass}_1 \div \text{Mass}_2 \text{ (or sp. gr.} = \text{Weight}_1 \div \text{Weight}_2\text{)}$$

Mass_1 is the weight of the mineral before placing in water.

Mass_2 is the weight of the mineral after placing in water.

A typical example is

$$(M_1)\ 497\text{ g} \div (M_2)\ 186\text{ g} = 2.67 \text{ (the specific gravity of the mineral tested)}$$

If you have access to chunks of several other minerals, go through the same procedures for determining the specific gravity of each. Record and compare your results.

• Conclusions

Regardless of the shape or volume of the mineral, when it is completely submerged, it replaces an equal volume of water. Specific gravity is a measure of the relative density of objects. But, unlike density (*m*/*v*), specific gravity is the ratio of the density of a specified material to the density of some standard material, for example, water. In other words, you can determine the density of an object in two ways, by dividing the mass of the sample by its volume, or by flotation to compare the weight of the sample with the weight of an equal volume of displaced water.

This project compares the given weight (mass) of a chunk of mineral to the weight (mass) of an equal volume of water that is displaced by the volume of the mineral being tested. Therefore, the example given for the sample mineral weighs 2.67 times as much as an equal volume of water. Since specific gravity IS the ratio of the density of a substance compared to the density of water, and since water's density is 1 gram/cubic centimeter (g/cm^3), therefore a mineral with a specific gravity of 2.67 also has a density of 2.67 g/cm^3. Most common minerals have a specific gravity between 2.5 and 3.5, or you can say most common minerals have a density between 2.5 g/cm^3 and 3.5 g/cm^3. Metals have significantly greater specific gravities (or densities) than do most minerals. Table A.1 gives examples of minerals and metals and their specific gravities (densities).

Table A.1
Specific Gravity of Some Minerals and Metals

Mineral	Sp. Gr. (density)
ice	0.9
water	1.0 (standard at 4°C)
graphite	2.2
granite	2.2
feldspar	2.6
quartz	2.7
calcite	2.7
mica	2.9
dolomite	2.9
pyrite	5.1
magnetite	5.2
galena	7.5
Metal	**Sp. Gr. (density)**
copper	8.9
silver	10.5
lead	11.3
mercury	13.6
gold	19.3

Geologists also use a subjective method to determine the approximate specific gravity or density of a mineral, rock, or metal. This is called the Heft method. In other words, you pick up an object of unknown specific gravity and by holding it in one hand, compare its heft with that of a known sample (of approximately equal size) in the other hand. The Heft method is not exact, but it is a quick-and-dirty way to determine the approximate density of a sample mineral.

An alternate method that can be used to determine the density, rather than the specific gravity, of a rock is to weigh the volume of water that is displaced by the rock's volume. Can you figure out a way to collect the water displaced by the rock? Once you do this, you can divide the mass (weight) of the rock by its equal volume of water ($d = m/v$). This is the method that Archimedes (287 B.C.E.–212 B.C.E.) used to ascertain whether his king's crown was solid gold or a mixture of gold and other metals.

Project 5: Mohs Hardness Scale

In 1822, Friedrich Mohs (1773–1839), a German mineralogist, devised a hardness scale, based on whether the sample was resistant to scratching or abrasion, to assist geologists in identifying minerals and rocks. His ten-point scale ranges from 1 for the softest mineral (talc) to 10 for the hardest (diamond). Since minerals with higher numbers can scratch other minerals with lower numbers, it is possible to classify minerals by the Mohs hardness scale.

• Materials

1. Samples of several minerals (use those from your collection or left over from Project 4)
2. A copper penny, an iron nail, a paper clip, a penknife, a small piece of broken window glass, sand paper, a steel file, grinding wheel

or corundum emery cloth, and a diamond ring (if possible)

3. Pencil and paper

• Procedures

1. Select the chunks of minerals you wish to test for hardness. Select one chunk and use it to scratch one of the other samples. Hold on to the harder mineral that scratched the softer one, and then do the same with each of the other samples. Record which ones were hardest (scratched the other samples) and arrange your samples and data from the softest to the hardest.
2. Next, using the other objects you collected, record which objects were able to scratch which sample minerals. Again, record your data and arrange from softest to hardest.

Figure A.5
The relative hardness of various minerals and rocks can be established by comparing the hardness of common objects with specific samples against Mohs hardness scale.

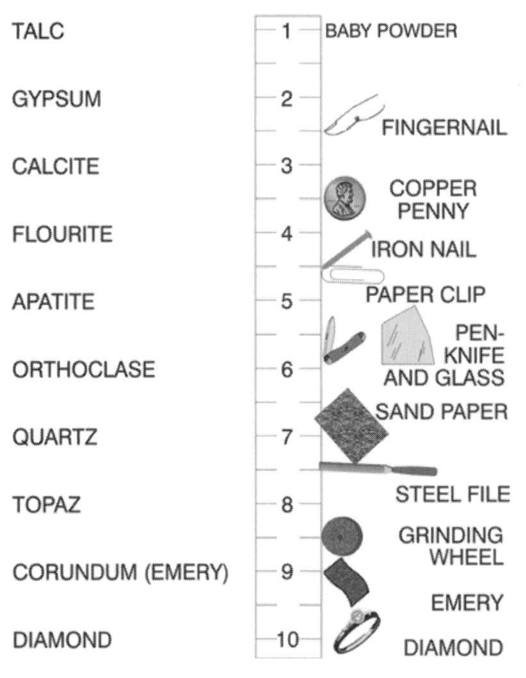

• Results

This is a somewhat subjective test because you must judge from your observations of which-scratched-which to compile a list using the Mohs scale. Compare your recorded results with Table A.2.

Project 6: The Streak Test

The streak test is related to the hardness test and is used to identify the color of the dusty powder for abraded minerals and rocks. Some of the minerals' streaks have colors with a metallic luster. The powder produced by the streak color test is often used by geologists, in conjunction with the hardness test, when working in the field. These tests provide a tentative identification for mineral and rocks in the field when a complete laboratory is not available.

• Materials

1. Several sample minerals and rocks that were used in projects 4 and 5
2. A piece of broken tile with an unglazed underside (white preferred)
3. Paper and pencil

• Procedures

1. Rub a corner of the mineral or rock across the unglazed tile. A streak of powder from some minerals leaves behind a powder that is not always the same color as the mineral.
2. Record your results correlating the colors with the mineral and rock samples.

• Results

The color of the streak is usually more reliable in identifying the mineral and rocks than just looking at a chunk of the sample. All minerals with the same composition produce the same color streak. Enter your data for the streak colors of the different minerals and rocks in a table similar to Table A.3. (Note: This is a good time to also check each

Table A.2
Mohs Hardness Scale

Minerals and Objects	Relative Hardness
*Talc ($Mg_3Si_4O_{10}(OH)_2$	1 softest
Baby powder	1
*Gypsum ($CaSO_4 \bullet 2H_2O$)	2
Fingernail	2.5
*Calcite (dolmite) ($CaCO_3$)	3
Copper penny	3.5
*Fluorite (fluorspar) (CaF_2)	4
Iron nail	4.5
Paper clip	4.5
*Apatite (Ca_5)(PO_4,CO_3)$_3$(F,OH,Cl)	5
Window glass	5.5
Pen knife	5.5
*Orthoclase (a feldspar) ($KAlSi_3O_8$)	6
*Quartz (a silica) (SiO_2)	7
Sandpaper (regular silicon type)	7
Steel file	7.5
*Topaz (gem stone) ($Al_2SiO_4(F,OH)_2$	8
Grinding wheel (used to sharpen knives)	8.5
*Corundum (emery) (Al_2O_3)	9
*Diamond (gem) (crystalline carbon)	10 hardest

Note: Samples identified with * are the minerals used originally by Mohs to establish his hardness scale.

rock and mineral for its magnetic properties.) Compare your recorded figures for the colors from the streak test with Table A.3.

• Conclusions

The heft, hardness, and streak tests are all subjective tests that geologists use in the field to aid in identifying minerals. They also identify minerals and rock in the field by observing their shapes and their textures (rough, dull, smooth, shiny), as well as the color of their ores and the geological formation in which they are found. Another test that can be used in the field is the acid test. More definitive chemical and physical tests are later conducted in the laboratory to arrive at more exact results.

Project 7: The Acid Test

Another way to identify and classify rocks and minerals is to test how they react to weak acids. As you recall, weak acids, such as carbon dioxide gas dissolved in water (carbonic acid: H_2CO_3), vinegar (acetic acid: CH_3COOH), and sulfur oxides dissolved in water droplets (weak sulfuric acid: H_2SO_4), react slowly with some types of rocks (i.e., limestone and dolomites) to form karsts (caves and sinkholes) and, over time, dissolve the surface of monuments and stone buildings. If available, and under supervision in a laboratory, you could substitute weak hydrochloric acid (HCl) or weak sulfuric acid (H_2SO_4) for the vinegar in this project.

Table A.3
Color Streak and Magnetic Properties Tests

Rock #	Mineral	Color	Magnetic?
1			
2			
3			
4			Etc.

Color Key:

Color	Mineral's Name	Magnetic?
light blue	azurite	
green	malachite & olivine	
greenish-yellow	pyrite	
bright red	cinnabar	
reddish-brown	hematite	
yellow-brown	limonite	
grayish	galena	
black	magnetite	
		Etc.

• Materials

1. Several small igneous, sedimentary, and metamorphic rocks from your rock collection (if you do not know which is which, just select several different looking rocks from your collection)
2. Marble chips, a small piece of limestone, and calcite (you may wish to include a small piece of chalkboard chalk)
3. A 16 or 24 ounce (oz.) bottle of distilled white vinegar (vinegar works well and is less hazardous than acids)
4. Several clear glasses large enough to hold one of the rocks covered with vinegar (4 to 6 oz. juice glasses are excellent, but make sure your rock samples are small enough to fit into the containers)
5. An eyedropper
6. A metal file or sandpaper
7. Paper and pencil

• Procedures

1. Number each glass.
2. Scratch the surface of each sample with a file or sandpaper to expose a fresh surface before placing it in a glass.
3. Place a different sample with the exposed scratched surface facing up in each glass.
4. Using the eyedropper, place several drops of vinegar or acid onto the exposed surface of each sample. Observe results.
5. Next, cover the surface of each sample with the fluid. Observe results.
6. Once the samples are completely covered, let them set for a few days.
7. Again, observe and record what occurred for each sample after one week.

• Results

Using your recorded results, organize the samples according to their reactions to the acid. Can you identify each sample as belonging to one of the three main groups of rocks, or types of minerals, or crystals structures? What conclusions can you make from your results?

Figure A.6
Use an unglazed tile or a piece of uncolored concrete to perform the streak test. Determining the color of the powder produced by scratching a piece of mineral or rock across the tile assists in identifying specific mineral samples.

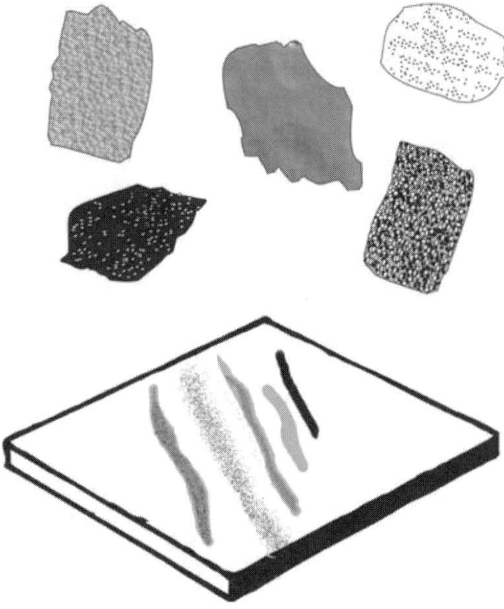

Figure A.7
Although dilute hydrochloric acid (HCl) works best, vinegar can be used for the acid test. Be sure to use caution when using acid. Construct a table indicating the reactions of each sample over specified periods of time.

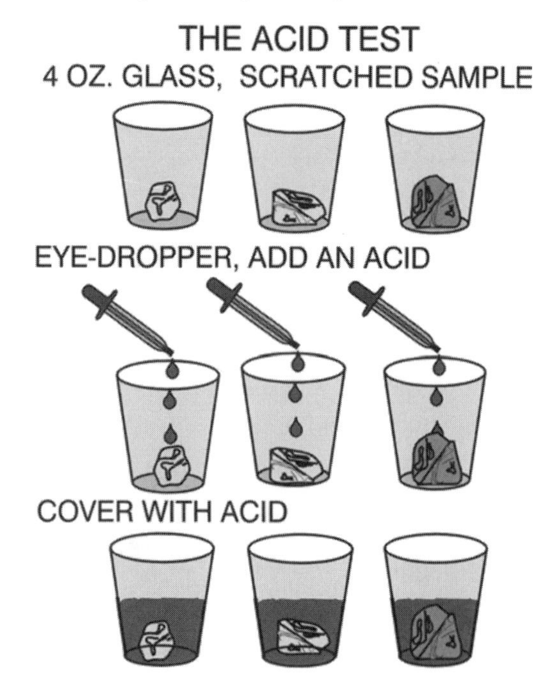

Project 8: Growing Crystals

Crystals can be composed of a single chemical element or a mixture of several elements whose atoms are arranged in such a way as to express different axes or surface planes. Crystal axes are three or four imaginary lines that pass through the crystal's center. These lines are used to describe the crystal's symmetry and structure. The plane structure of crystals determines how they might split or just fracture without splitting. Crystals are structured in orderly geometric arrangements of atoms (or molecules) that are repeated over and over, forming a crystal lattice with repeated space within the lattice. Crystals are divided into six major systems (see Figure 4.2 in chapter 4) that, in turn, consist of 32 possible combinations of symmetrical arrangements. Minerals (and rocks) exhibit particular crystalline types and shapes depending on the elements involved and the conditions of pressure and temperature that existed during the formation of the crystal. Some examples of minerals that grow as crystals are quartz, calcite, fluorite, halite (table salt), garnet, asbestos, kyanite, galena, and pyrite. Many gemstones are also crystals, for example, topaz, sapphire, and diamond. Many rocks are composed of an interlocking variety of mineral crystals formed from several different elements or compounds. Some examples are granite, marble, and limestone.

The following project involves growing a crystal of a mineral called alum, which is a group of minerals containing hydrous aluminum sulfates ($KAl(SO_4)_2 \cdot 12H_2O$).

• **Materials**

1. A 2 L (or 2 qt.) sauce pan that can be used to boil water
2. A stove or burner to boil the water
3. A 1–1.5 L (or 1–1.5 qt.) wide-mouth glass jar
4. 1 L (1 qt.) water (use distilled or bottled drinking water)
5. 90 g (3 oz.) of powdered alum
6. Small rock (about 30 g or 1 oz.)
7. String (about 25 cm or 10″ long)
8. A thin stick or pencil about 15–25 cm (6–8″) long
9. Magnifying glass or low-power microscope

• **Procedures**

1. Heat the 1 L (1 qt.) of bottled water in saucepan until it boils.
2. Remove from heat and slowly stir in the powdered alum until it completely dissolves.
3. Pour warm (not boiling) alum solution into wide mouth jar.
4. Tie one end of string securely around small stone.
5. Tie other end of string around the center of the stick or pencil. When placing the stick/pencil across top of jar, make sure the string is just long enough to suspend stone in water without touching the bottom or sides of jar (see Figure A.8). Ideally, the stone should be suspended at just below the center of the water in the jar.
6. Observe growth of crystals that collect on the string and the surface of the suspended stone as the liquid cools. Also, observe any growth over a period of several days. Keep a record of your observations. When the crystals seem to have stopped growing, remove the string and stone from the water.
7. Let crystals drain. Observe under magnifying glass or microscope.

• **Results**

Make a drawing of your crystals and compare with Figure 4.2 in chapter 4. Which of the six major types of crystals do the alum crystals resemble? Taste the crystal by just touching it to your tongue. Make a record of your observations of the crystals and how you would describe its taste.

This is a good time to examine the other solid minerals you used for the specific gravity and hardness tests with a magnifying glass or microscope. Can you identify the types of crystals by their structures?

Figure A.8
A number of salts in solution can be used to grow crystals. Try experimenting with table salt, sugar, and other materials. Not all solutions form crystals. Compare your crystals with the various types shown in Figure 4.2 in chapter 4.

FORMATION OF CRYSTALS

SOLUTION OF POWDERED ALUM WATER IN WIDE-MOUTH JAR

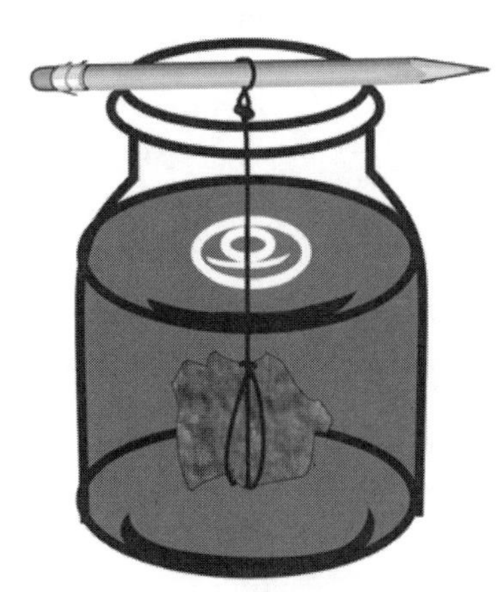

SMALL ROCK SUSPENDED ON STRING SECURED BY PENCIL

• **Conclusions**

Alum is a colorless or white mineral with a sweet-sour astringent taste. The astringent quality of alum aids in treating cold blisters by shrinking and drying up the lip sore. It is also used in some cooking and food preparations.

Can you think of other minerals in powder form that might form crystals in a similar manner? Collect a few samples and try them out. Hint: Try dissolving several tablespoonfuls of table salt (NaCl) in a small amount of water in a saucer-type dish. Set the dish aside and leave undisturbed until the water evaporates. Try the same experiment with magnesium sulfate, also known as Epsom salts ($MgSO_4 \cdot H_2O$), and common table sugar. Observe the resulting crystals with a magnifying glass and compare with crystal structures in Figure 4.2. Based on your observations and your data, make some conclusions about the formation of crystals during the formation of the Earth.

III. Earth's Forces

As mentioned in chapter 2, the dynamic nature of the Earth is driven by a multitude of natural chemical and physical forces that conform to universal scientific laws. Although we cannot see the results of many of these forces (tectonics) within a single lifetime, we can observe and experience the forces of nature that result in earthquakes and volcanic activity, as well as climate and weather. Plate tectonics is based on the power of these forces, over eons of time, moving continental plates across the face of the Earth so that the current geography of the globe is not the same as it was many millenniums ago.

Project 9: Continental Drift and Pangea

The Austrian geologist Eduard Suess (1831–1914) developed a theory that the southern continents were once combined as a supercontinent. He based his theory on the shapes of the coastlines of the continents in the southern hemisphere, along with ancient plant fossils found there. He named this supercontinent "Gondwanaland," after the Gonds of ancient India. About 50 years later the German meteorologist and mineralogist Alfred Lothar Wegener (1880–1930) advanced Suess's theory by including the northern as well as the southern continents into a more comprehensive arrangement of landmasses that he called "Pangea." Wegener's theory included the concept of continental drift, which was his explanation for the drifting of landmasses to form their present global continental configuration.

• Materials

1. About 6–8 pieces of card stock paper or light cardboard (if a jig saw is available, use 1/8″ thick plywood)
2. Scissors
3. Tracing paper, pencil
4. Jig saw (optional)

• Procedures

1. Trace the continents onto thin tracing paper and then transfer to heavy card stock or thin plywood (Use Figure A.9 or a map to trace a copy of individual continents.)
2. Cut out each continent.
3. Place the pieces on a table. Start by arranging the continents in the positions they were assumed to be to form Pangea. Make a diagram of this arrangement on a sheet of paper. (Use Figure A.9 or a reference book that provides a picture of Pangea.) Then slowly move the continents over the surface of a table to form the approximated present day configuration of the continents (make another diagram of this configuration). Label your diagrams appropriately.

• Results

Compare your two diagrams, and from what you know about continental drift and plate tectonics, try to think into the future and move each continental piece to where it might end up millions of years from now.

Figure A.9
Trace the outlines of the continents on a piece of paper. Cut out the shapes and then transfer each to heavier cardboard or thin plywood. When completed, rearrange the continents to resemble Pangea, the current configuration of the continents, and your prediction of the future positioning of landmasses.

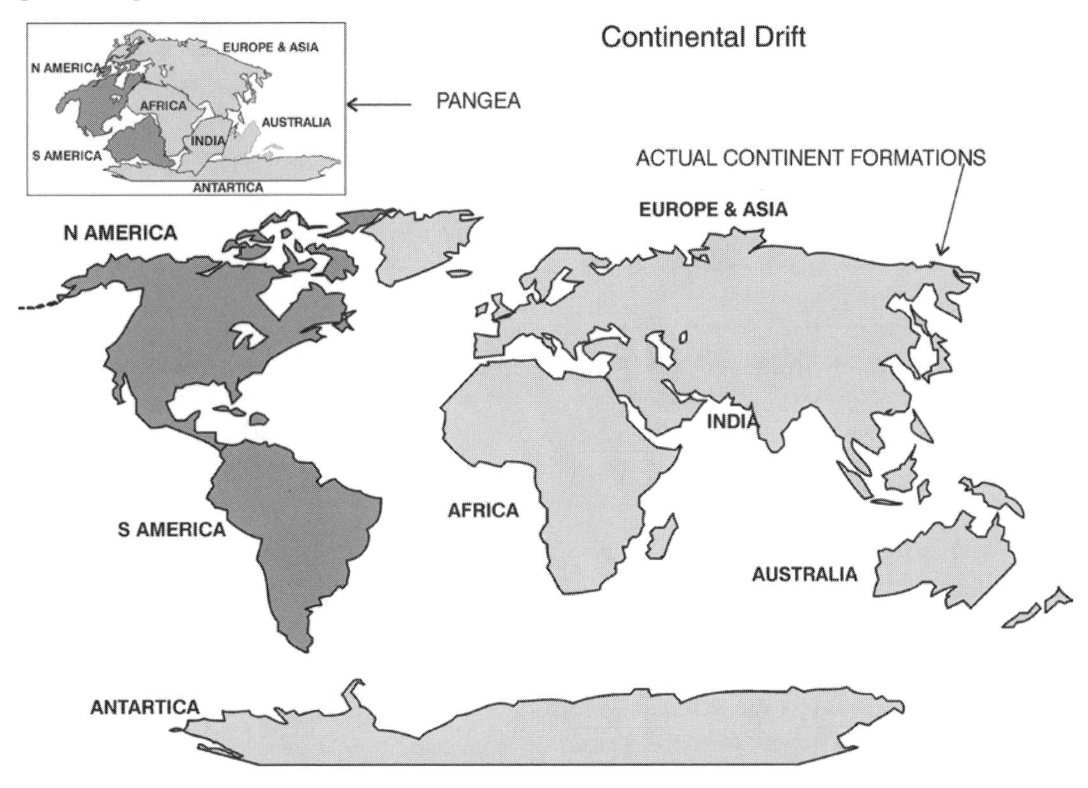

Make a third drawing predicting the future configuration of the continents on the surface of the Earth.

- **Conclusions**

Compare all three drawings that depict the movements of the continents. What conclusions can you draw about the dynamics of the Earth's oceanic and continental crusts?

Project 10: Mid-Atlantic Oceanic Ridge

As mentioned in chapter 2, as the tectonic plates deep in the Earth slide and move over the asthenosphere, they tend to run into each other, causing slow but steady disruption to the Earth's surface. In addition, as molten rock pushes upward, it tends to break through the thinner bottom of the Atlantic Ocean and forms a ridge as the new rock material spreads out both westward to the continents of North and South America and eastward toward Europe and Africa. In doing so, these continents have been pushed further apart over many thousands of years as the tectonic plates continue to move. These dynamic forces result in changes on the Earth's surface that cause long-term buckling to form mountains and, over much shorter time spans, earthquakes and volcanoes.

• Materials

1. An old shoebox, or similar rectangular box
2. One piece of legal size bond paper (8.5 × 14″)
3. Scissors

• Procedures

1. Turn the shoebox upside down. At the center of the box, cut a 5″ × 1″ slot across what was the bottom of the box (see Figure A.10 for direction of cut).
2. On one long side of the box cut an opening about 4″ × 6″, or large enough to stick your hand inside the box (see Figure A.10.1).
3. Cut legal size paper vertically in half to form two 4.25″ × 14″ strips.
4. Place one strip of paper on top of the other. Then staple just one end of the two strips together.
5. Place the stapled end of the two strips of paper down through the top slit, with each of the free ends extending toward and over an end of the box.
6. Put one hand in the side hole of the box, holding the stapled ends of the two strips down near the bottom of the box.
7. Slowly push the two strips of paper up through the slot in the top of the box so they bulge up but spread out toward the ends of the box.
8. Observe what happens to the paper strips as they emerge from the slit on the top of the box, and what happens to the ends of the strips that extend past the ends of the box.

Figure A.10
Part A demonstrates how magma rises and spreads on the ocean floor, forming a rift valley (the Mid-Atlantic Ridge). As you push the paper strips upward and outward (representing the rise of magma), the paper is pushed outward where it sinks over the ends of the box. Part B represents this continuing geological activity that pushes the North American and European continents farther apart.

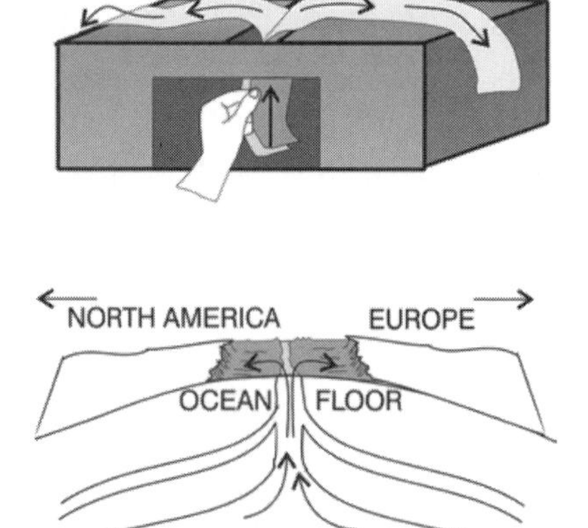

• Results

Write down what you observed. How and why did the paper strips first push up as you raised them out of the box through the slit? What does this represent? What happened to the paper strips as they extended past the ends of the box? What geological phenomenon does this represent?

• Conclusions

The paper pushing up through the slot on the top of the box represents the hot molten rock material that pushes up through the weak region of the Atlantic Ocean seafloor, forming the mid-oceanic ridge (the slot in the top of the box). The paper spreading outward from the top slot represents the new material that pushes older ocean crust material outward, spreading in both directions (east and west) perpendicular to the ridge. The spreading of the paper strips across the top of the box represents the older portions of the ocean floor that are slowly, but constantly, moving away from the central ridge, while the portion of the paper strips dipping over the ends of the box represent tectonic plates sliding under (subduction) the continental landmasses.

Project 11: Folding Rocks

As surface rocks erode, they are broken down into smaller and smaller pieces to form sand, clay, soil, and, in general, layer on layer of sediments in low areas and on ocean bottoms. The horizontal structure of these sediment layers is quite evident when deep cuts are made during the construction of mountain highways. Sometimes the horizontal layers are slanted, even at steep angles, indicating that something very powerful must have changed their horizontal orientation. The great temperatures and tremendous pressures within the Earth create heat and energy that force massive movement of the crust near the surface of the Earth. The movements of the tectonic plates are very slow, but steady, and can cause the surface to uplift, slip sideways, drop, or move in some other fashion.

There are two projects for Project 11, Project A and Project B.

• Materials for Both Project A and Project B

1. About 25–30 sheets of colored paper (try to use at least 4–6 different colors of paper)
2. Modeling clay of two or three different colors
3. A flat surface on which to work
4. A rolling pin or long glass jar

• Procedures for Project A

1. Stack several sheets of paper of one color onto several sheets of paper of a different color. Make sure the edges are even.
2. Place one hand at each end of the stack of mixed colored paper.
3. While pressing down slightly, slowly push the stack of paper toward its center to make it bulge (see Figure A.11). Observe and record what happens.

• Procedures for Project B

1. Use the rolling pin or jar to roll out and flatten each colored piece of clay into thin sheets.
2. Place a different colored sheet of clay alternately on top of each other. Press down on top layer to make sure colored layers adhere to each other.
3. Slowly push outer ends (long-wise) of clay sheet toward the center of the stack. Observe and record what happens.

• Results

Both Project A and Project B exhibit *stress* and *compression* and demonstrate *folding* of horizontal sedimentary rock layers. Make diagrams and describe your results.

Figure A.11
Projects A and B are two different ways to demonstrate the pressure on stratified rock layers that creates a great variety of land features by squeezing, bending, and folding of the lithosphere and crust.

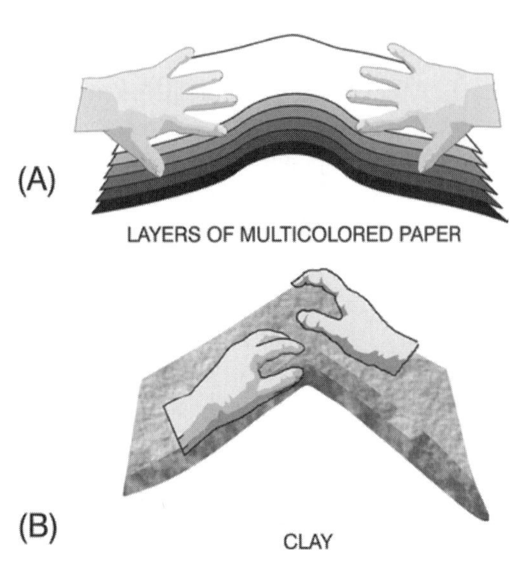

• Conclusions

What can you conclude from this activity? Can you manipulate the stack of colored paper or sheets of clay to create other formations? The bulge in the center of your paper or clay is called an *anticline.* Can you produce a downward *syncline?* Anticlines form ridges on the surface of the Earth as well as high mountains, which at one time must have been layers of sediments on the ocean bottom since some of them contain marine fossils.

Project 12: Earthquakes, Seismograph, and Seismogram

As the continental plates slowly shift and move over and under each other, or sideways to each other, they build up pressure that, in time, is released by a sudden shifting of their positions. During this shifting the joints between these adjacent plates may create a force great enough to disturb the surface, causing an earthquake (see chapter 2 for details). Geologists use an instrument called a seismograph that is sensitive to slight motions on and near the surface as well as deep inside the Earth. The seismograph produces a seismogram, which is a printed record that relates the degree of the Earth's motion to the Richter scale (see Table 2.2 in chapter 2 for a copy of this scale). Modern seismological instruments are much more sensitive than the ones that simply use a suspended pen over a piece of paper.

• Materials

1. A 3 or 4 foot strip of adding machine tape (about 2.5–3″ wide)
2. A round oatmeal box or some similar cylindrical, flat bottom container with an open top
3. Double-stick tape
4. A small piece of smooth plastic, glass, or metal that can be taped to the bottom of the box with double-stick tape (The exposed bottom surface of the sheet of plastic or glass or metal must be very smooth. Also, be sure the double-stick tape is only used between the bottom of the box and the surface of the plastic or glass plate that is attached to the box. It is important that the entire box slip and slide with little effort. Test it on a piece of paper on top of the table you plan to use.)
5. A marking pen, rubber bands, and sand or small stones
6. A small table that is not too steady (a card table works just fine)

• Procedures

1. Using rubber bands, fasten the marking pen upright on the outside of the round box. You may also wish to use some tape to secure the pen. Make sure the felt tip of the pen is touching the surface at the outside bottom of the box.
2. Fill the box partially full with sand or gravel. (You may need to adjust the weight in the box, depending on the smoothness of its bottom.)
3. Place box on top of table. One end of paper tape should be underneath the bottom of the box, with just a bit showing.
4. Very slowly pull this short end of the paper tape toward you. (Make sure the marker is making a straight line on the tape.)
5. As you continue to slowly pull the tape toward you, start shaking the table gently and then more vigorously. (Make sure the felt tip of the pen is recording the motion. If it is not making any marks, lower the pen a bit.) This is a difficult project and may require some adjustments and trial-and-error work on your part.

• Results

Note: The weighted box should not move as the table is jiggled. If it does, redesign your experiment so that the table

Figure A.12
This is a crude setup to demonstrate how actual seismographs record an earthquake's internal disturbances. Some adjustments to the proposed design for this project may be necessary for it to simulate seismic activity.

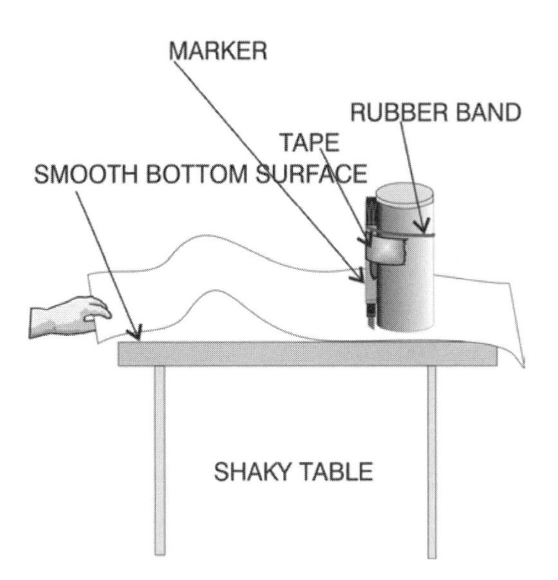

and paper tape move together while inertia keeps the box steady.

When everything is working correctly, the tape should show a straight line, followed by a curvy line, which becomes more zigzag as the vibration of the table increases.

• Conclusions

It should be obvious that a real seismograph requires a design that can maintain its at rest inertia as the Earth moves. Today, it is possible to detect and distinguish between the different types of vibrations as they pass through these inner layers of the Earth at different speeds. This aids in determining the density and chemical/physical composition of these inner spheres. Even earthquakes, as well as underground nuclear explosions on the other side of the Earth, can be detected.

Project 13: Volcano

Most volcanic activity on the surface of the Earth occurs as new magma forms mid-oceanic ridges, resulting in seafloor spreading as well as deep-sea volcanoes. Volcanoes also result at the edge of subduction zones as ocean plates slide under continental plates. These weak spots on the boundaries of tectonic plates act as routes for the molten magma from deeper spheres to escape through the surface. Magma consists of molten rock under high pressure and temperatures, and thus it is also of less density than more solid rock material. Therefore, plumes of magma follow the path of least resistance as they are pushed upward toward the surface of the Earth, where they may erupt in a violent explosion of rocks, ash, and gases or seep out as molten lava. Plumes of magma also form deep in the mantel before they push upward through cracks in the lithosphere and crust.

There are several ways to construct a model volcano. Project 13 provides instructions for two different models, Volcano A and Volcano B.

• Materials for Volcano A

1. Large pan in which to place volcano
2. A second pan in which to mix ingredients
3. Paper strips torn from newsprint to make papier-mâché
4. A test tube about 1–1.5″ in diameter and 6″ long (a small juice can open at one end also works well)
5. Baking flour, water
6. Rubber hose about 1/4″ in diameter and 2–3 feet long
7. Two or three cups of Puffed Rice

• Procedures for Volcano A

1. Mix flour with water to make paste, slurpy but not too thick.

2. Form a cone using paper strips into the shape of a volcano with the test tube, or small juice can open at one end, down the center to form a vertical hole in the cone. Let dry, then remove tube from center of cone.
3. Make a horizontal hole near the bottom of the cone with a pencil into the central hole that was formed when you removed the larger tube.
4. Insert rubber tubing through this 1/4″ hole into the bottom of the hole in the cone.
5. Mix a small amount of the dry flour with Puffed Rice in a bowl.
6. Fill the opening in the top of the volcano about 3/4 full with the flour/Puffed Rice mixture.
7. Blow into the exposed end of the 1/4″ rubber/plastic tubing, gently at first, then vigorously.

• Results for Volcano A

The explosion of the lighter flour first, and then the heavier Puffed Rice, out of the cone is somewhat like the eruption of a volcano but without the gases and molten lava.

• Materials for Volcano B

1. Several sticks of modeling clay
2. A test tube about 1″ or 1.5″ in diameter about 6–8″ long (a small juice can open at one end also works well)
3. About a 1/2 cup of baking soda and 8 oz. of vinegar

• Procedures for Volcano B

1. Form a cone using clay with the test tube or juice can in the center of the cone with the top of the tube or can even with the top of the cone. When cone is completed, DO NOT remove the test tube or juice can from the clay volcano.
2. Place cone in larger pan with newspapers spread out over the table for protection.
3. Fill tube about half full with baking soda.
4. Pour about 1 oz. of vinegar into tube and stand back.

• Results for Volcano B

The reaction of vinegar (acid) with the baking soda (base) causes a chemical reaction releasing carbon dioxide, which rapidly expands and pushes the contents of the volcano over the sides of the cone.

• Conclusions

Although the models of the two volcanoes do not accurately depict a real volcano, the first one gives an idea of how rocks and particles are spewed out of the cone. The second volcano provides an example of how gases force molten lava to flow over the sides of the cone.

Figure A.13
These are two different designs for constructing a model volcano. Volcano A uses papier-mâché made out of newspaper and flour paste; volcano B uses modeling clay. You may find it necessary to adjust the design to the materials available.

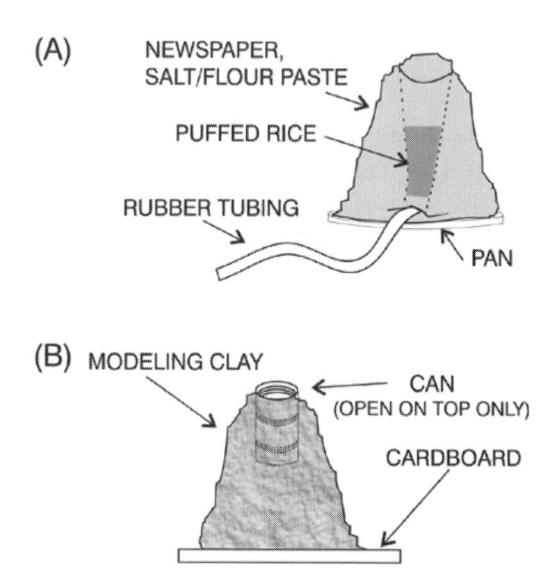

IV. Convection and Magnetic Fields

The molten magma that forms deep in the Earth's mantle is constantly being circulated toward the surface. As it cools, it sinks back deeper into the mantle to again be heated by energy produced by fission of radioactive elements, thus driving the convection cycle. The main sources of Earth's internal high temperatures are the result of fission of the radioactive elements uranium and thorium.

The temperatures, pressures, and chemical composition of the inner spheres are in a constant state of flux. The physical and chemical composition of the material that makes up the central core and lower mantle results in different rotation speeds for these inner spheres. The rotating motion of these inner spheres is one theory of how electromagnetic forces are created at their boundaries. The result is a huge magnetic field that extends far out into space from the Earth's surface.

Project 14: Convection

Convection is defined as the transport of heat energy or mass by the movement of currents, creating a diffusion of the heat or mass. Convection takes place in water and air, as well as in molten magma deep inside the Earth. The mantle acts like a huge convection cell where the rock material picks up heat from the core and becomes less dense (and thus lighter) as it rises toward the surface. If the temperature and pressure are great enough and the magma finds a weak spot in the lithosphere and crust, a volcanic eruption occurs. Most of the magma, however, cools as it approaches the lithosphere and crust and becomes denser (and thus heavier), causing it to sink back down to be heated again, thus continuing the convection current.

• Materials

1. A 1 or 2 L Pyrex beaker, or a 1 or 2 qt. Pyrex clear glass container (if clear heat-resistant glass is not available, use a metal pot)
2. About 1.5 L or 1.5 qt. of water
3. Hot plate (or similar heating device)
4. Several raw carrots and black peppercorns

• Procedures

1. Cut about 1/2 cup of raw carrots into very fine chunks (about 1/8″ or less), and mix with a tablespoon of peppercorns.
2. Place water in pot, set on hot plate, and turn on high heat. Just before boiling takes place turn the controls to a lower heat setting.
3. After water starts to move in the pot (before boiling takes place), add cut-up carrots and peppercorns.
4. Observe movements of particles in the water currents.

• Results

Make a diagram of what you observed.

• Conclusions

What conclusion can you draw from your observations? Convection provides a means for the heat energy generated by the fission of radioactive elements in the core to be diffused and then dissipated near the Earth's outer surface. As the hot magma rises it cools, thus becoming denser, causing it to again sink toward the core to be reheated, continuing the dynamics of convection deep in the Earth. Remember that convection is the movement of energy (heat) that is diffused in a medium, which can be air or water or magma.

Figure A.14
You can use any type pot and any ingredients that do not dissolve in water to demonstrate convection. Heat is one of the important internal forces of the Earth that drives convection, resulting in the movement of magma, continental drift, earthquakes, and volcanic activity.

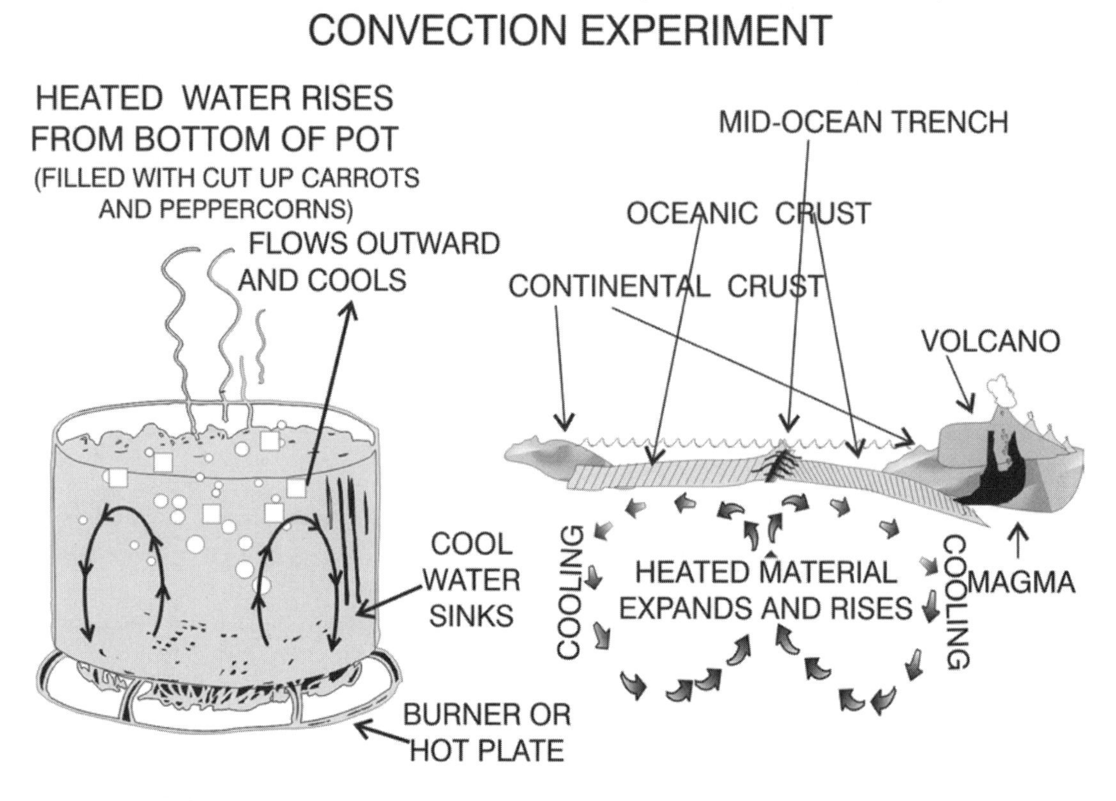

Project 15: Magnetic Fields

The inner core of the Earth consists of a mass of iron, nickel, and other metals. It is maintained in a solid state by the tremendous pressure surrounding it, even though the temperatures are high enough to melt these metals. The high temperatures are the result of radioactive fission of the unstable elements found in several of the inner spheres. The tremendous pressure exerted by the outer spheres squeezes the inner core, which increases its density, resulting in solid rather than liquid metal. The outer core portion of the central sphere is thought to be a combination of metals in a liquid state. As the Earth spins, the different cores react with each other as well, at different speeds of rotation at the boundary between the core and the lower mantle.

William Gilbert (1544–1603), the English physicist and physician, was the first to consider the Earth as a large magnetic lodestone. He constructed a globe from a lodestone that he used to demonstrate how a compass needle behaved in relation to the Earth. The closer a compass needle was to the poles of his lodestone earth, the deeper it dipped. The dipping of a compass needle had been previously noticed by sailors, but it was Gilbert who related this phenomenon of the position of the magnetized compass needle to Earth's latitude. More recent theory suggests that the different densities of the inner and outer cores of the Earth and the lower mantle cause them to spin at slightly different rates. This phenomenon and the convection of the liquid core are believed to result in the generation of a large magnetic field. This field causes the Earth to act as a giant electromagnet with north and south poles. This magnetic field surrounds the Earth, extending well above the Earth's atmosphere. The two magnetic poles, although aligned closely to the geographic North and South Poles, have changed position and even reversed themselves over eons of time. Similar magnetic fields and patterns have been detected surrounding other planets, suggesting that they, too, are composed of different spherical layers.

Magnetic fields not only surround the Earth and other planets but also the sun, which has its own inner spheres. The different rates of movement between internal spheres can create electromagnetic forces that can extend far out into space. These electromagnetic fields assist astrophysicists in determining the structures of other planets and the sun.

• Materials

1. A bar magnet (If a regular 4″ or 6″ bar magnet is not available, use the largest rectangular refrigerator magnet you can find.)
2. Iron filings (If you can't find a supply, you may be able to collect some fine iron filings from the bench in a shop where they grind iron metal.)
3. A thin sheet of window glass about 10″ or 12″ square (A thin sheet of firm clear plastic can also be used.)
4. A glass marking pen

• Procedures

1. Using the marking pen, draw a circle about the same diameter as the length of the bar magnet you are using in the center of the glass or plastic square. Then draw a straight vertical line through it representing the earth's axis, and mark the ends of the line as N and S.
2. Place some newspaper or other protection on top of the table you are using.
3. Place the glass over the bar magnet. Center the magnet under the glass with the marked N and S poles in line with the bar magnet.
4. Hold the iron filings in a container about 8–10″ above the center of the glass and sprinkle lightly to cover most of the glass.
5. Gently tap the glass with the eraser end of a pencil.

Figure A.15
The Earth, as well as some of the other planets, acts as a giant electromagnet. The magnetic particles in some rocks indicate that the Earth's magnetic poles have changed over past eons of time. Earth's magnetic field extends much further into space than is indicated in the scale of this figure.

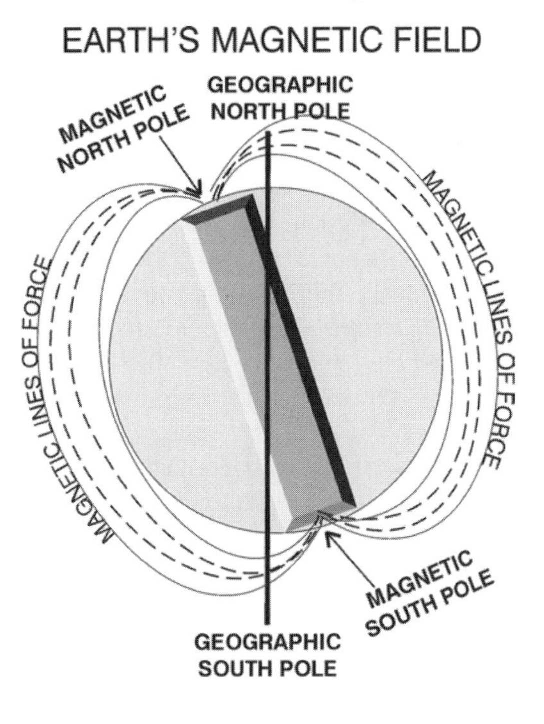

• **Results**

As you gently tap the glass (or plastic) the iron filings should align themselves with the magnetic field. Turn the glass (or plastic) to almost align the magnetic field with the poles but keep them just a bit off, since the magnetic and geographic poles are not exactly in the same position.

• **Conclusions**

The magnetic field that surrounds the Earth at great altitudes captures high-energy particles that are emitted by the sun and from outer space. These particles then follow the magnetic field and are led around the Earth to converge at the magnetic polar regions, resulting in the aurora borealis, also known as the northern lights.

Note: Can you repeat this project using a large iron nail; several feet of thin, covered wire; and a dry-cell to form an electromagnet, which is more representative of the Earth as a magnet than as a bar magnet?

V. Weathering and Erosion

In chapter 6 the physical and chemical forces responsible for weathering and erosion of rocks are discussed. One of the main results of weathering is that larger rocks are broken or abraded into smaller pieces that become gravel, sand, and silt. After deposition and burial of these sediments, and over eons of time, these become sedimentary rocks. The sizes of weathered and eroded rock particles determine how they are distributed in the ocean and streams, rivers, and lakes. The size of eroded rocks and gravel also determines how glaciers distribute them in rivers or on land as till to form moraines.

Project 16: Settling Rates

• **Materials**

1. A tall (16–20″) glass cylinder or jar (A five-gallon clear glass jug can be used if a more narrow, tall, cylindrical container is not available.)
2. Water
3. Small paper cups filled with each of the following:
   (a) Large pebbles or small stones (1/2″ to 3/4″)
   (b) Large gravel (crushed 1/4″ to 1/2″ pea gravel works well)
   (c) Fine gravel (fish tank gravel works OK)
   (d) Clean dry sand
   (e) Pulverized soil (with no or little organic matter)
   (f) Silt or mud
4. A stopwatch or clock with a large second hand
5. Pencil and graph paper

Note: You may want to collect more than a cupful of each of the materials in item 3 for use in subsequent projects.

• **Procedures**

1. Fill tall glass cylinder or jug with water.
2. Have stopwatch set and ready to start.
3. Take one tablespoon full of the large pebbles or small stones (a) and dump them gently (but all at once) into the water.
4. Make sure you start the stopwatch at the moment the rocks enter the water.
5. Also make sure you stop the stopwatch at the moment the rocks come to rest at the bottom of the container.
6. Repeat step 3, and again record the sinking time for another sample of the same material (i.e., a). If the second time recorded is much different from the first time, repeat step 3 for the third time, and then calculate an average time required for the (a) sample to descend to the bottom of the container.
7. Do this procedure for each of the 6 types of material you have collected. Make a list specifying, for each sample, the type of material used and the time required for it to reach the bottom of the container.

8. Keep a record of your results, then enter your results on the graph chart.

• Results

Organize your list of (a)–(f) samples and the time required for each to settle to the bottom of the container into a graph, with time as the vertical axis and particle size the horizontal axis. Also make a drawing depicting the bottom of the container and the layers of rocks, pebbles, gravel, and so forth, as deposited on its bottom.

• Conclusions

What conclusion can you make from both your observations and the graph, comparing the size of the particles to the time required to descend to the bottom of the water? Why is this project referred to as the *rate* of settling? Why were you asked to repeat the experiment with each sample three times and calculate an average when your first two measurements did not agree? Write up your results and conclusions.

The bed of a steep, fast-flowing river entering a lake has boulders and larger rocks deposited on the down slope of the riverbed before the river reaches the lake. Next, are pebbles and gravel extending just into the lake. And finally, there are sand and silt (as soil and mud) deposited on the bottom of the lake at a greater distance from where the river entered the lake. This is referred to as "horizontal sorting," which is due to both the sizes of the particles and the velocity of the flowing water. Underwater landslides can result in the sorting of rocks and particles into alternate layers of particles of different sizes. Your project is an example of vertical sorting, where the roundest, largest, heaviest, densest particles sort out fastest and thus become deposited deeper than are the less round, smaller, lighter, and less dense particles (silt) that form layers on top of the more dense material.

Figure A.16
Select six samples of rocks, gravel, sand, and dirt with different particle sizes for this project. Keep a record of the rates compared to particle size for the samples and use these data to construct a graphic chart.

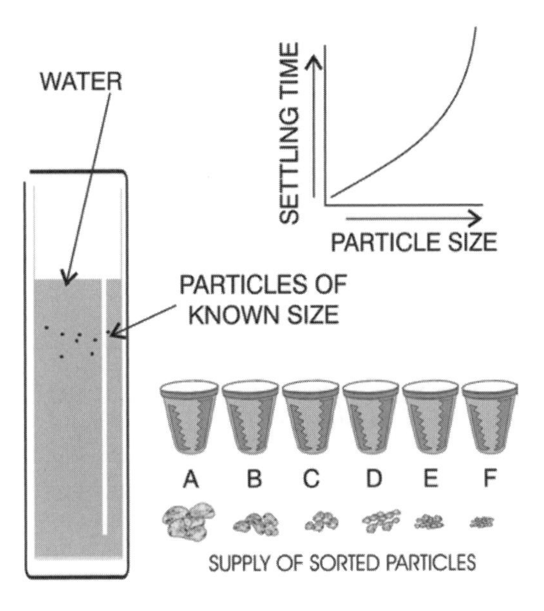

Project 17: Sedimentary Rock

The vertical sorting described in the previous project occurs in nature and, in time, becomes solidified to form sedimentary rock. Again the larger, denser materials usually settle to the bottom to form the lower layers of sedimentary rock.

• Materials

1. The unused rocks, pebbles, gravel, sand, silt, and soil [samples (a)–(f)] saved from Project 16.
2. A large paper cup (at least a 10–12 oz. size)
3. About 1 pound of plaster of Paris (Note: Save some plaster for a later project.)
4. Water, mixing spoon
5. A bucket, pan, or bowl in which to mix the plaster, water, rocks, and so forth

• **Procedures**

1. Fill the cup with dry plaster of Paris. Pour the plaster into the bucket (or bowl) and add just enough water to make a thick liquid. Be sure it is not too firm, somewhat like the consistency of pancake syrup. Keep it liquefied.
2. Pour the liquefied plaster of Paris into the large paper cup. Fill it to about 1″ from the top of the cup.
3. Drop about 1–1.5 tablespoons of *each* of the above rocks, gravel, sand, and so forth into the cup containing the liquefied plaster mix. Start with the heavier rocks and pebbles, then finish with the lightest samples.
4. Do not shake or stir. Let the plaster mix settle and dry until hard.

Figure A.17
The key to the success of this project is to have all materials ready before you start. Once the plaster of Paris is poured, it sets up quickly. After you remove your sedimentary rock from the cup, you may wish to cut or break it in order to examine the internal layering of the particles.

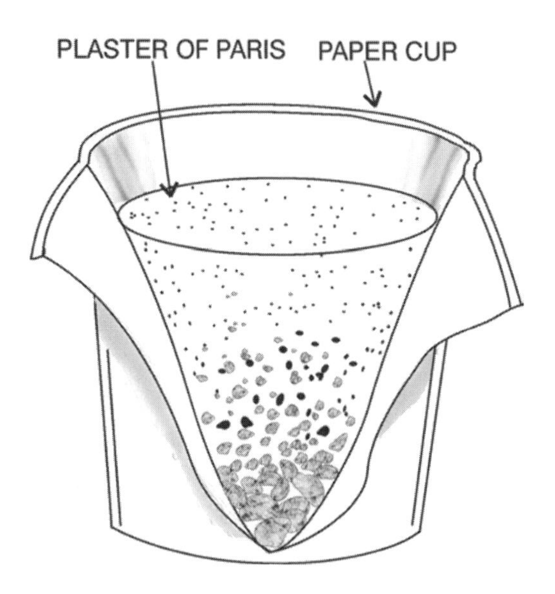

5. Cut open the paper cup and expose the solidified mixture.

• **Results**

You might notice that not all the particles were layered on the bottom of the cup as most of them were in the bottom of the container in Project 16. Some of the lighter samples were suspended in the plaster. Make a drawing of your sedimentary rock.

• **Conclusions**

Particles that settle according to size, shape, weight, and density form a great variety of types of sedimentary rocks, but all such rock structures are formed in layers under great pressures and over long periods of time. If there are no larger pebbles or gravel, but just silt, it may become shale. If the final sedimentary rock consists of more sand than gravel, it may end up as sandstone. It could also consist of particles of various sizes and become a conglomerate type rock.

VI. Hydrology

Hydrology is the science that deals with water, its properties, its distribution, and its circulation on or below the Earth's surface. The hydrological cycle (water cycle) refers to the constant circulation of water from the surfaces of lakes and seas through evaporation as water vapor in the atmosphere and its eventual return as precipitation to Earth. Through the processes of transpiration and evaporation, water is again recycled into the atmosphere. Water on the surface flows from higher altitudes to lower levels, while at the same time it permeates the soil and percolates into vast underground areas of stored water.

Project 18: Soil Permeability and Percolation

The amount of rain and surface water that can carry nutrients into and through the ground is dependent on soil permeability. The seeping and flowing of groundwater through the permeable soils is known as percolation. Farmers are concerned with how well their soil retains and drains water. If the soil retains too much water, crops rot. If the soil is extremely porous (sandy) and water passes through it too fast, plants may dry up. Soil permeability and percolation are especially important in areas where crop irrigation is practiced. Soil permeability is also of concern to home gardeners who raise potted plants. They can mix soils (and water supply) to the requirements of the plants they intend to grow much more easily than farmers are able to do.

• Materials

1. About one pound of each of the following:
   - (a) Gravel (fish tank gravel works OK)
   - (b) Clean, dry sand
   - (c) Dry mud or dry potter's clay
   - (d) Clean, dry potting soil or topsoil (from garden shop)
2. Five large paper cups (10 to 12 oz. size)
3. Five paper coffee filters
4. Water, and measuring cup
5. Pan or bucket to mix soil types, and mixing utensil
6. Ballpoint pen and marking pen
7. Five large plates or 8″ cake pans
8. Ten small sticks or pencils

• Procedures

1. Punch eight small holes in the bottom of each of the five paper cups. Use a knitting needle or the point of a ballpoint pen.
2. Set a cup on each filter paper, trace around the cup's outside bottom, then cut the circle out of each filter paper and fit one into the inside bottom of each cup.
3. Label (mark) the outside of each paper cup according to the following mixtures, but do not fill the cups just yet.
   - (a) Mix 1/2 gravel with 1/2 dry sand for cup A
   - (b) Mix 1/4 gravel, with 1/4 sand and 1/2 dry potting soil for cup B
   - (c) Mix 1/4 sand, 1/4 dry mud (or potter's clay), and 1/2 dry potting soil for cup C
   - (d) Mix 1/2 dry mud (or clay) with 1/2 dry potting soil for cup D
   - (e) Mix 100% dry mud or potter's clay for cup E
4. Fill cup A 3/4 full with the (a) mixture, then fill cup B 3/4 full with the (b) mixture, and so on for each of the five cups.

Figure A.18
There are a variety of ways to set up this project. If you do not have the exact materials suggested, be creative and design your own experiment. You may also wish to design a graph chart to display your findings.

5. Set each labeled paper cup across two sticks or pencils resting on a plate or cake pan.
6. Slowly pour 1.5 cup (12 fluid oz.) of water into each cup.
7. Wait one hour. Pour the water from the plate/pan under cup A into the measuring cup. Record the amount of water that percolated through mixture A. Repeat for cups B–E and record how much water drained through each cup.

• Results

Write down your observations and results for each of the labeled cups. Be sure to keep a record of how much water drained through each cup, and the type of mixture that was in each cup. Either set up a list of your results or try to plot the results on a graph.

• Conclusions

Both the retention of water (and dissolved nutrients) and the drainage of excess water from the soil is important for the growth of healthy plants. The supply of subsurface and flowing underground water in aquifers is dependent on the percolation rate of surface water. If the soil is too coarse and allows too much water to rapidly drain into deeper levels of the crust, however, surface plants do not receive adequate amounts of water. Gravity is responsible for percolation as it pulls water deeper into the subsoil until it reaches an impermeable layer that stops the seeping of water. If this layer is a hardpan close to the surface, the soil may retain too much water. Most crops cannot survive in standing water. The water table for an area is the upper level of underground water. It may also be defined as the boundary between the upper region where aeration takes place and the lower region where saturation occurs. This boundary for the water table can be near the surface, as with hardpan, or much deeper, as with aquifers. Can you conclude from your data which mixture would be the best for raising crops? And which would be most suited for recharging aquifers?

Project 19: Water Pressure

Water has weight, just as do gases and solids. As the depth of water increases, so does the weight of the water above it, which results in an increase in water pressure as the depth increases. Pressure (for water or gas) is defined as the unit of force exerted per unit of area, which is equal in all directions. Therefore, at any given depth, the water pressure exerted on a submerged surface is equal in all directions.

Project 19 provides instructions for two different water pressure projects, Project A and Project B.

Project A

• Materials

1. A 2 L or 2 qt. clear plastic soda bottle
2. Large, sharp nail or other sharp instrument with which to make smooth holes about 1/8″ diameter in the plastic bottles
3. Water
4. A sink or area with a drain
5. Ruler, paper, and pencil

• Procedures

1. Use a marking pen to place four holes in a straight vertical line on one side of the plastic bottle. Use a ruler to make sure each mark is evenly spaced from 1/2″ off the bottom to 2″ off the top of the bottle.
2. Use the nail to punch a round hole in each of the four marks.
3. Set the bottle over the drain or in a large bucket.
4. Fill the bottle rapidly with water, and keep the water level as near to the top of the bottle as possible for several seconds. Do not cap the bottle.

5. While pouring in the water have someone else try to measure the horizontal length of the water spouting out of each hole.
6. Record your results.

• Results

The spout of water from the lower hole will be longer than the one emerging from the upper holes. Make a diagram of the apparatus, and construct a graph depicting your data.

• Conclusions

Is it the volume of water or the height of water in the bottle that determines the pressure on the water at the level of each hole? Why do water storage tanks for towns and villages have a narrow pipe rising many feet with a globular water holding tank sitting on top of the thinner structure? Why don't they build water storage tanks with the large spherical container of water at the bottom? Use your diagram and data (graph) to answer these questions.

Project B

Set up another experiment to demonstrate if the volume or the height of the column of water determines water pressure. Hint: Use a narrow plastic bottle and a wide plastic bottle, both about the same height (see Figure A.19). Punch just one hole on the side, near the bottom of each bottle, and fill with water to the same level in each bottle. (Be sure the holes are the same diameter and at the same height up from the bottom of the bottles.)

Observe the length of the spouts of water from each. Does the spout of water from the wider bottle, which holds a greater volume of water, differ in length from the spout of water from the narrow bottle, which holds less water, although the spouts are at the same starting height?

Figure A.19
Projects A and B demonstrate that water pressure is determined by the height of the water column, not the volume of water.

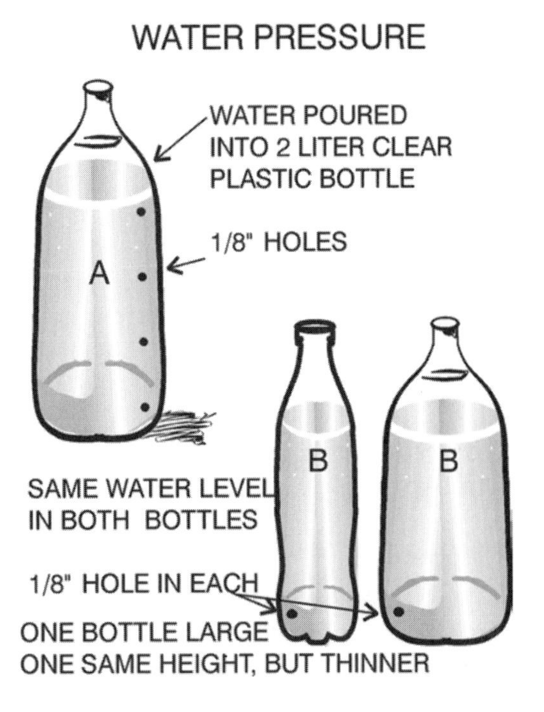

VII. Atmosphere

Our global atmosphere is a mixture of many gases and other chemical matter that exist in several physical states. The carbon cycle, nitrogen cycle, oxygen cycle, water cycle, and, to some extent, nutrient cycle are dependent on the atmosphere as the medium that enables these complex systems to function.

The projects for this section are of a different nature. They require some research to complete the tasks.

Project 20: Constructing Atmospheric Cycles

• Procedures

Use reference materials (try libraries) that provide information on the following atmospheric cycles to construct your own diagrams. Use a circle with arrows as indicators to track one phase to the next in the direction in which the cycle proceeds.

1. The carbon cycle
2. The nitrogen cycle
3. The oxygen cycle
4. The water cycle

• Results

When your diagrams are completed, identify the role of the atmosphere in each of these cycles. Next, describe the importance of each cycle.

Project 21: Structure of the Atmosphere

• Procedures

Draw horizontal lines on a piece of paper that represents the various levels of the atmosphere, starting at the bottom with the surface of the Earth at sea level. Include the following:

1. Troposphere and tropopause
2. Stratosphere, including the ozone layer and stratopause
3. Mesosphere and mesopause
4. Thermosphere

For each of the atmospheric layers identified in your diagram, add the following (averages or extremes):

1. The altitude of each layer
2. The temperatures for each layer
3. The air density or air pressure for each layer
4. Types of clouds for each layer

• Results

Identify the layers where most weather takes place (wind, rain, snow, hail, etc.), jet airplanes fly, meteors break up, chemical reactions take place (e.g., ozone), and cosmic radiation causes the aurora borealis.

Conclusions

As you do your library research and complete the project, it will be obvious that the atmosphere is not the same at all altitudes. Your final diagram should include these differences.

VIII. Climate and Weather

In a sense, the climate of a region, or for the entire globe, is a composite of a great number of interactive chemical and physical phenomena that cause what is known as *weather.* Measuring the factors that produce weather, and correlating them, is the basic means of predicting weather. Weather is a complex of systems, and because the factors that cause weather keep changing, the extent of their interactions can only be measured in the short-term. Anticipating weather conditions and predicting what the weather will be in just the next few days, even with the best instruments, is a crap shoot. Most weekly weather forecasts are no more than 50% accurate. The following simple projects are designed to demonstrate how some of these physical phenomena that affect weather can be measured.

Project 22: Air Weight/Density

• Materials

1. Two medium size paper bags (use grocery store or paper lunch bags, but not grocery store plastic bags)
2. A 36″ dowel rod, or yard stick, or meter stick
3. Small candle or alcohol lamp
4. Double-stick tape
5. String, and the back of a chair for support

• **Procedures**

NOTE: It is advisable to conduct this project in an area where the risk of fire is minimal, or outside where no flammable material is nearby.

1. Attach the bottom of each opened paper bag to the end of the stick with the double-stick tape (see Figure A.20).
2. With the open bags facing downward, tie a string around the center of the stick so that the two bags on the ends of the stick balance each other. Adjust the string until the stick is level.
3. Find a support for the string, or balance the stick over the back of a chair so that the bags are balanced even with each other (level).

Figure A.20
Make sure the two paper bags are evenly balanced and there is no air movement in the room in which you are conducting this experiment. Use caution when using a candle. Will a 100 watt light bulb do as well as the candle?

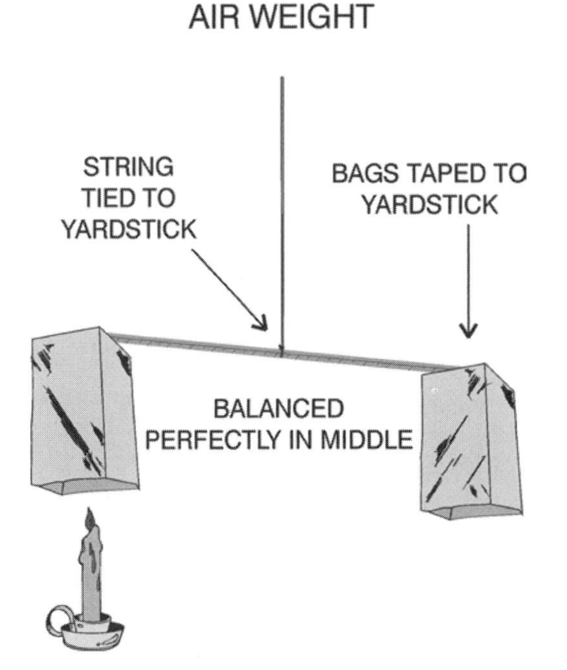

4. Place a lighted candle (or other source of heat) under one bag, making sure that the heat flows into the open mouth of the bag. The candle should be a few inches from the opening, but not close enough to catch the paper on fire.
5. Keep the candle under the bag until you see some change in the level of the two bags.
6. Observe what happens.

• **Results**

Make a diagram of before and after you lit the candle. Write up what you think happened, and why.

• **Conclusions**

A gas (air), like anything else that has weight and occupies space, expands when heat increases molecular motion, thus the volume of a given mass increases. Thus, the heated air expanded, and since the volume inside the bags is the same, the expanded air inside the heated bag weighed less than before it was heated. In other words, the heated air expanded, became less dense than the air inside the bag on the other end of the stick, and thus was lighter than the air in the other bag.

Project 23: Air Pressure

Air is stacked up above the Earth. Therefore, the lower the altitude from the outer atmosphere toward the Earth's surface, the more upper layers of air are pushing down on the lower layers. After all, air does have weight. In other words, wet can visualize that for every one square inch of surface at sea level, a column of air extends up as far as air can be found. This column weighs about 14.7 pounds for each square inch of surface. Since the height of a column of air would be inconvenient to measure, meteorologists use the equivalent weight of a column of mercury as compared to the column of air, which is 29.92 inches of mercury. In other words, a 30-inch column of mercury is equal in weight to an equal column of air. The previous project demonstrated that air's density is affected by changes in temperatures, and thus its weight per unit volume is also changed. Using this information, it is possible to construct a simple barometer to measure minor changes in the density (weight of air above us), which is one of the indicators for predicting weather.

Project 23 provides instructions for constructing two different types of simple barometers. Project A, the balloon barometer, uses a rubber dam to represent an aneroid barometer. Project B, the water barometer, substitutes water for a mercury barometer.

Project A: Balloon Barometer

• Materials

1. A 1 qt. or 2 qt. jar with medium wide-mouth opening
2. A torn balloon, or similar piece of thin rubber
3. Rubber bands
4. Drinking straw
5. Pin or needle
6. Small amount of glue
7. Wooden or plastic 12″ ruler
8. Chunk of modeling clay

• Procedures

1. Stretch the rubber balloon or thin rubber sheet over the mouth of the jar. Fasten it tightly around the mouth with rubber bands.
2. Glue the needle onto the tip of one end of the straw so that the needle extends past the end of the straw.
3. Using just one drop of glue, fasten the other end of the straw horizontally onto the rubber that is stretched over the mouth of the jar. Be sure the glued end of the straw extends only to the center of the stretched rubber.
4. Stick one end of the ruler into the clay for support. The ruler should be vertical and stable. Place the glass jar and ruler in an area where they won't be disturbed.
5. Adjust the ruler so that the pin or needle on the end of the straw is close to, but does not touch, the fractional inch marking on the ruler.
6. Regardless of the height of the jar, the needle should be set on the ruler's scale (but not touching the ruler). Record the level of the needle at the 1/32″ ruler markings.

Figure A.21
This project is considered a balloon barometer because it uses a rubber sheet over a wide-mouth jar to react to the changing air pressure. The key is to arrange the stick glued to the rubber sheet so that the free end can move to a greater extent than does the fixed end. This movement can then be measured on a vertical scale. This balloon barometer works similarly to an aneroid barometer.

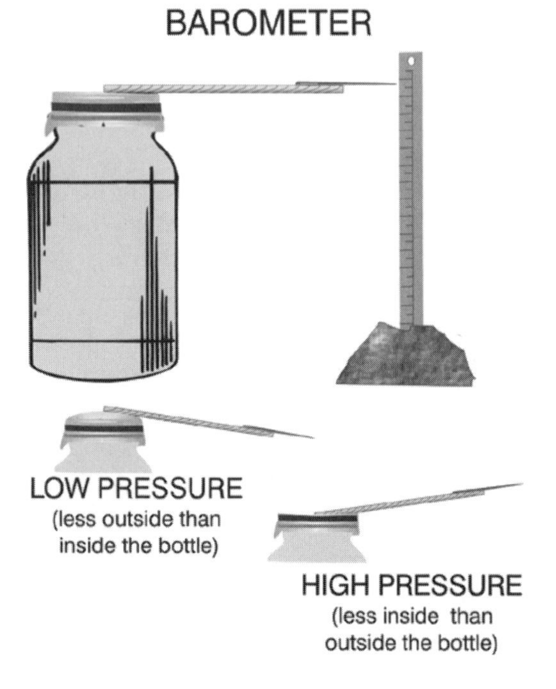

• Results

Take a daily (or morning and evening) reading of the needle's position by 1/32″ markings on the ruler. Record your results.

• Conclusions

Can you establish any relationship between your daily readings and the barometric pressure reported in the local newspaper or on television? Why does the needle move from time to time? Why does it rise sometimes or drop at other times? Do changes in the room's temperature affect the reading? If so, why?

Project B: Water Barometer

• Materials

1. A three-foot piece of clear plastic tubing 1″ in diameter (available in most hardware stores)
2. A cork or stopper that fits tightly in the end of the tubing
3. Paraffin or candle wax
4. A bucket or pitcher of water, and a funnel
5. Food coloring (darker colors work best)
6. Marking pen

Figure A.22
The U-shaped tube can be a piece of clear plastic tubing long enough to shape into a U. Be sure the U is smoothly shaped without a kink in it. If this project is attempted outside during freezing weather, you may wish to add some antifreeze, salt, or alcohol to the water. This water barometer works similar to a mercury barometer.

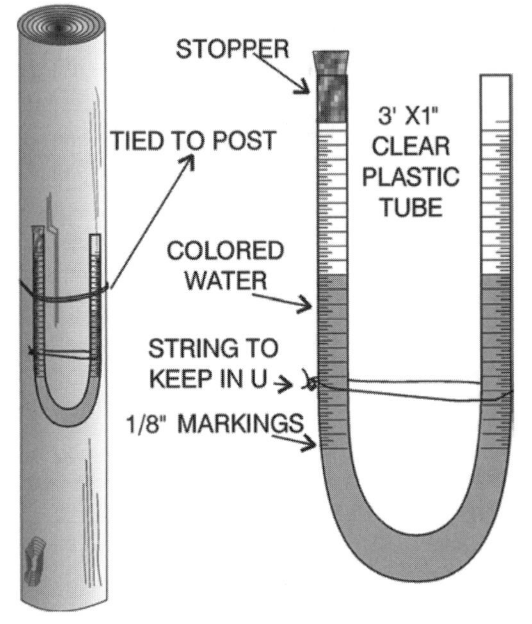

• Procedures

1. Mix food coloring and water in the pitcher.
2. Bend the plastic tubing into U shape. (Hold it in this position by tying a string around it to maintain the U shape.)
3. Place the funnel in one end of the tube and fill the tube with colored water until it is just about half full, making sure there is about an 8–10″ air space at each end.
4. While keeping the tube level in its U shape, place the cork or stopper in one end of the tube. Make sure it is firmly in place. Seal with wax if possible. No air should leak in or out of this end of the tube.
5. Find a place outside where you can fasten (tie) the U tube. Find a post, tree, or some other support and fasten the U tube to the support with the twine at eye level.
6. Using the marking pen, make a mark of the water level at each side of the U. Then, make several more markings, about 1/8″ apart, both above and below the present water level marks at both ends.
7. Make a daily (or morning and evening) reading of the level of the water.

Note: If you place your barometer outside in freezing weather, use salt water or antifreeze to prevent the water from freezing in the tube.

• Results

Correlate your reading with the daily barometric weather report in the local newspaper or television weather news. Which way does the water level change in the end of the tube without the cork when a storm approaches? Why? Does it work just as well indoors? Do temperature changes affect the readings?

• Conclusions

Which direction of movement of the water level in the tube indicates an increase in air pressure? How does a decrease in pressure affect the water level, and what does that tell you about the weather? Why? After keeping records for several days, can you determine when a storm is approaching? Are there other types of barometers in use? How can a barometer be used as an altimeter?

Project 24: Hygrometer

A hygrometer measures the amount of moisture (water vapor) in the local atmosphere. The relative humidity is the ratio between the amount of water vapor in the air at the present time and the total amount of water vapor that the air can hold at the current temperature. This ratio is expressed in percent, for example, 50% relative humidity means that the current amount of moisture in the local air is just half of the amount of water vapor the air could possibly hold at the current temperature. The higher the percentage, the more moist (humid) the air is, and, if the humidity increased enough or the temperature dropped enough, it rains or snows. Why?

Note that several different setups can be devised using the same general principles described for this project.

• Materials

1. A long piece of hair, about 10–12″ in length (Note: If a long hair cannot be found, use a piece of pure silk thread)
2. A flat board about 18″ long and 3″ wide (Add some means to support the board in vertical position, and place it so that it can be viewed without moving it.)
3. A 1–1.5″ nail, and a 1″ flat machinist's washer (not too heavy)
4. A flat toothpick and small amount of glue
5. A plain 3″ × 5″ index card, and pen

• Procedures

1. Pound the nail part way into the center of the board about 1″ from the top.
2. Glue the flat toothpick across the center of the washer (see Figure A.23).
3. Tie or glue the hair or thread to the washer.
4. Tie the other end of the hair or thread around the nail. Be careful not to break the hair or thread.
5. Draw fine horizontal markings on the 3 × 5 card.
6. Glue, or tape, the card to the board behind where the washer is hanging from the hair or thread with the horizontal toothpick pointing to the central mark on the card (see Figure A.23).

Note: The washer/toothpick should hang freely by the hair or thread, about 1/4″ from the board. It is OK if the hair/thread slightly twists the washer/toothpick so that they touch the card, but the washer should be free to move without much friction. The card should be aligned with the toothpick pointing to a central mark you made on the card.

• Results

Once the hygrometer is set up and calibrated, make daily or twice daily readings to see if the toothpick moved up or down the scale of marks on the index card mounted

Figure A.23
This model of a hygrometer is very sensitive, and care must be taken during construction to ensure the hair or thread is long enough and free to rotate slightly. The hygrometer should be used in an area where there is little or no air movement.

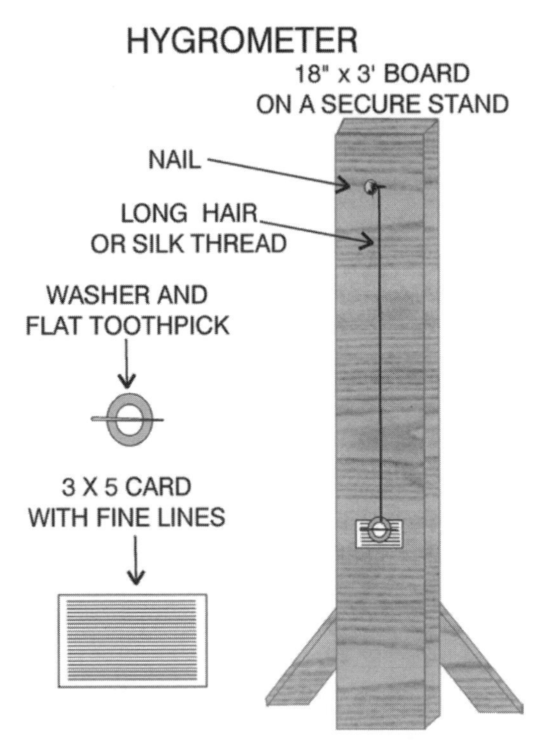

behind the washer. Check your figures with the daily weather reports. Do your readings correlate with the reported relative humidity figures?

• Conclusions

The hair or silk thread is hygroscopic and thus affected by the changes in humidity, as many women can attest to when their hair becomes frizzy on humid days. Professional meteorologists use more accurate instruments, including a psychrometer that uses both wet bulb and dry bulb thermometers that compare the relative humidity in moving air. Are there other types of hygrometers? How do they work?

Project 25: Wind Speed (Anemometer)

The speed at which air moves over the surface of water and land is an important aspect of weather systems. In 1806 Francis Beaufort (1774–1857), a British sea captain, devised a wind scale to determine the effects of wind on the sails of ships. Later the scale was revised as to the wind's effects on the sea's surface, and still later it was adapted to the conditions that resulted from different wind speeds on land. The scale ranges from 0, which is calm, to 13, which is above the speed at which wind is associated with a hurricane or tornado (i.e., over 100 miles per hour). An example of the scale, for your use with your anemometer, is given in Table A.4.

• Materials

1. Four small (3 oz.) paper cups
2. Two thin sticks or dowel rods about 12″ long and 1/4″ in diameter
3. An empty wooden or plastic thread spool
4. A piece of wood about 6″ square (or a large block of clay)
5. Glue, thin wire, penny, punch or nail set, hammer, colored marker, sharp knife, a pencil with rubber eraser, and a large sewing needle (about 1.5–2″ long)

• Procedures

1. Mark the center of each stick or dowel. Cut out small U-shaped areas at the centers so that you can cross the sticks, and then glue and secure with a small, tightly bound wire.
2. Set the penny on a firm surface that will not be damaged when you pound on it. Place the point of the nail set or punch at the center of the penny and hit hard with the hammer. You are making a dimple to act as a bearing to allow your wind gauge to rotate freely.
3. At the point where the sticks cross, attach the dimpled penny with the concave side

Table A.4
Wind Speeds

Beaufort Scale	Wind Speed MPH	Wind Speed Description	Wind Speed Clues
0	less than 1	Calm	No air movement.
1	1–3	Light air movement	Smoke drifting with wind.
2	4–7	Light breeze	Wind felt, leaves rustle, flags stir.
3	8–12	Gentle breeze	Leaves move & flags move.
4	13–18	Moderate breeze	Dust or loose dirt moves, flags flap.
5	19–24	Fresh breeze	Small trees sway, flags ripple.
6	25–31	Strong breeze	Branches move, flags beat.
7	32–38	Moderate gale	Whole trees move, flags extended.
8	39–46	Fresh gale	Twigs and small branches break.
9	47–54	Strong gale	Minor damage to houses, awnings.
10	55–63	Whole gale	Trees uprooted, house damage.
11	64–75	Storm	Widespread damage to trees, etc.
12	75–100	Hurricane	Great damage to structures, floods.
13	over 100	Hurricane & tornado	Excessive, serious damage.

facing away from the sticks so that the needle pivot fits inside the dimple.

4. Cut about 1/2″ off the outer rim of the paper cups to shorten them and make them lighter in weight.
5. Glue or otherwise fasten the bottom of each cup to the flat side at the end of each stick. Make sure each cup is facing in the same clockwise direction.
6. Color the outside of just one of the cups with the marking pen.
7. Glue the wooden spool onto the center of the 6″ × 6″ block of wood.
8. Place a pencil in the hole in the spool with the eraser pointing up. Glue or otherwise secure the pencil inside the spool.
9. Stick the blunt end of the needle into the pencil eraser. Then set the crossbar with the cups onto the pointed end of the needle, making sure the needle fits into the penny dimple bearing. If your crossbar with cups does not set level, adjust with tiny bits of clay attached to the wooden stick until you get it level. Make sure the crossbars spin smoothly and easily on the needle.
10. Carefully move your anemometer outside. Place it in an area where it can catch some unobstructed air movement and yet be partially protected from being disturbed.

• Results

During different times of the day, and on different days, count the revolutions made by the cups by the number of times the colored cup passes a certain point each minute. This gives you a basic rate of the spins. Try to relate your figures to the wind descriptions of the Beaufort scale. Record and write up your results.

• Conclusions

More elaborate anemometer models can be made. Perhaps you can design one that will automatically do the counting for you. You can design a wind vane for your model to determine the direction of the wind. After some practice your records may be checked with the local weather forecasts for wind

Figure A.24
The key to success in constructing this model anemometer is to balance the cross-arms with the cups and to have a very easily moving pivot for the rotation of the arms. Try to find a way to count the number of turns of the arms (color one cup so that you can count revolutions), and then compare your figures with known wind speeds.

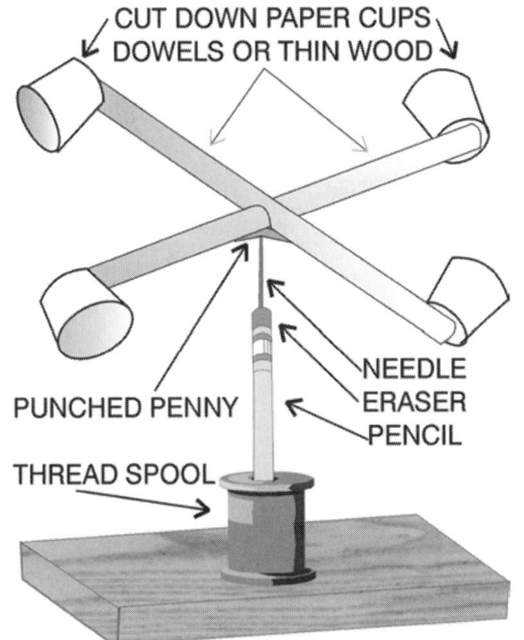

speeds at different times of the day. Many local weather forecasters use the Beaufort scale when reporting the speed and damage caused by winds.

Project 26: Simulated Tornado

This project uses water to imitate the wind and rain action of a real tornado. Temperature and pressure differences are responsible for the formation of actual tornadoes, but we use the force of swirling water to simulate the funnel-shaped winds of a tornado.

• Materials

1. Two empty 2 L (or 2 qt.) clear plastic soda bottles
2. Duct tape
3. Pencil, scissors, towel, water

• Procedures

1. Fill one plastic bottle 3/4 full of water.
2. Place a 3″ strip of duct tape over the mouth of the empty bottle.
3. Punch a clean round hole in the tape over the mouth of the empty bottle. The hole should be smooth and slightly larger than a pencil.
4. Dry the necks of both bottles.
5. Place the empty bottle over the one with water and seal the connection with duct tape. Make the seal as strong and waterproof as possible.
6. Turn your tornado simulator upside down with the water bottle on top.
7. While holding the section where both bottles are joined, rotate the apparatus clockwise a few times until the water starts draining from the top bottle to the lower one in a spiraling fashion.

• Results

Once the descending water begins to swirl, a whirlpool-type of vortex forms in the upper, water-filled bottle. This is not the same as a real tornado, but the water simulates the air motion of a tornado.

• Conclusions

The water funnel of your model is similar to the air funnel of a real tornado but does not have a tornado's great wind speeds and destructive force. Can you come up with other designs to demonstrate the formation of a tornado? Try to construct one that uses

Figure A.25
It may take some practice to create a good water tornado. Can you make the water rotate in both clockwise and counterclockwise directions?

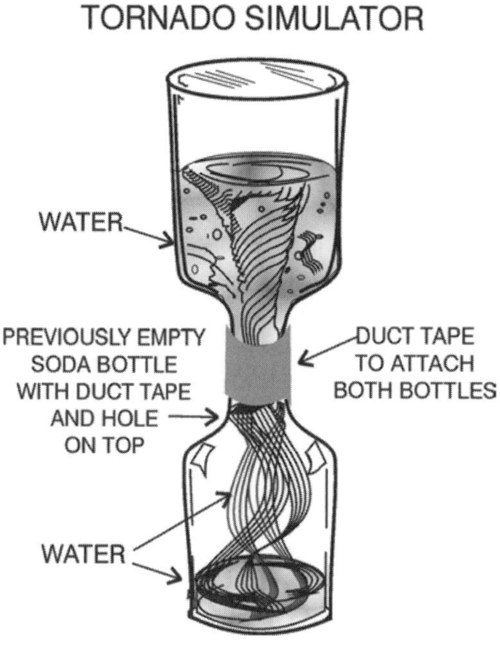

different levels of water in the top bottle, then vary the speed of rotation and note the results.

IX. Geological Time: Fossils

Geology and biology complement each other as a means for determining the age of the Earth, minerals, and rocks, as well as ancient plants and animals. The levels or strata of rocks, mainly sedimentary rocks, in which fossils of prehistoric animals and plants are found aid geologists, as well as biologists, in dating our past.

Project 27: Making a Fossil

• Materials

1. An assortment of moderate size sea shells (about 1.5″ or 2″ across)

Figure A.26
There are a number of ways you can make molds of shells and other objects that can be filled with plaster of Paris to produce casts. You can start a collection of your own fossils.

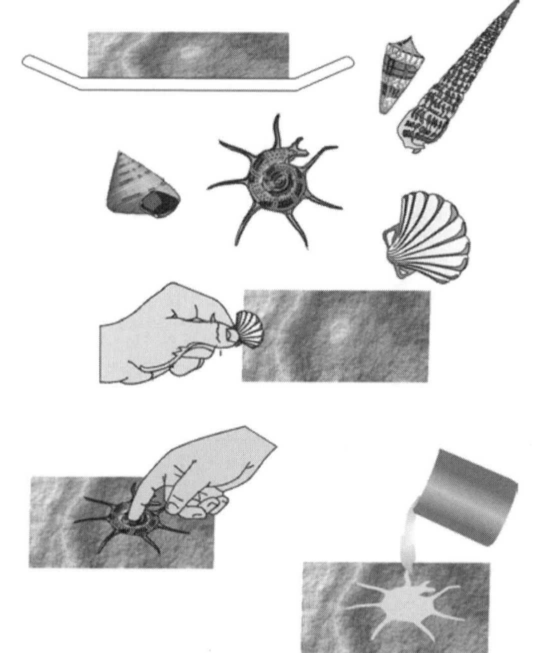

2. Several chunks of modeling clay
3. Clear Vaseline
4. Plaster of Paris
5. Several paper plates
6. A disposable container to mix the plaster, a mixing stick, and water

• Procedures

1. Form several pieces of modeling clay into 3–3.5″ squares, about 3/4″ thick.
2. Place each square of clay in a paper plate to catch any spills.
3. Lightly cover the convex (bulging) side of each shell with clear Vaseline. Use only enough to make just a very thin film.
4. Press the convex side of a shell firmly into the clay to make an impression of the shell.

Make sure you press hard enough to make a good mold. Repeat with other shells, using as many as desired.

5. Remove the shells from the clay molds.
6. At this time, mix the plaster of Paris in a disposable container with just enough water to make it flow, but so it is not runny.
7. Before the plaster in the container sets up, pour just enough plaster into each clay imprint or mold of each shell.
8. Wait for the plaster to thoroughly dry before removing from the clay.

• Results

After you lift your plaster fossil out of the clay mold, clean off any stray plaster from the edges. Although the shells you used are not old, the molds you made simulate the natural process of making fossils over long periods of time.

• Conclusions

Most fossils are found in sedimentary rocks that at one time were mud or silt. The structure of plants and animals that died and were buried in these sediments left an impression even if the bones, wood, and leaves eventually disintegrated. Over time the hard parts of the plants or animals are replaced to form exact hard copies of the original animal, plant or shell. Check the Internet for information on the fossilization processes of invertebrates, which are the most common, or how ancient plants are carbonized.

Glossary

Abiotic Not biological. Inorganic environmental factors, for example, climate, geology, and geography, that influence ecological and environmental systems. Some examples are earthquakes, chemical pollution, and volcanoes.

Absolute Zero Also known as "absolute temperature." In physics and chemistry, on the Kelvin scale, absolute zero is 0° Kelvin, which is –273.16°C or −459.60° F. It is the temperature at which all matter possesses no thermal energy and at which all molecular motion ceases. (Although there is no positional movement, molecules do vibrate at absolute zero.)

Absorption The penetration of one substance into the inner structure of another.

Accretion Refers to the increase or growth in the size of an object through a process of external addition of matter. It is one theory for the creation and growth of larger bodies as they accumulated smaller particles to form the present planets.

Acid Rain A controversial form of air pollution that is blamed for local environmental damage to ponds, lakes, and forests in the northeastern United States and southeastern Canada, England, Europe, and Africa. It is both a natural and a man-made condition in which gaseous sulfur and nitrogen compounds combine with atmospheric moisture. These compounds then create a slightly acidic rain, snow, hail, or fog.

Adiabatic Refers to a thermodynamic process in which there is no transfer of heat into or out of a closed (isolated) system.

Adsorption Adherence or collection of atoms, ions, or molecules of a gas or liquid on the surface of another substance, called the "adsorbent." It is the opposite of **absorption,** which is the penetration of a substance into the bulk of another solid or liquid. Adsorption is an important process in the dyeing of fabrics.

Aggradation A process by which the level of a streambed or other surface is raised through the deposition of sediment or the spread and/or growth of permafrost.

Allotropic Also "allotropism." Refers to the formation of two or more different structures by a specific element, where the different forms usually exhibit different stabilities when experiencing different temperature ranges, for example, the different crystalline forms of carbon—charcoal, graphite, and diamonds.

Amplitude A variation in magnitude of some characteristics (e.g., the variations in brightness of the sun). The amplitude of an electromagnetic wave is the measurement of the wave's height.

Angular Momentum Related to *linear* momentum of a body (p), which is the vector product of its mass (m) times its velocity (v) [i.e., $p = mv$]. Angular momentum is similar, except that it is associated with the rotation of a mass around an axis. It is important in the structure, shape, and evolution of celestial bodies, for example, planets and moons.

Anthropic Related to human activities within their environment, particularly during the period of human civilization (i.e., the Holocene epoch).

A Priori From the Latin, meaning "from the former." In science, it generally refers to something that is *not* based on an experiment or factual study but is instead an assumption related to a known cause.

Autopoiesis The self-organization and self-maintenance of a living organism.

Bathysphere A spherical diving chamber constructed as to be strong enough to withstand water pressure while studying ocean depths.

Big Bang A cosmological theory that posits that at the beginning of the universe an incredibly dense, small ball, a point-source of mass, containing all the matter and energy in the universe, exploded and expanded rapidly outward, producing a higher temperature than our sun's. The explosion was forceful enough to overcome the gravity of the particles that expanded and that are still expanding after ~14 or ~15 billion years. The particles and energy are what make up everything in the universe. Where the original matter for the Big Bang came from, or exactly of what it consisted, still has not been determined.

Biotic Refers to all living or biological organic factors in the environment that affect other organisms (e.g., bacterial infection, predation of one organism on another).

Board-Foot (Feet) A unit of lumber measurement that equals 144 cubic inches (2,360 cubic centimeters) and is equal to the volume of a board one foot square by one inch thick.

Carcinogenic Refers to a substance that causes cancer.

Catalyst Any substance that affects the rate of a chemical reaction without itself being consumed or undergoing chemical change. Platinum/palladium pellets in automobile catalytic converters are chemical catalysts. A biological catalyst (e.g., an enzyme) affects chemical reactions in living organisms.

Catastrophism In biology and geology, the idea that catastrophic events that have occurred in the past (e.g., earthquakes and volcanoes, meteor impacts, and major climate changes) radically altered the Earth's surface and/or biological processes.

Centrifugal Force An apparent outward force on a body (mass) in motion along a curved line. The force appears to be directed away from the center of curvature or axis of rotation. It is an apparent force because it is fictitious but is created by the angular momentum of a mass. Historically it was considered a balance to centripetal force.

Centripetal Force A force directed toward the center or axis of a body that causes it to move in a circular path. In the case of orbiting artificial satellites, their mass is held in orbit by the constant gravitational force of the Earth at particular heights above the Earth based on the velocity of the satellites. Thus, the height of an artificial satellite can be determined by its velocity and mass. Or, if the satellite's mass is known, the speed that it must reach to attain a particular orbit about the Earth can be determined.

Chaos Theory A theory in the new science of chaoplexity that deals with the breakdown of ordered systems into random, disordered, or chaotic systems. A chaotic system is very sensitive to the initial conditions of that system. Small differences in initial starting points, no matter how minute, combined with other small changes cause the system to progress in very different ways. In other words, depending on changing environmental conditions, the system will progress differently in the final phases depending on the slight differences in its early phase.

Chemical Bond Electrostatic force that holds together the elements that form molecules or compounds; known as ionic bonds and covalent bonds. Van der Waal bonds are less strong, but all are important for chemical reactions. This attractive force between atoms is strong enough to hold the compound together until a chemical reaction causes the substance to either form new bonds or break the bonds that form the molecule.

Colloidal Refers to finely divided substances or particles, 1–1,000 nanometers in size, that are suspended in a continuous medium and will not diffuse through a membrane. Some examples are (a) oil in water; (b) gelatin; (c) aerosols that are dispersions of liquid or solid particles in a gas, as in fog or smoke; and (d) foams that are dispersions of gases in liquids or solids.

Convection The process of transferring heat from one part of a fluid to another by the movement of the fluid itself. *Natural convection* occurs as a result of gravity when the hottest part of the fluid expands, becoming less dense and thus rising. At the top of the fluid its density increases as it cools, thus sinking to again be heated and continue the cycle. *Forced convection* occurs when the hot fluid is transferred from one region to another using a fan or a pump, for example, as in a convection oven.

Coriolis Force Refers to the inertial force that acts on rotating surfaces at right angles to the rotating body. The Earth rotating on it axis causes surface water and air to follow a curved path opposite the direction of the rotating Earth.

Corundum An extremely hard mineral (aluminum oxide, Al_2O_3) that is found in various colors and that crystallizes in a hexagonal system. Some examples are rubies, sapphires, and the common gray, brown, and blue forms that are used in abrasives. Corundum is 9 (out of 10) on the Mohs hardness scale.

Cosmology The scientific study of the entire physical universe.

Covalent Bonding Sharing of electrons by two or more atoms. In a nonpolar covalent bond, the electrons are not shared evenly, whereas in a polar bond, they are shared evenly.

Cracking A process used in the refining of petroleum by which large hydrocarbon molecules are broken down into smaller molecules to form more useful petrochemicals.

Critical Mass The minimum mass of fissionable material (U-235 or Pu-239) that will initiate an uncontrolled fission chain reaction, as in a nuclear (atomic) bomb. The critical mass of uranium 235 is about 33 pounds, and of plutonium 239, about 10 pounds.

Crystallization The growth of crystals that occurs from (a) condensation from a gaseous state, (b) precipitation from a solution, or (c) cooling of a melt.

Degradation The lowering of the land's surface by erosion.

Dendritic Having a branching or treelike structure. Dendritic patterns similar to spreading fingers from the palm of a hand are formed by depositions of sediments at the mouths of rivers as they enter oceans, (e.g., deltas).

Density Mass per unit volume ($d = m/v$). The density of a solid or liquid is compared to the density of 1 milliliter (ml) or cubic centimeter (cm^3) of water at ~4°C. The weight of 1 cm^3 of water is exactly 1 gram, thus the density of water = 1. Therefore, water is used as a relative standard for comparison of the densities. Specific gravity is the same as density for liquids and solids, but it is not exactly the same as the density for gases, which is expressed as grams per liter.

Depolarizer A substance or agent that counteracts or prevents polarization. For example, manganese oxide is added to the electrolyte of a primary cell (flashlight battery) to oxidize the hydrogen bubbles by chemically combining with the hydrogen gas as it forms.

Dew Point The temperature at which water vapor condenses and forms droplets of water.

Diffraction The bending of light and/or other electromagnetic waves as they pass through an obstacle or aperture.

Echo Sounders Devices that record and determine the depth of oceans by measuring the time needed for sound waves to travel to the seafloor and return.

Electrode A conductor that allows an electric current to pass through an electrolyte (solution).

Electroplating Sending an electrical charge from one electrode through an electrolyte to deposit a thin coating of a metal on an object at the opposite electrode.

Entropy A measure of the disorganization or disorder within a system, or the lack of information about a situation, thus the uncertainty associated with the system. In thermodynamics, it is the function of the system where the amount of heat transfer introduced in a reversible process is equal to the heat absorbed by the system from its surroundings, divided by the absolute temperature of the system.

Eon The longest unit of geological time, having two or more eras; for example, the Phanerozoic eon includes the Paleozoic, Mesozoic, and Cenozoic eras.

Equilibrium A state or condition in which the influences of energy forces and related reactions are canceled by each other, the result of which is a balanced, stable, and unchanging system. *Thermal equilibrium* is said to occur when no heat exchange has taken place within a body or between the body and its surroundings.

Eukaryote A eukaryotic cell contains a definitive nucleus, in which nuclear material is surrounded by a membrane and cytoplasm-containing organelles. Eukaryotic cells comprise all living things except **prokaryote** bacteria and blue-green algae, which have no well-formed nuclei.

Exfoliation A weathering process by which concentric shells, slabs, sheets, and/or layers of rock, ranging in thickness from less than a centimeter to several meters, are successively stripped from the surface area of a large rock mass.

Feedback In systems, negative feedback is the control of an input to the system that affects the output to the extent that a portion of the output is then returned to alter the original input. Positive feedback is the opposite and reinforces the original input.

Felsic A shorthand adjective derived from *fe*ldspar + *l*enad (feldspathoid) + *si*lica + *c,* which is used to describe light-colored igneous rocks such as quartz and feldspars. It is a complementary term to **mafic**.

Fission The splitting of an atom's nucleus with the resultant release of enormous energy and the production of smaller atoms of different elements. Fission occurs spontaneously in the nuclei of unstable radioactive elements, such as U-235 and Pu-239, which are used in the generation of nuclear electrical power, as well as in nuclear bombs.

Fluorescence The emission of electromagnetic radiation from a substance resulting from some form of energy flowing into that substance. The emission (light) will stop when the external energy stops striking the substance (e.g., the phosphor coating inside fluorescent light tubes only glows when excited by flowing electrons when electricity to the tube is turned on.)

Foraminifer An order of marine protozoa having a secreted shell enclosing its body.

Friable Describes extremely brittle rock that crumbles easily, for example, thinly bedded shale.

Gluons Hypothetical, massless, neutral elementary particles that carry the strong force (interactions) that binds quarks, neutrons, and protons together.

Guyots Seamounts with flat tops.

Gyre The circular or spiral movement of seawater in each of the major ocean basins. Gyres are located on a subtropical, high-pressure region. The gyre's motion is caused by (a) the transmission of warm surface water that flows toward the poles, (b) the rotation of the Earth on its axis, and (c) the prevailing winds. (See also **Coriolis force.**) The water in the gyre moves clockwise in the Northern Hemisphere and counterclockwise in the Southern Hemisphere.

Hexagonal Close-Packed Refers to a *crystal system* that is distinguished by one axis of either threefold or sixfold symmetry. The perpendicular axis is unequal in length to the three identical axes that intersect at 120° angles.

Homogenous Refers to substances that have the same structures and consistency throughout their dimensions.

Humus The brown or black, generally stable, organic substance in soil that consists of decomposed vegetable matter and other unidentified decayed sources.

Hydrate A mineral compound containing water combined within its crystal structure.

Hydrocarbon An organic compound whose molecules consist of *only* carbon and hydrogen. Hydrocarbons can form chains of molecules known as "aromatic compounds" or form stable rings known as "aliphatic compounds." All petroleum products and their derivatives are hydrocarbons.

Hydrogenation Also known as "hardening." Hydrogenation involves the conversion of an unsaturated compound into a saturated compound by the addition of hydrogen. For example, liquid oils are converted to solids by adding hydrogen, such as hydrogenating corn oil to form margarine.

Hydrolysis The chemical reaction of a compound with water; for example, salts of weak acids or bases hydrolyze in aqueous solutions. In geology, it refers to the reaction between silicates and either water or an aqueous solution.

Hypotenuse In a right triangle, the side opposite the right angle.

Igneous Refers to rock formed by the cooling of magma (molten rock).

Inorganic Involving neither organic life nor the products of organic life or carbon compounds in living tissues (i.e., nonorganic). Many inorganic compounds contain carbon but are not organic, for example, CO_2, soot, diamonds, cyanide (CN).

Insolation The combined solar and sky radiation striking the Earth or other planets, and the geological effect of those rays on the surface material of the Earth or planet, namely, the weathering of rocks.

Intermontane Lying between mountains, for example, valleys. Also referred to as "intermountain."

Ions Atoms or molecules that have gained or lost electrons and thus have acquired either a negative or positive electrical charge. If the ion has a net positive charge in solution, it is a *cation.* If it has a net negative charge in solution, it is an *anion.*

Isostasy A theory that states that there is a condition of equilibrium among the outer spheres of the Earth, or when Earth's theoretical gravity is in balance with the continental and oceanic crusts.

Isotopes Atoms of the same element that have the same atomic number (protons) but have different atomic masses (number of neutrons). Isotopes of a given element all have the same chemical characteristics (electrons and protons), but they have slightly different physical properties.

Kapok The silky fiber from the silk-cotton tree that is native to the tropical regions of South America (rainforests). The fibers are harvested and used for padding in mattresses, pillows, and life preservers.

Karst A topographical area of irregular, dissolved limestone, dolomite, or gypsum characterized by fissures, sinkholes, underground streams, and caverns.

Kinetic Energy Energy associated with motion (e.g., molecular motion).

Knap To flake or chip rocks, such as flints, quartz, and obsidian, with a sharp blow.

Latent Heat The quantity of heat that is absorbed or released when a substance undergoes a change in its physical state, for example, from ice to water or from water to steam.

Lava Molten or fluid rock that flows from a volcano or fissure in the Earth's surface.

Leached Removed or dissolved, as the soluble components from rock or ores by the percolation of liquid, for example, water.

Legumes Plants that take nitrogen (N_2) from the atmosphere and incorporate it into nodules on their roots, for example, bean and pea plants.

Lenticular Refers to the shape of an object that resembles a convex lens, for example, the lens-shaped rock found in an anticline.

Mafic A term derived from *ma*gnesium + *f*erric + *ic* to describe dark igneous rocks and ferromagnesian minerals. A complementary term of **felsic**.

Magnotometry The science that studies and measures the magnitude and direction of the Earth's magnetic field using an instrument known as a magnetometer.

Moment of Inertia A measure equal to an object's resistance to angular acceleration. It is the sum of all the products of each mass element of a body multiplied by the square of its distance from an axis.

Nascent Means "newborn" and refers to something coming into existence. In chemistry, it refers to an atom or simple compound that, at the moment of its liberation from chemical combination, may have greater activity than during its usual state.

Neutrons Fundamental particles of matter. Each neutron has a mass of 1.009 (of a proton) and has no electrical charge. It is a part of the nucleus of all elements except hydrogen.

Omnivore In zoology, it refers to an animal that feeds on both animal and vegetable substances.

Opaque Describes the characteristics of a material or a liquid that is impervious or impenetrable to the passage of light.

Ores Naturally occurring rocks and/or minerals or composites from which valuable components can be extracted for profit, for example, iron ore, sulfur ore.

Outcropping A geological formation that juts out on the Earth's surface above the soil, for example, outcropping of bedrock.

Oxidation The process by which an element (a metal or nonmetal) combines with oxygen and forms an oxide, for example, magnesium oxide.

Panspermia An evolutionary theory, first proposed by the Swedish chemist Svante August Arrhenius, that states life came to Earth as a bacterial spore or other simple form from outside the Earth's solar system.

Particulates Fine, solid particles that remain individually scattered in gases and other emissions. Particulates, such as bacteria, pollen, and dust, can be irritants and are often present in rain, fog, smog, fumes, and so forth. Very fine particulate matter less than 2.5 microns is found in ambient (nonpolluted) air.

Pedology The scientific study of the origin, structure, classification, and use of soils.

Percolation In geology, percolation is the process by which the nonturbulent flow of water infiltrates the small openings of a porous material.

Perturbation A variation or departure of an electron or celestial body from its orbit as a result of the influence of one or more external bodies or forces acting on that electron or celestial body.

Photosynthesis Process by which chlorophyll-containing cells in plants and bacteria, in the presence of sunlight, convert carbon dioxide and water into carbohydrates, resulting in the simultaneous release of energy and oxygen.

Phytoplankton Microscopic plants (algae) that float on or near the surface of the sea. They are the basis of the food web for a variety of aquatic life and the plant equivalent of **zooplankton**.

Plano-Convex Refers to an optical lens that is flat on one side and convex (or curved outward) on the other.

Plasma In physics, the electrically neutral, highly ionized gas composed of ions, electrons, and neutral particles. In biology, the clear, yellowish, liquid portion of the blood. In geology, the part of soils that has been moved and reorganized by natural processes.

Pluton Deep-seated igneous rock that formed beneath the Earth's surface during the cooling off and hardening of magma. Also referred to as "plutonic rock."

Potable Refers to water that is safe to drink.

Precession Refers to the wobbling or circling of the Earth's orbit. It is a complex motion of a rotating body (Earth) subject to a torque acting upon it as a result of gravity.

Prokaryote A primitive cell or organism where genetic material (DNA) is not enclosed in a nucleus encapsulated within a membrane. Bacteria and blue-green algae are examples. All other organisms are **eukaryotes**, in which the DNA is enclosed within a nucleus.

Protons Positively charged particles found in the nuclei of atoms. The atomic number is equal to the number of protons in the element's nucleus.

Pumice A light-colored, glassy, abrasive, porous volcanic rock.

Pyrolysis Chemical change or decomposition caused as a result of extreme heat.

Quantum Theory In 1900, the German physicist Max Planck developed the quantum action theory, which states that energy does not flow in an unbroken stream but instead proceeds or jumps in discrete packets or quanta (singular "quantum"). A photon is a quantum of light, which was described by Albert Einstein as a "packet of electromagnetic particle/energy." Today, many people mistakenly consider a "quantum leap" to be a great stride or large advancement of events or accomplishment, rather than the tiny quantity of energy or mass it describes.

Quarks Hypothetical subnuclear particles that have an electric charge one-third to two-thirds that of the electron.

Radicals Also known as "free radicals." A group of atoms that can act as a single unit during a chemical reaction and that have one unpaired electron.

Radiometric Refers to the essence of kinetic energy of gas molecules measured by a radiometer, a device designed to measure the motion of molecules in a rarefied atmosphere.

Reducing Refers to an agent (reductant) that chemically induces reduction, that is, acceptance of one or more electrons by an atom or ion. It is the opposite of **oxidation,** although the reducing agent itself is oxidized (loses electrons).

Rheology The study of the deformation (alteration) and flow of matter.

Seamount An undersea mountain with an elevation of at least 1,000 meters above the seafloor.

Seismic Refers to vibrations or tremors on the Earth, including earthquakes and artificially induced fluctuations, that is explosions.

Seismic Profiler An instrument used to assemble a profile of the surface features of the ocean floor, specifically of the stratigraphy of rocks and sediment beneath the ocean floor.

Speciation The evolutionary process in the formation of new species. "Species" is the lowest ranking in the classification of organisms. It is the distinguishable group with a common ancestry, able to reproduce fertile offspring, and being geographically distinct.

Spectrometer A kind of spectroscope designed to measure, either directly on a scale or by a photometer, the electromagnetic wavelengths and their radiation intensities given off by materials.

Standard Model In physics, a collection of established experimental knowledge and theories that summarizes the concepts of field. It includes the three generations of quarks and leptons, the electroweak theory of weak and electromagnetic forces, and the quantum chromodynamic theory of strong forces.

Subduction A geological process whereby one edge of a lithospheric (crustal) plate descends below an adjoining plate.

Sublimation The direct passage of a substance from a solid to a vapor without going through the liquid state.

Submersible A compact vessel or vehicle that is self-propelled and capable of operating underwater for the purpose of exploration and investigation.

Sunspot Giant loop of magnetic flux around the sun that causes a temporary dark area on the surface of the sun. The sun has a high magnetic field that is wound around the sun rather than aligned north to south, as is the Earth's magnetic field. Sunspots range from about 3,000 to 62,000 miles in diameter and become solar flares that emit radio waves that leap out from the sun, often affecting electronic communications on Earth. The sun has a 22-year cycle, whereas sunspots exhibit two 11-year cycles during this period. The periods for sunspot cycles have varied over history, and their effects on Earth's climate are still not fully understood.

Superposition Pertains to the order of sedimentary rocks, that is, where the youngest sedimentary strata are at the top and the oldest at the bottom.

Symbiotic Refers to the relationship between two different organisms in close association with one another that may be beneficial to at least one of the organisms but does not cause harm to either.

Syncline In geology, a fold in the bedrock that forms a troughlike area where the rocks incline inward from opposite sides. The core of the syncline contains the younger rocks. The opposite of an anticline.

Temperature Inversion The abnormal increase of air temperature with height, that is, the upper portion of the atmospheric layer is warmer than the lower. This increase is opposite to the usual state of the troposphere, the lowest level of the Earth's atmosphere, in which the temperature decreases with height. Inversions occur when the lower air layer is cooled by an upper layer of air that has abated or sunk to a lower level. For example, during nighttime hours outgoing radiation cools the surface of the Earth, which subsequently cools the lower air layers and creates a nocturnal surface inversion, ranging from a few inches to several hundred feet high. The effect is the curtailment of the vertical movement of the air. Inversions can also trap pollutants in the air that can be potentially hazardous or irritating to portions of the population who suffer from respiratory conditions.

Tetrahedron A solid geometric shape seen in some crystals that has four faces, each with equal intercepts on all three axes.

Thermocline The region in a thermally stratified body of water in which the temperature drops rapidly with depth.

Thermodynamics The study of energy and the laws governing the transfer of energy from one form to another, particularly relating to the behavior of systems in which temperature is a factor (i.e., the flow of heat and availability of energy to perform work).

Thermohaline Refers to the joint activity of temperature variations and salinity of the oceans that affects the vertical movements of seawater, which are generated by differences in density.

Thermosphere The layer of air of steadily increasing temperatures that extends from the top of the mesosphere to the exosphere and onto outer space. It ranges from about 70–80 kilometers (40–50 miles) to the outer edge of the atmosphere.

Tillite A sedimentary rock formed by the lithification of unstratified and unsorted glacial sediment known as glacial till.

Tomographic Refers to the procedure that uses computer-aided x-ray-type equipment to photograph a predetermined plane of a solid body by eliminating the images of the other planes of that solid body. This is the principle of CAT (computer-aided tomography) scans of the human body that are used for medical diagnosis.

Tritium A radioactive isotope of hydrogen with an atomic mass of 3 and a half-life of 12.5 years. It is one form of heavy hydrogen; the other is deuterium, which has an atomic mass of 2. Both make up very little of all the hydrogen compounds found on Earth.

Tube Worms There is a great variety of tube worm species. Some live near the surface of salt or brackish water and foul piers and bottoms of boats. More exotic red and white tubelike species live near the hydrothermal vents that appear at the deepest levels on the seafloor and from which superheated water spews forth and provides the chemicals necessary to sustain these and other organisms in an environment completely devoid of sunlight.

Ultramafic Igneous rocks, such as olivines and augites, which are composed chiefly of mafic minerals (see **mafic**).

Ultraviolet (UV) The radiation wavelength in the electromagnetic spectrum from 100 to 3900 angstroms (Å), between x-rays and visible violet light.

Uniformitarianism A principle devised by the British geologist Sir Charles Lyell that states "The present is the key to the past," meaning that currently observed geological changes and processes are adequate to explain geological history. He based much of his concept on classification of the strata of ancient marine beds. He observed that the layers of sediment closest to the surface contained the shells and remains of animals species still living in modern times. Conversely, the deeper, older strata contained fossils of extinct species.

Vector A quantity specified by magnitude *and* direction whose components convert from one coordinate system to another in the same manner as the components of a displacement. Vector quantities may be added and subtracted. For instance, in the equation $F = ma$, the F stands for force, which is a vector quality that has both magnitude and direction and is calculated by the product of the mass (m) and acceleration (a) related to the push or pull on an object.

Viscous Refers to heavy, molasses-like fluids that have a high resistance to flow.

Wane Synonymous with "decline." For example, it refers to a gradual *decrease* or decline in the illuminated visible surface area of the Earth's moon.

Wax Opposite of **wane**. It refers to a gradual *increase,* for example, in the illuminated visible surface area of the Earth's moon.

Xerophytic Refers to desert plants that have adapted to and grow in an environment with little moisture, for example, cactus.

Zooplankton The animal corollary of **phytoplankton**. Zooplankton are microscopic aquatic animals that float on or near the surface of the sea.

Selected Bibliography

Print Media

Allen, Myles, et al. "Uncertainty in the IPCC's Third Assessment Report." Policy Forum: Climate Change. *Science,* 20 July 2001, 430–33.

Alley, Richard B., and Michael L. Bender. "Earth From the Inside Out: Greenland Ice Cores: Frozen in Time." *Scientific American,* special publication, 2000, 4–9.

Alvord, Valerie. "Hard Times Pull Down the 'Mine in the Sky.'" *USA Today,* 4 October 2000.

Asimov, Isaac. *Isaac Asimov's Guide to Earth and Space.* New York: Fawcett Crest, 1991.

———. *Asimov's Chronology of Science and Discovery.* New York: Harper and Row, 1989.

———. *Isaac Asimov's Beginnings: The Story of Origins—of Mankind, Life, the Earth, the Universe.* New York: Berkley Books, 1987.

Bailey, Ronald. *Earth Report 2000: Revisiting the True State of the Planet.* New York: McGraw-Hill, 2000.

———. *EcoScam: the False Prophets of Ecological Apocalypse.* New York: St. Martin's Press, 1993.

Bair, Jeffrey. "Biker Runs Harley with Soybean Oil." *Houston Chronicle,* Houston, Tex., 27 February 2001.

Balter, Michael, and Ann Gibbons, "A Glimpse of Humans' First Journey Out of Africa." *Science,* 12 May 2000, 948–50.

Barnes-Svarney, Patricia, ed. *The New York Public Library Science Desk Reference.* New York: Macmillan, 1995.

Bates, Robert L., and Julia A. Jackson. *Dictionary of Geological Terms.* New York: Anchor Books/Doubleday, 1984.

Battle, M., et al. "Global Carbon Sinks, and Their Variability Inferred from Atmospheric O_2, O_3, and C." *Science,* 31 March 2000, 2467–70.

Bobick, James E., and Margery Peffer. *Science and Technology Desk Reference.* 2d ed. New York: Gale, 1996.

Bochinski, Julianne Blair. *The Complete Handbook of Science Fair Projects.* New York: John Wiley and Sons, 1996.

Bower, Miranda. *Experiments with Weather.* Minneapolis: Lerner, 1994.

Brooks, Harold. "National Severe Storms Laboratory." Weather Focus. *USA Today,* 17 May 2001.

Crawford, Mark, J. *Physical Geology.* Lincoln, Nebr.: Cliffs Notes, 1998.

Chernicoff, Stanley. *Geology: An Introduction to Physical Geology.* New York: Houghton Mifflin, 1999.

Crowley, Thomas J., and Robert A. Berner. "CO_2 and Climate Change." *Science,* 4 May 2001, 870–72.

Daintith, John, Sarah Mitchell, Elizabeth Tootill, and Derek Gjertsen. *Biographical Encyclopedia of Scientists.* 2d ed., 2 vols. Bristol, England, and Philadelphia: Institute of Physics Publishing, 1981–1994.

de Menocal, Peter B. "Culture Responses to Climate Change during the Late Holocene." *Science,* 27 April 2001, 667–73.

Denecke, Edward J., Jr. *Let's Review: Earth Science.* Hauppauge, N.Y.: Barron's Educational Series, 1995.

Dobbin, Ben. "Small Town Gets Big Payoff." *Valley Morning Star*, Hurlingen, Tex., 22 October 2000.

Dombroski, Lori. "The Grain Dilemma: Feeding Livestock or Feeding the World." *Everybody Eats: Fort Collins Food Co-op,* September/October 1995.

Erickson, Jon. *Plate Tectonics: Unraveling the Mysteries of the Earth.* New York: Checkmark Books, 2001.

Ernst, W. G., ed. *Earth Systems, Processes and Issues.* New York: Cambridge University Press, 2000.

Falkenberg, Lisa, "Air Pollution Could Trigger Heart Attacks." *Valley Morning Star,* Hurlingen, Tex., 12 June 2001.

Flavin, Christopher, Harry French, and Gary Gardner, et al. *State of the World 2002: Worldwatch Institute.* New York: W. W. Norton, 2002.

Fredrickson, James K., and Tullis C. Onstott. "Microbes Deep Inside the Earth." *Scientific American,* "Earth from Inside Out," special publication, 2000, 10–15.

Goldstein, Natalie. *Earth Almanac: An Annual Geophysical Review of the State of the Planet.,* 2d ed. Westport, Conn.: Oryx Press, 2002.

Gonzales, Frank, I. "Tsunami." *Scientific American,* May 1999, 56–65.

Gorbachev, Mikhail. "Out of Water: The Distant Alarm Comes Closer." *Civilization,* October/November 2000, 82–83.

Gunn, Angus M. *The Impact of Geology on the United States: A Reference Guide to Benefits and Hazards.* Westport, Conn.: Greenwood Press, 2001.

Hamblin, Kenneth W., and Eric H. Christiansen. *Earth's Dynamic Systems.* Upper Saddle River, N.J.: Prentice Hall, 1998.

The Handy Science Answer Book. New York: Carnegie Library of Pittsburgh, 1997.

Hazen, Robert M. "Life's Rocky Start." *Scientific American,* April 2001, 77–85.

Herzog, Howard, Baldur Eliasson, and Olav Kaarstad. "Capturing Greenhouse Gases." *Scientific American,* February 2000, 72–79.

Hodell, David A., et al. "Solar Forcing of Drought Frequency in the Maya Lowlands." *Science,* 18 May 2001, 1367–70.

Isaacs, Alan, John Daintith, and Elizabeth Martin, eds. *Concise Science Dictionary.* 3d ed. New York: Oxford University Press, 1996.

Jones, Lorraine. *Super Science Projects about Weather and Natural Forces.* New York: Rosen Publishing Group, 2000.

Kardos, Thomas. *75 Easy Earth Science Demonstrations.* Portland, Me.: J. Weston Walch, 1997.

Kious, Jacquelyne W., and Robert I. Tilling. *This Dynamic Earth: The Story of Plate Tectonics.* Denver, Colo.: U.S. Geological Survey; Washington, D.C.: U.S. Government Printing Office, 1999.

Kolbert, Elizabeth. "Comment: Hot and Cold." *The New Yorker,* 13 August 2001.

Kramers, Jan. "The Smile of the Cheshire Cat." *Science,* 27 July 2001, 618–19.

Krebs, Robert E. *Scientific Laws, Principles, and Theories: A Reference Guide.* Westport, Conn.: Greenwood Press, 2001.

———. *Scientific Development and Misconceptions through the Ages.* Westport, Conn.: Greenwood Press, 1999.

———. *The History and Use of Our Earth's Chemical Elements: A Reference Guide.* Westport, Conn.: Greenwood Press, 1998.

Kuntz, Margy. *Adventures in Earth Science.* Parsippany, N.J.: Fearon Teacher Aids, 1988.

Lewerenz, Dan. "Grain Furnace Usage Heats Up: More Planters Save Money by Burning Crops Instead of Fuel." *Houston Chronicle,* Houston, Tex., 24 February 2001.

Lewis, Richard L. *Hawley's Condensed Chemical Dictionary.* 12th ed. New York: Van Nostrand Reinhold, 1993.

Lloyd, Alan C. "The Power Plant in Your Basement." *Scientific American,* July 1999, 80–86.

Lomborg, Bjørn. *The Skeptical Environmentalist: Measuring the Real State of the World.* New York: Cambridge U. Press, 2001.

Lovelock, James. *Gaia: A New Look at Life on Earth.* New York: Oxford University Press. 1995.

———. *The Ages of Gaia: A Biography of Our Living Earth.* New York: Bantam Books, 1988.

Mandell, Muriel. *Simple Weather Experiments with Everyday Materials.* New York: Sterling Publishing, 1990.

Margulis, Lynn, and Dorion Sagan. *Slanted Truths: Essays on Gaia, Symbiosis, and Evolution.* New York: Springer-Verlag, 1997.

Mason, Stephen F. *A History of the Sciences.* New York: Macmillan, 1962.

Marcus, David L. "Charles Darwin Gets Thrown Out of School." *U.S. News and World Report,* 30 August 1999.

McGuire, Thomas. *Reviewing Earth Science: The Physical Setting.* New York: Amsco School Publications, 2000.

McNeil, John S. "Peeking Behind the Sun: New Ways to Predict Stormy Solar Weather." *U.S. News and World Report,* 5 June 2000.

Millar, David, Jan Millar, John Millar, and Margaret Millar. *The Cambridge Dictionary of Scientists.* New York: Cambridge University Press, 1996.

Mongillo, John, and Linda Sierdt-Warshaw. *Encyclopedia of Environmental Science.* Phoenix, Ariz.: Oryx Press, 2000.

Mussett, Alan, E. and M. Aftab Khan. *Looking Into the Earth: An Introduction to Geological Geophysics.* New York: Cambridge University Press, 2000.

Nemecek, Sasha. "Trends in Archeology: Who Were the First Americans?" *Scientific American,* September 2000, 80–87.

Newton, David, E. *Oryx Frontiers of Science Series: Recent Advances and Issues in Chemistry.* Phoenix, Ariz.: Oryx Press, 1999.

Ogawa, Hiroshi, et al. "Production of Refractory Dissolved Organic Matter by Bacteria." *Science,* 4 May 2001, 917–20.

Oreskes, Naomi, ed. *Plate Tectonics: An Insider's History of the Modern Theory of the Earth.* Boulder, Colo.: Westview Press, 2001.

Oxlade, Chris, Corinne Stockley, and Jane Wertheim. *The Usborne Illustrated Dictionary of Physics.* Tulsa, Okla.: EDC Publishing, 1996.

Parker, Sybil P., ed. *McGraw-Hill Concise Encyclopedia of Science and Technology.* 3d ed. New York: McGraw-Hill, 1994.

———. *McGraw-Hill Dictionary of Scientific and Technical Terms.* 5th ed. New York: McGraw-Hill, 1994.

Ransom, Barbara, and Sonya Wainwright. *Recent Advances and Issues in the Geological Sciences.* Westport, Conn.: Oryx Press, 2002.

Reilly, John, et al. "Uncertainty and Climate Change Assessments." Public Forum: Climate Change. *Science,* 20 July 2001, 430–33.

Reynolds, Ross. *Cambridge Guide to the Weather.* New York: Cambridge University Press, 2000.

Robinson, Edwin Simon. *Basic Physical Geology.* New York: John Wiley and Sons, 1982.

Rudin, Norah, *Dictionary of Modern Biology.* Hauppauge, N.Y.: Barron's Educational Series, 1997.

Santos, Miguel. *The Environmental Crisis.* Westport, Conn.: Greenwood Press, 1999.

Scientific American Science Desk Reference. New York: John Wiley and Sons, 1999.

Seitz, Frederick. *Hot Talk, Cold Science: Global Warming's Unfinished Debate.* Oakland, Calif.: The Independent Institute, 1997.

Singer, Charles. *A History of Scientific Ideas: From the Dawn of Man to the Twentieth Century.* New York: Barnes and Noble, 1959.

"Spacecraft Finds Water beneath Ganymede." *USA Today,* 18 December 2000.

Suess, Erwin, Gerhard Bohrmann, and Jens Greinert. "Flammable Ice." *Scientific American,* November 1999, 76–83.

"Sun's Magnetic Flip Affects Cosmic Ray Penetration." Science and Education. *USA Today,* 10 April 2001.

Suplee, Curt. "Studies May Alter Insights Into Warming." Global Warming Information Page. *Washington Post,* 15 March 1997.

Thain, M., and M. Hickman. *The Penguin Dictionary of Biology,* 9th ed. London, Penguin Books, 1994.

Thompson, Graham R., and Jonathan Turk. *Earth Science and the Environment.* New York: Harcourt Brace, 1999.

Tolman, Marvin N. *Hands-On Earth Science Activities: For Grades K-8.* West Nyack, N.Y.: Parker Publishing, 1995.

Tomikel, John. *Basic Earth Science: Earth Processes and Environments.* New York: Allegheny Press, 1981.

"USA Today Weather Focus." *USA Today,* 17–18 July 2001.

VanCleave, Janice Pratt. *Janice VanCleave's A + Projects in Earth Science.* New York: John Wiley and Sons, 1999.

———. *Janice VanCleave's Earth Science for Every Kid.* New York: John Wiley and Sons, 1991.

Vecchione, Glen. *100 Amazing Make-It-Yourself Science Fair Projects.* New York: Sterling Publishing, 1994.

Vergano, Dan. "Ancestor Theory Falls From Grace." *USA Today,* 15 January 2001.

Watson, Traci, and Jonathan Weisman. "6 Ways to Combat Global Warming." *USA Today,* 16 July 2001.

Weinberg, Steven. *Facing Up: Science and Its Culture Adversaries.* Cambridge, Mass.: Harvard University Press, 2001.

———. "A Unified Physics by 2050?" *Scientific American,* December 1999, 68–75.

Wennberg, P.O., et al. "Hydrogen Radicals, Nitrogen Radicals, and the Production of O_3 in the Upper Troposphere." *Science,* January 1998, 49–53.

White, Larry. *Air: Simple Experiments for Young Scientists.* Brookfield, Conn.: Millbrook Press, 1995.

Winchester, Simon. *The Map That Changed the World: William Smith and the Birth of Modern Geology.* New York: Harper Collins, 2001.

Wright, Robert. "The Accidental Creationist." *New Yorker,* 13 December 1999, .

Zachos, James, et al. "Trends, Rhythms, and Aberrations in Global Climate 65 Million Years Ago to Present." *Science,* 27 April 2001, 686–93..

Digital Media

CD-ROMs

Academic Press Dictionary of Science and Technology. San Diego: Academic Press, 1996. CD-ROM.

Crolier Multimedia Encyclopedia. Danbury, Conn.: Grolier Interactive, 1998. CD-ROM.

Encyclopedia Britannica. Chicago: Britannica, 2001. CD-ROM.

GETIT (Geoscience Education through Interactive Technology). The Learning Team. New York: Cambrian Systems, 2000. CD-ROM.

Microsoft Encarta. New York: Funk and Wagnalls, 1994. CD-ROM.

Our Earth. Tucker, Ga.: Megasystems, 1999. CD-ROM.

Oxford English Dictionary. London: Oxford University Press, 2000. CD-ROM.

Internet Web Sites

Alden, Andrew. *A Guided Tour to Today's Mantle:* About Geology. http://www.geology.about.com/science/geology.libr, accessed 2001.

Alean, Jurg, and Marco Fulle. Etna Hoops it Up: BBC News. http://news.bbc.co.uk/hi/english/sci/tec/newsid_696000/696953.stm, accessed 2001.

Atmospheric Chemistry in the Mesosphere and Thermosphere. http://www.sprl.umich.edu/SPRL/research/atmo_chem.html, accessed 2001.

The Atmospheric Oxygen Cycle. http://www.agu.org, accessed 2001.

Babin, Steven M. *Water Vapor Myths: A Brief Tutorial.* Penn State University 1998. http://www.fermi.jhuapl.edu/people/babin/vapor/index.html, accessed 2001.

Biogeochemical Cycle: Nutrient Cycle. http://www.xrefer.com, accessed 2001.

Burning Issues. *Particle Sources.* http://www.buriningissues.org, accessed 2001.

Carlowicz, Michael. *Inge Lehmann Biography.* http://www.agu.org/inside/awards/lehmann2, accessed 2001.

Clark, Greg. *Extraterrestrial Water: Water from the Heavens.* http://www.space,com/science/astronomy/meteorite_water.html, accessed 2001.

Clickable Geologic Time Scale. http://www.geol.ucsb.edu/Outreach/tTimeScale, accessed 2001.

Cohen, Ronald, and Lars Stixrude. *Crystals at the Center of the Earth: Anisotropy of the Earth's Inner Core.* http://www.psc.edu/science/Cohen_Stix/cohen.stix.html, accessed 2001.

Columbia Encyclopedia. 6th ed. "Atmosphere." http://www.bartleby.com, accessed 2001.

Columbia University. http://www.columbia.edu.html, accessed 2001.

Common Ozone Depletion Myths. http://www.members.aol.com/jimn469897/common.html, accessed 2001.

Creation/Flood Myths of the World. http://www.templar.bess.net/Comp.names/book of, accessed 2001.

The Daily Revolution: Ancient Weather. http://www.dailyrevolution.org, accessed 2001.

"The Deadliest Eruptions." In *A Sourcebook on the Effects of Eruptions,* 1984. www.volcano.und.nodak.edu/vwdocs/vw_hyperexchange/deadly_volcs, accessed 2001.

Debates, Gatherings and Court Decisions. http://www.talkorigins.org/origins/faqdebates, accessed 2001.

Department of Geology and Geophysics, University of Alaska. http://www.uaf.edu/geology/geo_time.html, accessed 2001.

Earth Day 1998 Fact Sheet. *Myths and Facts About the Environment.* http://www.nationalcenter.org/EarthDay99Myths, accessed 2001.

Earth's Cycles. http://www.rainbow.ideo, accessed 2001.

"Earth's Magnetism." *Scientific American:* Ask the Experts: Geology. http://www.sciam.com/ask expert/geology/geology4.html, accessed 2001.

The Edwards Aquifer Homepage. http://www.edwards aquifer.net, accessed 2001.

Environmental News Service. http://www.ens.lycos.com/ens/may99/1999L-05-06-06, accessed 2001.

Evolution of Life on Earth. http://(www.scibridge, sdsu.edu/coursemats/introsci/evolution_of_life, accessed 2001.

Find Articles.com. *Science News.* http://www.findarticles.com/m1200/20-15658037882, accessed 2001.

The Franklin Institute Online. *Investigation Wind Energy.* http://www.sln.fi.edu/tfi/units/energy, accessed 2001.

Geo-2000: Freshwater: State of the Environment. http://www.grid2.cr.usgs.gov/geo2000/english/0046, accessed 2001.

Global Climate Maps. http://www.fao.org/waicnt/faoinfo/sustdev/eldirect/climate/Elsptext, accessed 2001.

Greening Earth Society. http://www.greeningearth society.org/Articles, accessed 2001.

Hans, Junginger. *The Human Skin: Your Largest Organ.* http://www.arubaaloe.com.humanskin, accessed 2001.

Hipschman, Ron. *Lightning.* http://www.explora torium.edu/ronh/weather, accessed 2001.

History of Climate Change: *How do we Know About Past Climate.* http://www.athena.ivv.nasa.gov, accessed 2001.

"How Earthquakes Cause Damage: *Tsunamis.*" http://www.aolsvc.worldbook.aol.com, accessed 2001.

How to Forecast Weather by Nature. http://www.stalkingthewild.com/weather.html, accessed 2001.

Koermer, Jim. *PSC Meteorology Program Cloud Boutique.* http://www.vortex.Plymouth,edu/clouds, accessed 2001.

Köppen Climate Classification. http://www.huizen.dds.hl?~gvg/ctkoppe2, accessed 2001.

Louie, J. *Mercalli Earthquake Damage Scale.* http://www.seismo.unr.edu/ftp/pub/louie/class/100/mercalli.html, accessed 2001.

Louie, J. *Richter Earthquake Damage Scale.* http://www.seismo.unr.edu/ftp/pub/louie/class/100/magnitude.html, accessed 2001.

Low Level Clouds (pictures). http://www.met.hu/cloudalbum/alfelhok.htlm, accessed 2001.

Massett, Larry. *The Weather Notebook: Ancient Weather Theory.* http://www.Mountwashington.org, accessed 2001.

Metamorphic Rocks (Metamorphic Settings). http://www.science.ubc.ca/~geol202/meta/metaset, accessed 2001.

Metamorphic Rocks (Mr. Burns Science Page). http://www.home.epix.net/~tgburns/metarock, accessed 2001.

Monastersky, Richard. "The Mush Zone: a Slurpy Layer Lurks Deep inside the Planet." *Science News.* http://www.findarticles.com, accessed 2001.

Most Deadly Earthquakes in History. http://www.aolsvc.worldbook.aol.com, accessed 2001.

Myths of 1999. http://www.sepp.org/NewSEPP/environmyths, accessed 2001.

National Aeronautics and Space Administration. http://www.nasa.gov, accessed 2001.

National Oceanic and Atmospheric Administration. http://www.ngdc.noaa.gov, accessed 2001.

T*he Nitrogen Cycle.* http://www.clab.cecil.ccmd.us, accessed 2001.

Nitrogen Cycle: CHM 110. http://www.elmhurst.edu, accessed 2001.

Nutrient Cycling: High School Biology. http://www.homeworkhelp.com, accessed 2001.

Oxygen Cycle. http://www.xrefer.com, accessed 2001.

Ozone Reality Check. http://www.foe.org/ozone/into, accessed 2001.

Parsons, Robert. Re: *Critical Analysis of* "Environmental Overkill, Ch. 3 & 4." http://www.tp.alternatives.com, accessed 2001.

Peel the Planet: Atmosphere. http://www.discovery.com, accessed 2001.

Rockdoctors Guide to Metamorphic Rocks. http//www.cobweb.net/~bug2/rock5, accessed 2001.

The Science of Ozone Depletion. http://www.ucsusa.org/resourses/ozone.science, accessed 2001.

SciTech Daily News (Science News Sources). http://www.scitechdaily.com, accessed 2001.

Sedimentary Rocks. http://www.home.expix.net, accessed 2001.

Singer, Fred S. *Environmental Soil Resources on the Web.* http://www.swcs.org, accessed 2001.

10 Tips for Water Conservation. http://cleanoceanaction.org/COATips/COASTips/H2Oconservel, accessed 2001.

The Terrible Power of Earthquakes. http://www.aolsvc.worldbook.aol.com, accessed 2001.

Texans Can Be Water Wise by Making Simple Changes at Home. http://www.tnrcc.state.tx.us/water/wu/drought/nycu.html, accessed 2001.

This Dynamic Earth. U.S. Geological Survey. http://www.pubs.usgs.gov/publications/text/understanding, accessed 2001.

This Week In Science. *The Longest Wake.* http://www.labonweb.com, accessed 2001.

Three Mile Island 2 Accident Report: Health Effects. http://www.nrc.gov/OPA/gmo/tip/tmi/htm, accessed 2001.

Ultralow Velocity Zones (ULVZ). *Investigating Earth's Lower Mantle with Seismology.* http://www.geo.berkeley,edu/seismo/annual-report/ar97, accessed 2001.

University of Alaska. *Geologic Time Scale.* Department of Geology and Geophysics. http://www.uaf.edu/geology/geo_time.html, accessed 2001.

University of California, Berkeley. *UCMP Web Time Machine.* http://www.ucmp.berkeley.edu, accessed 2001.

University of Connecticut. http://www.uct.ac.za, accessed 2001.

University of Idaho. *Climate Classification.* http://www.snow.ag.uidaho.edu, accessed 2001.

University of Illinois. *Online Meteorology Guide.* http://www.2010.atmos.uluc.edu, accessed 2001.

"USA Today Weather Focus." *USA Today,* 17–18 July 2001.

U.S. Geological Survey. http://www.usgs.gov/pubs.html, accessed 2001.

U.S. Geological Survey. *The Geologic Time Scale.* http://www.geology.er.usgs.gov/paleo.geotime.html, accessed 2001.

U.S. Geological Survey. *Major Divisions of Geologic Time.* http://www.pubs.usgs.gov/gip/geotime/divisions.html, accessed 2001.

U.S. Geological Survey. *Radiometric Time Scale.* http://www.pubs.usgs.gov/gip/geotime/radiometric, accessed 2001.

U.S. Geological Survey. *Relative Time Scale.* http://www.pubs.usgs.gov/gip/geotime/relative, accessed 2001.

"Viewpoint: Global Warming Natural, May End Within 20 Years." *Ohio State University News Research Archive.* http://www.osu.edu/research news/archive/nowarm.htm, accessed 2001.

Villard, Ray. *The Milky Way Will Never Be the Same.* 21 September 1999. http://www.space.com, accessed 2001.

Vitousek, Peter M. *Human Alteration of the Global Nitrogen Cycle: Causes and Consequences.* http://www.esa.sdsc.edu, accessed 2001.

A Walk Through Time. http://www.hysics.nist.gov/genint/Time/Ancient.

Water Myths. http://www.1.stpaul.gov/depts/water/pages/myths.htm, accessed 2001. http://www.awwa.org/ bluethumb/mythsrealities.htm, accessed 2001. http://www.hashville.org/ws/drinking1.htm, accessed 2001.

Weather Basics: More About Thunderstorms. http://www.usatoday.com, accessed 2001.

Weather Folklore. http://www.members.aol.com/ Accustiver/wxworld_folk.html, accessed 2001.

Weather: The Wind Chill Formula, Its Development. http://www.usatoday.com/weather/wchilfor, accessed 2001.

Whitfield, John. "Sink Hopes Sink." Science Update, *Nature.* http://www.nature.com/msu/010524/010524-14.html, accessed 2001.

Subject Index

About the Author

ROBERT E. KREBS is retired Associate Dean for Research at the University of Illinois Health Sciences Center. He is also a former science teacher, science specialist for the U.S. Government, and university research administrator.